Plant Mutagenesis and Crop Improvement

Crop improvement using classically induced mutagenesis is now well standardized. Many new promising varieties of different crops have been successfully developed worldwide using both physical and chemical mutagens. Voluminous literature is now available on basic and applied aspects of mutagenesis. The mutation technique has been refined and holds the promise of generating much wider and more desirable variability than classical breeding. Recent advances in technology combined with classical mutation breeding offer new and exciting challenges for the development of new varieties. A global inventory of induced mutagenesis activities for crop improvement is required. This book covers both basic and applied aspects of mutation and its impact on various crops: it is extremely well prepared and contains a huge volume of information accumulated using classically induced mutagenesis on different crops in different countries.

Three key features:

1. Describes the importance of induced mutation in crop plant research and its application to production
2. Highlights new advances in the understanding of plant mutagenesis in crop improvement
3. Contains contributions from major leaders in the field of plant mutation research

This volume brings together all the important and relevant literature in the field. It provides a complete account of the mutation breeding of crops, presenting conclusions about the value of the method, its possibilities, limitations, and shortcomings, and the possible difficulties of further application in various crops. The initial chapters deal with the interactions between mutagenic treatment and plant material, such as aspects of mutagenic treatment, postirradiation behavior of shoot apices, and adventitious bud techniques. All available literature is then discussed crop by crop and critically evaluated. This will serve as an extremely comprehensive guide for researchers, teachers, students, and individuals who are interested in using induced mutagenesis as a tool for crop improvement.

Plant Mutagenesis and Crop Improvement

Edited by
Nitish Kumar

CRC Press
Taylor & Francis Group
Boca Raton London New York

CRC Press is an imprint of the
Taylor & Francis Group, an informa business

First edition published 2024
by CRC Press
2385 NW Executive Center Drive, Suite 320, Boca Raton FL 33431

and by CRC Press
4 Park Square, Milton Park, Abingdon, Oxon, OX14 4RN

CRC Press is an imprint of Taylor & Francis Group, LLC

ISBN: 9781032492629 (hbk)
ISBN: 9781032492636 (pbk)
ISBN: 9781003392897 (ebk)

DOI: 10.1201/9781003392897

Typeset in Times
by Newgen Publishing UK

Contents

Preface

Sustainable food production will remain a preeminent challenge in the decades to come. Today, more than 800 million people do not have enough food to meet their daily needs. By 2050, the global population is expected to increase from 7 billion people to 9.8 billion. To feed everyone, farmers will by then have to produce 50% more food. To overcome this increasing demand for food and proper nourishment, an improvement in food production is urgently needed. There are different mechanisms for harnessing the heritable variations encoded in the genetic makeup of existing crop plants so as to use them in crop improvement programs. The incorporation of desired traits from nonadapted landraces/crop wild resources can speed up crop improvement. Putative parental material can also be induced to mutate so as to obtain new genes that control desired traits for new crop variety development. Among the different strategies to enhance crop improvement programs, induced mutagenesis has contributed immensely by creating mutant varieties with improved and desirable genetic changes in agronomically important traits of the crop plants. Mutagenesis has become more efficient in combination with advanced molecular biology techniques and in vitro culture methods that result in enhancement of crop improvement/breeding programs, particularly under global climate change. Such induced mutagenesis also helps in the mining of new gene alleles that do not occur in the germplasm. Different types of physical, chemical, and combined mutagens have been used by various breeders to induce genetic variability in various crops. Mutation breeding improves several qualitative and quantitative characteristics of crop plants and has been successfully applied in several cereals, grains, legumes, oil seeds, vegetables, fruits, medicinal plants, ornamental plants, and fodder crops. With the advancement of various plant breeding, genetics, and biotechnological tools, mutation breeding contributes toward the increase in global food and agriculture production, which ultimately can be used to overcome global hunger and improve the nutritional status of the globe.

This book discusses the mutation induction mechanisms, different types of mutagens, the role of mutation breeding in crop improvement, and some highlights of mutant varieties, the mutation breeding strategy for obtaining mutants, and the economic impact of a new mutant variety. Featuring general discussions and research papers by leading international experts, it is a valuable resource for scientists, teachers, students, and industrialists working in the field of agriculture.

Nitish Kumar
Gaya, Bihar, India

Contributors

Geetika Bhandari
School of Agricultural Biotechnology, Punjab Agricultural University, Ludhiana, Punjab, India

Aanchal Bhatia
School of Agricultural Biotechnology, Punjab Agricultural University, Ludhiana, Punjab, India

Nihar Ranjan Chakraborty
Department of Genetics and Plant Breeding and Crop Physiology, Institute of Agriculture, Visva-Bharati University, Sriniketan, West Bengal, India

Gautam Chhabra
School of Agricultural Biotechnology, Punjab Agricultural University, Ludhiana, Punjab, India

Sibananda Darjee
Indian Agricultural Research Institute (IARI), New Delhi, India

Sandip Debnath
Department of Genetics and Plant Breeding and Crop Physiology, Institute of Agriculture, Visva-Bharati University, Sriniketan, West Bengal, India

Vinod Janardan Dhole
Nuclear Agriculture and Biotechnology Division, Bhabha Atomic Research Centre, Trombay, Mumbai, Maharashtra, India

Shilpa Gupta
Department of Biochemistry, Punjab Agricultural University, Ludhiana, Punjab, India

Souframanien Jegadeesan
Nuclear Agriculture and Biotechnology Division, Bhabha Atomic Research Centre, Trombay, Mumbai, Maharashtra, India

Krishna Sai K
School of Agricultural Biotechnology, Punjab Agricultural University, Ludhiana, Punjab, India

Achal Kant
Narayan Institute of Agricultural Sciences, Gopal Narayan Singh University, Jamuhar, Sasaram, Rohtas, Bihar, India

Gaganpreet Kaur
School of Agricultural Biotechnology, Punjab Agricultural University, Ludhiana, Punjab, India

Gurwinder Kaur
School of Agricultural Biotechnology, Punjab Agricultural University, Ludhiana, Punjab, India

Jessica Kaur
School of Agricultural Biotechnology, PAU, Ludhiana, Punjab, India

Maninder Kaur
School of Agricultural Biotechnology, Punjab Agricultural University, Ludhiana, Punjab, India

Ramanjot Kaur
School of Agricultural Biotechnology, Punjab Agricultural University, Ludhiana, Punjab, India

Shabneek Kaur
School of Agricultural Biotechnology, Punjab Agricultural University, Ludhiana, Punjab, India

Gurbachan S. Miglani
School of Agricultural Biotechnology, Punjab Agricultural University, Ludhiana, Punjab, India

Pratikshya Mishra
Odisha University of Agriculture and Technology (OUAT), Bhubaneswar, India

Ayesha Mohanty
Department of Soil Science and Agricultural Chemistry, Odisha University of Agriculture and Technology (OUAT), Bhubaneswar, Odisha, India

Wajhat Un Nisa
Department of Plant Breeding and Genetics, Punjab Agricultural University, Ludhiana, Punjab, India

Smruti Ranjan Padhan
Indian Agricultural Research Institute (IARI), New Delhi, India

Kaushik Kumar Panigrahi
Department of Plant Breeding and genetics, Odisha University of Agriculture and Technology (OUAT), Bhubaneswar, India

Umesh Preethi Praba
School of Agricultural Biotechnology, Punjab Agricultural University, Ludhiana, Punjab, India

Biswajit Pramanik
Department of Genetics and Plant Breeding, Palli Siksha Bhavana (Institute of Agriculture), Visva-Bharati, Sriniketan, West Bengal, India

Dhanasekar Punniyamoorthy
Nuclear Agriculture and Biotechnology Division, Bhabha Atomic Research Centre, Trombay, Mumbai, Maharashtra, India

Surinder K. Sandhu
Department of Plant Breeding and Genetics, Punjab Agricultural University, Ludhiana, Punjab, India

Rajveer Singh
School of Agricultural Biotechnology, Punjab Agricultural University, Ludhiana, Punjab, India

Satwinderpal Singh
College of Agriculture, Punjab Agricultural University, Ludhiana, Punjab, India

Diksha Singla
Department of Biochemistry, Punjab Agricultural University, Ludhiana, Punjab, India

Muskan Tiwari
School of Agricultural Biotechnology, Punjab Agricultural University, Ludhiana, Punjab, India

Yogesh Vikal
School of Agricultural Biotechnology, PAU, Ludhiana, Punjab, India

Editor Biography

Nitish Kumar has been Senior Assistant Professor at the Department of Biotechnology, Central University of South Bihar, Gaya, Bihar, India, for the last 12 years. Dr. Kumar completed his doctoral research at the CSIR (Council of Scientific and Industrial Research)—Central Salt and Marine Chemicals Research Institute, Bhavnagar, Gujarat, India. He has research experience in a wide variety of areas in the field of crop improvement using plant biotechnology and mutation breeding techniques. He has published more than 70 research articles in leading international and national journals, more than 20 book chapters, and 7 books with Springer and Taylor & Francis. Dr. Kumar received the Young Scientist Award from the Science and Engineering Research Board (SERB) in 2014. He has received many awards/fellowships/projects from various prestigious Indian government organizations such as CSIR, DBT, ICAR, SERB-DST, and BRNS-BARC. He is a reviewer for various international journals and serves as an associate editor of the journal *Gene* (Elsevier).

Acknowledgement

First and foremost, I would like to praise and thank God, the Almighty, who has granted countless blessings, knowledge, and opportunity to accomplish the book project work.

Thanks to all the authors of the various chapters for their contributions. It was quite a long process, from the initial outlines to developing the full chapters and then revising them in light of reviewers' comments. We sincerely acknowledge the authors' willingness to go through this process. I also acknowledge the work and knowledge of the members of our review panels, many of whom were called on at short notice. Thanks to all the people at CRC Press, India, especially Ms. Renu Upadhyay and Ms. Jyotsna Jangra, with whom I corresponded, for their advice and facilitation in the production of this book. I am grateful to my family members, Mrs. Kiran (wife), Miss Kartika Sharma and Laavanya Sharma (daughters), and my parents for their unconditional, selfless, and loving support all the time.

Nitish Kumar
Gaya, Bihar, India

1 Mutagenesis as a Tool in Plant Genetics, Functional Genomics, and Breeding

Pratikshya Mishra,[1] Smruti Ranjan Padhan,[2] Sibananda Darjee,[2] Ayesha Mohanty,[1] and Kaushik Kumar Panigrahi[1]*

[1]Odisha University of Agriculture and Technology (OUAT), Bhubaneswar, India

[2]Indian Agricultural Research Institute (IARI), New Delhi, India

*Email: kaushikpbg@gmail.com

INTRODUCTION

Plant breeding emerged during the first agricultural revolution (neolithic period), when hunter-gatherer societies adopted a more sedentary and agricultural lifestyle, starting as early as 10,000 B.C. Domestication took place concurrently in a number of subtropical areas during this time, including southeast Asia, western and South America, central Africa, and the Mediterranean (Gepts, 2002). The most viable samples from each harvest were presumably only used in the earliest plant breeding studies, which significantly affected crop productivity (Evans, 1993). Given that human selection usually clashed with natural selection, this changed the plants in unique ways (Hillman and Davies, 1990). In 1859 the term "artificial selection" was coined by Charles Darwin to emphasize the difference between natural and artificial selection, as the plants that have been domesticated could not be termed as "natural" (Darwin, 1859). Domesticated plants have undergone systematic selection to the extent that their wild cousins are now routinely categorized as wholly different taxa. Greater crop yields allowed for the development of settlements, an increase in human population density, and the specialization of employment outside of food production within these communities. Humanity suffered several negative consequences as a result of the switch from foraging to agriculture, including the spread of new infectious illnesses and epidemics brought on by greater population density and commerce, as well as a decline in the variety of available foods (Gepts, 2002). However, it is safe to say that the very cornerstone of contemporary civilization is plant breeding. Only a small share of the roughly 2 lakh plant species on Earth have endured domestication

DOI: 10.1201/9781003392897-1

throughout history due to the high human demand for desirable features and productivity (Panigrahi et al., 2023). Only about 200 species have been fully domesticated; however, many more species may have been used for food, feed, and fodder at some point in time. The entire world's food supply is now produced by 15–20 species (Chrispeels and Sadava, 2003; Balick, 1997).

1.1 GENETIC VARIATION INDUCTION

A crucial way to address the problem of food security and nutrition component of the world is crop genetic engineering (Ronald, 2011). It is anticipated that to be able to fulfill the demands of an exploding population by the year 2050, food production will need to at least double (Ray et al., 2013; Tester and Langridge, 2010; FAO, 2009). Heritability of trait of any variety is necessary for the genetic advancement of crops. Insufficient natural variation can be created through random or deliberate activities. Besides recombination, treating plant materials with mutagens (chemical/physical) is the most common method for creating new varieties that has been most frequently described. Although diverse effects on plant genomes have been found through different mutation causing agents and positional biases have been discovered, physical and chemical mutagenesis are frequently seen as to have caused more random mutagenesis (Greene et al., 2003). The type and concentration of the mutagens affect how different mutagens affect the DNA sequence. After sufficient genetic variation has been produced, selecting materials with the necessary changed features is the next stage (Figure 1.1).

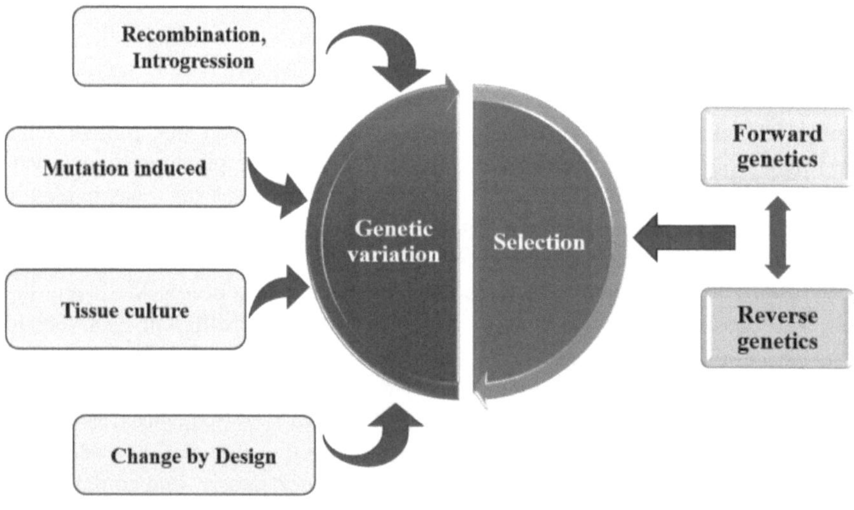

FIGURE 1.1 Techniques for crop development on the basis of generation and deployment of genetic variety.

1.2 PRACTICAL ASPECTS OF INDUCED CROP MUTAGENESIS

The mutation breeding process involves broadly three phases. The first step is the induction of mutation, followed by screening of potential mutant candidates, which is further followed by testing and releasing the mutants (Figure 1.2). The last phase is often standardized countries; thus research and development cannot (easily) increase efficiency there. The induction of mutations has been successfully used in many animals while being far from straightforward. The most time-consuming stage continues to be variant selection and mutant screening. Though phenomics has made incredible progress over the past five years, phenotyping is still more complex and time-consuming than genotypic selection (Cobb et al., 2013; Fiorani and Schurr, 2013). The species' historical accomplishments, as well as accessibility, affordability, and infrastructure, are typically taken into consideration when deciding which type of mutagen to be used for such programs

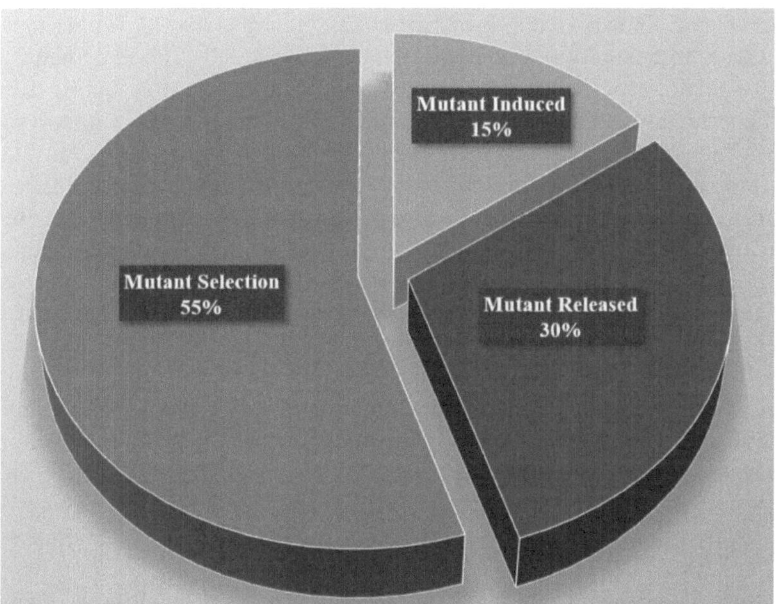

FIGURE 1.2 A method for release of enhanced crops through mutation breeding. Each component is drawn in proportion to the projected amount of time required for the development of a cereal grown from seeds (7–10 years). The initial phase, known as mutation induction, might take up to a year. The longest and most challenging step is mutant selection. It normally takes several years to find beneficial features that endure propagation cycles. The final step, the introduction of mutant varieties, adheres to the nation's standardized practices. Multilocational experiments with farmer participation are frequently needed for this. Although the duration of this phase can vary, it normally lasts less time than the testing and selection phase. If the chosen mutants are employed as parents in hybridizations, the process lengthens and becomes more difficult.

(Bado et al., 2015; Mba, 2013; MVD, 2016). Most of the mutations in mutant varieties are caused by ionizing radiation, notably gamma rays (MVD, 2016). This may be primarily attributable to the Food and Agriculture Organization of the United Nations and the International Atomic Energy Agency (FAO/IAEA) Joint Programme's active promotion of gamma irradiation, but it may also have biological significance because larger genomic aberrations are generally caused by physical mutagens as compared to chemical mutagens, and are more dominant, are easily detectable, and may be produced at a higher frequency (Jankowicz-Cieslak and Till, 2015). The process to employ standard operating procedures and general recommendations to generate mutations in plants' explant (seed/vegetatively propagated) by using mutagen like gamma rays and ethyl methane sulfonate (EMS) has been articulated by some (Lee et al., 2014; Bado et al., 2015; Till et al., 2006; Mba et al., 2010). In order to overcome a substantial bottleneck in plant mutation breeding, it is crucial to develop and analyze large mutant populations in order to enhance the possibility of detecting a desired variant. When cells of different genotypes coexist in the tissues of the same mutant plant, a phenomenon known as the dissociation of chimeras, also known as mosaics, takes place. In sexually reproducing plants, solitary cells called gametes serve as the foundation for the following generation, making chimeras simple to eradicate. For vegetatively propagated plants to yield homohistonts or genotypically uniform material, several regeneration cycles may be necessary (Van Harten, 1998; Mba et al., 2009). Using cell suspensions or (embryogenic) callus, certain cells that exhibit totipotency can be altered to stop chimerism in species that are propagated vegetatively (Van Harten, 1998). Less is known about the likelihood of chimerism at the DNA sequence as a result of these techniques' less frequent use than that of multicellular organs and tissues. It is interesting to observe what happens in a cell after mutation has been induced and the genetic material has changed. For instance, EMS mutagenesis causes alkylation, where the original base is not physically changed and the mutation is only corrected as a result of a replication error in the afflicted base. Two daughter cells with different genotypes could be created in this situation.

1.2.1 Using Induced Mutations to Create New Crop Varieties

The succeeding stages of the mutation breeding process parallel those of conventional breeding once a mutant population has been created. The generation in which selection for desired putative mutations might start must be considered (Figure 1.3). Selecting consistent phenotypes in the M2 population may be challenging depending on the frequency of mutations. This is due to the potentially complicated consequences of combining harmful lesions with epistasis, which influence how different proteins operate. The observable trait might disappear in later generations if phenotypes are selected too early because nonlinked alleles segregate separately. The researcher has the option to consider this risk and choose everything in the first nonchimeric generation that is of interest for further characterization (M2 for seed). To maximize the finding of novel mutations when contemplating reverse-genetic techniques, it is

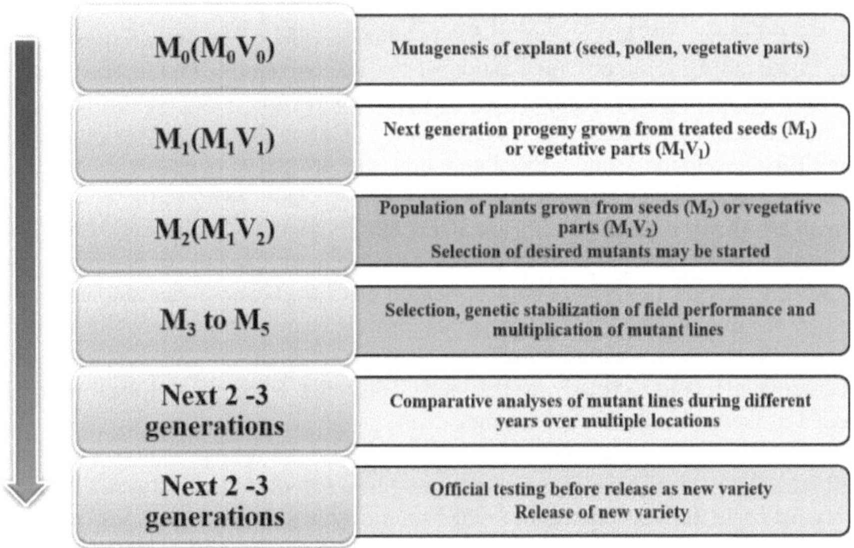

FIGURE 1.3 The conventional mutation breeding scheme. The stages for a specific generation are described in each row. For seed or pollen mutagenesis, the generation nomenclature begins with M0, and for vegetative organs, it begins with M0V0, where M refers to the meiotic generation and V for the vegetative generation. Prior to mutagenesis, all materials are labeled with a "0," and after mutagenesis is complete, they are labeled with a "1." When multicellular material is mutagenized, the first generation is not acceptable for evaluation since the ensuing plants will be chimeric. The M2 is the first homohistont (nonchimeric) generation in a seed-propagated material. To stabilize the inheritance of mutant alleles and make an asexually propagated material genotypically homogeneous, it may take multiple cycles. The initial nonchimeric generation can be used for screening and selection. To guarantee that the features are reproducible, succeeding generations often involve the assessment and selection of mutant phenotypes. The materials can then begin varietal release trials after this is finished.

generally better to perform molecular screening on the first nonchimeric generation (Jankowicz-Cieslak and Till, 2015). Additionally, the process used to choose acceptable phenotypes is quite significant. Even though phenomics strategies have been quickly growing in past years, the advancement of systematic species-independent protocols, as is possible with most genomic screening tools, is complicated by the diversity of morphological variations, disease responses, and physiological measures that varies from species to species (Cobb et al., 2013).

1.2.2 Elite Crop Varieties Created by Artificially Inducing Mutations

The use of this method began to develop new varieties by the plant breeders as soon as they found out that mutations could be induced while dealing with *Drosophila* (Muller, 1927) and cereals (Stadler, 1928a, b). A mutant tobacco variety known as

"Chlorina" was created in the 1930s as a result of the X-ray irradiation of flower buds (Tollenaar, 1934; Konzak, 1957; Coolhaas, 1952). The Joint Programme of the Food and Agriculture Organization of the United Nations and the International Atomic Energy Agency (Joint FAO/IAEA) in Vienna, Austria, maintains the Mutant Varieties Database, which contains information on more than 3220 crop varieties that have been developed using induced mutations and are grown in numerous nations all over the globe (MVD, 2016). Cereals make up over half (48%) of these crop types, which are grown from seeds. Ahloowalia et al. (2004) and Kharkwal and Shu (2004) examined the advantages of these mutant crop kinds for food security, nutrition, and economic success (2009). Examples include different types of rice that are widely grown in India, Pakistan, China, Australia, and Thailand; sunflower and peppermint in the US; barley in various European countries; sorghum in Mali; and a variety of decorative plants in Germany, Netherlands, and India. Millions of dollars have been added to the brewing and malting industries' profits thanks to the high-yielding and dwarf mutant barley cultivars "Diamond" and "Golden Promise" and their offspring (Ahloowalia et al., 2004). Some examples are several durum wheat types farmed in Italy for pasta that are exported to other countries, the Rio Star grapefruit variety in the United States, and the "Gold Nijesseiki" pear variety in Japan. The use of novel alleles produced through mutagenesis in the global production of superior agricultural varieties is difficult to quantify exactly. They are basically considered the same way as any other allele that a breeder can unintentionally or intentionally introduce into a new, enhanced cultivar in most parts of the world.

1.3 PHENOTYPIC SCREENING

The examination of plant features chosen by scientists, such as yield, quality, and tolerance to biotic/abiotic stressors, is the general definition of plant phenotyping. Depending on the need and the inquiry, the list may need to be enlarged. The five primary areas of the Mutant Variety Database are given in Figure 1.4 and Table 1.1. These mutant plants have been released and registered. The fact that 3222 officially registered mutants have 5569 improved characteristics suggests that many mutants have several improved qualities. For example, when a plant's ability to withstand biotic or abiotic stressors is improved, its yield rises. As a result, such a mutant will possess several enhanced features. It is still difficult to even guess how many altered genes and alleles are responsible for produced variation. The discovery of mutant alleles in charge of altered phenotypes will be made possible by genomic technology. The preponderance of the mutant varieties (48%) found in the Mutant Variety Database exhibits enhanced agronomic and botanical features, which is an interesting discovery. This might be because most botanic and agronomic features can be detected visually and can be screened without the need for specialized tools. Stresses caused by biotic and abiotic factors have the least mutations within their categories. It is noteworthy that these traits are crucial breeding objectives despite being complex and challenging to screen for. Techniques and procedures must be created to increase the effectiveness of mutant breeding. Based on the parental genotype,

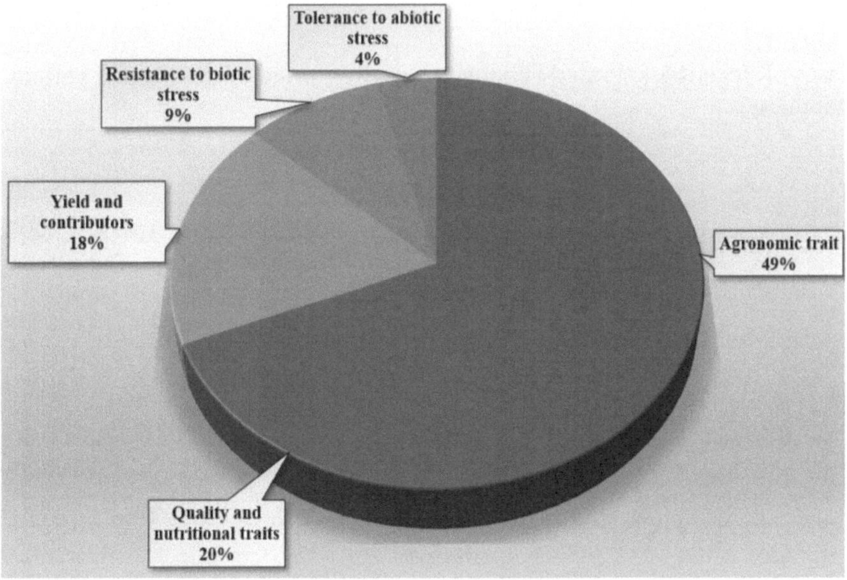

FIGURE 1.4 Registered mutants in the MVD are categorized based on their improved characters (traits). For 3222 types, enhanced characters are described a total of 5569 times. Five broad categories are used to group them: "agronomic trait" (49%), "quality traits" (20%), "yield and contributions" (18%), " biotic challenges" (9%), and " abiotic stresses" (4%). Maturity, blooming period, and plant structure are examples of agronomic and botanical features.

Source: MVD, 2016.

the method of propagation (seed vs. vegetatively), the desired improvement in the characteristic of interest, and the resources available, each phase of the process may be different. For instance, a screening technique for seed composition is called near-infrared reflectance spectroscopy (NIRS). Conventional methods used a disparaging approach, which is appropriate for identifying a sophisticated mutant line with a lot of seeds. The ability to quickly screen huge mutant populations and nondestructive techniques that measure the entire seed allow NIRS to be employed in this way. To assess seed constituents like protein concentration, calibration standards can be used using NIRS spectra. Detailed characterization can start once intriguing mutants have been located through a quick preliminary screen. Proteomic analysis can completely document the impact of genetic variability on the collection of proteins produced in grains or tissues. Another nondestructive technique is digital imaging that can be utilized for phenotypic assessment of morphological alterations brought on by mutagen treatment. For example, the response of roots to abiotic stress conditions like drought is quite important.

TABLE 1.1
Cases of Recently Released Enhanced Varieties Listed in the Mutant Variety Database

Attribute	References	Species and mutant variety name	Explanation	Growth type
Yield and contributors	Sigurbjoernsson and Micke (1974)	*Prunus armeniaca* L. Var – Early Blenheim	Early maturing, high-yielding, larger fruits and self-compatible pollen	Thermal neutron treatment of dormant scions (thN)
Agronomic and botanic traits	Newhouse et al. (1992)	*Triticum aestivum* L. Var – Above	White glumed, early maturing, awned	Treatment of seed with sodium azide (NaN_3)
Abiotic stresses tolerance	MVD (2016)	*Lycopersicon esculentum* M. Var – Maybel	Vigorous performance under drought conditions	Gamma rays induce seeds
Biotic stresses resistance	MVD (2016)	*Fragaria x ananassa* Var – Akita Berry	Increased resistance to *Alternaria alternata*, the causal organism of black leaf spot disease	Meristem culture induced with somaclonal mutation
Quality and nutrition traits	Kunter et al. (2012)	*Prunus avium* L. Var – Aldamla	Improved fruit quality, long petioles, condensed growth habit (70–80%)	Gamma rays irradiated dormant buds

1.4 GENOTYPIC SCREENING OF MUTANT PLANTS

Plant genotyping, broadly speaking, can be characterized as any experimental test intended to assess variations in the nucleotide sequence in species. Since nucleotide variation is the main cause of heritable phenotypic variation, this approach is extremely effective. Nucleotide variation identification techniques also enable choices that avoid the confusing effects of genotype by environment (G×E) interactions and provide crucial information on the evolution of plants (Annicchiarico, 2002). Several plant genomes can now be sequenced thanks to improvements in the protocols for collecting and analyzing genomic DNA (Weigel and Mott, 2009; Panigrahi et al., 2021).

1.4.1 Cheaper Techniques for Mutation Discovery and Genotyping

The issue with such inventions, though, is that they are frequently pricey and necessitate a high level of technical know-how. Because of this, not all laboratories have access to modern equipment. However, a variety of efficient, affordable techniques that work in labs with various infrastructures can be created. The first step in every genotyping experiment is the extraction of DNA. Before DNA extraction dehydrating and preserving leaf tissue in silica gel at room temperature eliminates the need for liquid nitrogen and −80°C freezers for long-term storage (Till et al., 2015). High-quality genomic DNA extraction from leaf tissue is normally carried out using pricey kits or more labor-intensive procedures using hazardous organic compounds, such as the CTAB method. In order to extract DNA from silica, chaotropic salts must bond to the material. This costs around 10% less but is comparable to the chemistry found in pricey kits. Therefore, it is significant that high-quality genomic DNA can be recovered without the need for specialized tissue grinding equipment or hazardous organic waste disposal methods. The end of low-cost techniques is not brought on by genomic DNA extraction. In reverse-genetics, a gene's expression or activity is changed in order to assess its function, which begins with the phenotype and finishes with the gene sequence. While endogenous transposons have been used to break genes, the advent of TILLING, a method that utilizes induced mutations, marked a significant advancement in the field of reverse genetics (Hunter et al., 2014; Conrad et al., 2008; Hirochika, 2001; McCallum et al., 2000a; Meeley and Briggs, 1995). According to some findings (Jankowicz-Cieslak et al., 2011; Kurowska et al., 2011; Greene et al., 2003) via the application of mutagens, TILLING produces a high density of induced mutations that are dispersed randomly throughout the genome. It is possible to produce a population of 3000–6000 mutant lines that have numerous mutations in every gene in the genome. For many years, a DNA and seed library can be built up and used as a resource. PCR and enzymatic mismatch cleavage are used in conventional TILLING to screen the DNA library for mutations in the targeted target genes. It is possible to make TILLING entirely affordable. Mutation detection can be carried out using regular agarose gels.

1.4.2 Techniques for High-Throughput Genotyping and Mutation Discovery

The next-generation sequencing technologies significantly outperform low-cost methods in terms of screening throughput. A three-dimensional pooling technique and simultaneous mutation identification in several gene targets are provided by the TILLING by sequencing protocol (Tsai et al., 2011). There are other uses for contemporary tools outside reverse genetics. The bulk of mutant crop types that have beenfficiallly released are the product of forward-genetic screening of radiation-damaged plant material (MVD, 2016). Genomic approaches have the potential to significantly improve the classic forward mutation breeding method, which has been used for 70 years. Determining whether a population contains a high density of advantageous mutations is the remaining challenge. It is acknowledged that differences seen in M1 plants do not imply heritable DNA changes, even though the abovementioned visual examination of M1 plants is desirable due to its quickness and relatively inexpensive (Preuss and Britt, 2003). Hence, M1 phenotypic changes do not necessarily correlate with mutation density. Thanks to next-generation sequencing technology, it is now possible to quickly evaluate the mutation densities and spectra of the M2 generation. The size of numerous plant genomes makes whole-genome sequencing of the necessary number of plants impractical for any but the most financially secure laboratories. A solution is provided by the sequencing of genomes with reduced representation. For a valid calculation of mutation density in this situation, tens of millions of base pairs from each mutant plant can be sequenced.

1.4.3 Improved Traits Caused by Cloning Mutant Alleles

Another significant obstacle to forward mutation breeding is finding and cloning the mutations responsible for the enhanced phenotype. This problem can be solved by utilizing the previously discussed techniques. By combining cosegregation of genotype and phenotype, it is expected to sequence entire genomes in plants with reduced genome sizes (Schneeberger et al., 2009; Cuperus et al., 2010). A technique called MutMap has been reported for cloning EMS-induced alleles in rice, utilizing a bulked segregant strategy. The technique has been refined to allow for the cloning of alleles without outcrossing (Abe et al., 2012; Fekih et al., 2013). Because of the throughput and cost limitations of whole-genome sequencing, this is far more challenging in crops with larger genomes. Reduced representation genome sequencing of researcher-designed regions is made possible by targeted capture-resequencing methods. Coding sequences are a great option when looking for mutations that alter gene function. Exome capture techniques are described by Henry and colleagues for recovering EMS-induced mutations in rice and wheat (Henry et al., 2014). If causative mutations are present in the regions the researcher chose for sequencing, this strategy permits huge enhancement of functional portions of the genome and makes software like MutMap viable for large genomes like wheat. Most attempts have so far been focused on recovering point mutations, including those brought on by exposure to the chemical mutagen EMS. Recent research indicates that mutagens primarily result in significant genome deletions. For instance, deletions of 1.2 million and 232,000 base pairs were found in *Zea mays* exposed to gamma radiation (Yuan et al., 2014). Whole-genome sequencing evaluation of rice exposed to rapid neutron radiation shows a greater spectrum of mutations than what is seen in gamma-irradiated rice (Li et al., 2016). Large genome deletions may significantly ease the process of cloning. For instance, a diploid plant like sorghum, which has a 730 Mbp genome, may contain 3000 induced mutations, making it challenging to pinpoint the exact mutation that causes a given feature. The same gamma-irradiated genome might only be able to develop a restricted number of big genomic indels. Therefore, it is much easier to pinpoint the mutation that caused the trait. Using this technique, poplar pollen's gamma-induced mutations have been catalogued (Henry et al., 2015). Cloning big indel mutations and SNPs will probably become more common soon as sequencing technology develops. Breeders will be able to use desirable mutant alleles for marker-aided introgression into superior germplasm.

1.5 MUTAGENESIS AND TILLING

Crop evolution has been slowed down throughout time as a result of breeders' increased attention to "elite" cultivars. Finally, this genetic deterioration reached a point of no return, and a variety of techniques for inducing mutations and artificial ways to boost variation emerged (Smartt and Simmonds, 1995). The first used mutagen by scientists was X-ray radiation because it was easily accessible. Muller demonstrated in 1927 that exposure to X-rays could raise the rate of mutation in a population of *Drosophila* by 15,000% (Muller, 1927), and Stadler found that exposure to X-rays and radium caused sterility in maize tassels and significant phenotypic diversity in barley seedlings (Stadler, 1928a; Stadler, 1928b). Subsequently, more sophisticated techniques were created by freshly founded nuclear research centers, including gamma and neutron radiation. Chemical mutagens were used in addition to radiation-based treatments during and immediately after World War II because they were less damaging, more widely available, and simpler to use. Pioneers in the field, Auerbach and others showed that *Drosophila* mutation frequency increased following exposure to mustard gas (Auerbach and Robson, 1946; Auerbach, 1949). Methane sulphonates and other still used chemical mutagens came after this work a few years later (Westergaard, 1957). Breeding through mutagenesis aims to maximize genomic diversity while minimizing viability loss. Among radiation-based approaches, X-ray and fast neutron bombardment have eclipsed X-ray in many applications. Fast neutron bombardment, which results in translocations, chromosome losses, and massive deletions, is more destructive than X-ray bombardment, which only produces point mutations and tiny deletions. Both forms of radiation greatly impair viability and inflict more extensive damage than chemical mutagens (Leung et al., 2001; Wu et al., 2005). Chemical mutagens are increasingly prevalent since they can cause a high rate of mutations, are simple to employ, and do not need expensive equipment. Chemical mutagens primarily cause single base-pair (bp) mutations as opposed to deletions and translocations. The most often employed chemical mutagen at the moment is EMS (ethyl methanesulfonate). When DNA-polymerase places a thymine residue over a cytosine residue during DNA replication as a result of EMS's preference for alkylating guanine bases, a random point mutation is produced. Transitions from GC to AT base-pairs account for 70–99% of alterations in EMS-mutated populations (Till et al., 2004, Till et al., 2007). Coding-area mutations may be silent, missense, or meaningless. Gene transcription can be up- or downregulated as a result of mutations that change the promoter sequences and other regulatory elements in noncoding areas. Additionally, mutagenesis may affect protein translation, mRNA stability, and incorrect splicing of the mRNA. Az-MNU solutions and other mutagens like sodium azide (Az) and methylnitrosourea (MNU) are often used. Genetically, Az-MNU mainly promotes transitions from GC to AT or AT to GC. Therefore, a shift can happen either way, unlike in EMS (Till et al., 2007). As was to be expected, all three chemical mutagens are extremely carcinogenic and need to be handled carefully. MNU is more difficult to manage than EMS since it is unstable above 20°C and sensitive to shock. Az is a solid dust in its ground state as opposed to EMS and MNU, which are both liquids. Because this makes it more difficult to handle than EMS and MNU, which are both liquids, it is the acutely poisonous and volatile material that needs to be dissolved before application. Over the years, mutagenesis has contributed significantly to plant breeding programs all over the world by creating a significant amount of genetic variety. Accounts collected by the joint FAO/IAEA Division in Vienna show that over the past four decades states, 2965 crop cultivars with few advantageous characters deriving from induced mutations have been distributed throughout the world (FAO-IAEA, 2011). Examples of crops with a significant positive economic impact include several wheat types (particularly durum wheat, which is used to create pasta), barley, rice, sunflower, including malting barley, grapefruit, and cotton. The introduction of TILLING technology has led to a resurgence in the application of chemically induced mutagenesis over the past decade. Chromosome-level DNA is extracted from each mutant line and utilized in TILLING to conduct a DNA-level population screening. TILLING seeds are subjected to a powerful mutagenic chemical that causes random mutations in genome, like conventional mutagenesis. Most researchers start by constructing a "kill curve" for their selected mutagen, in which the concentration of the mutagen is plotted against the seed viability, before creating the TILLING population. A 30–80% survival rate is the standard guideline (Wang et al., 2010; Chawade et al., 2010). To create a new generation of seeds following mutagenesis, the M1 seeds are planted and allowed to self-pollinate (M2). The M2 population is typically generated by sowing one seed from each line, and each M2 plant's DNA is separated. It is likely that each gene in the genome has a mutant allele somewhere in the population if the number of mutations per genome and the population is enormous. The ploidy of the desired crop must be taken into consideration when determining the ideal size of a TILLING population. The frequency of induced mutations and ploidy level appear to be strongly correlated. It has been shown that hexaploid plants like oat and wheat can have a mutation frequency as high as one per 25 Kb without dying or becoming infertile, whereas the maximal mutation frequency of diploid plants like barley and rice is substantially lower (Table 1.2). So, more than 5000 lines are infrequently needed for a hexaploid TILLING population. Contrarily, diploid populations often require tens of thousands of individuals (Chawade et al., 2010; Caldwell et al., 2004). It is advisable to think about the TILLING logistics before mutagenesis because TILLING is a critical and time-taking task in plants. It can take a lot of time and space to harvest and clean individual lines without cross-contamination, as well as to prepare, store, and organize thousands of bags of seed and the DNA samples that go with them. If kept in poor circumstances, many seeds quickly lose viability, making careful storage crucial. Furthermore, the creation of a database and barcoding system makes it much easier to manage a TILLING population and related data throughout several generations and to keep track of seed availability figures. It has been demonstrated that TILLING may be used to identify single mutations in particular genes in model systems like *Drosophila* and *Arabidopsis* (McCallum et al., 2000a; Winkler et al., 2005). TILLING is now effectively useful for a variety of crops, including those for wheat, rice, pea, maize, barley, oats, and soybean. Breeders now have a novel and sophisticated tool for crop enhancement thanks to this technology.

TABLE 1.2
Mutant Populations in Various Plant Species

Screening process	Mutagen	Species	Reference
Li-Cor	EMS	Maize, Durum wheat, Pea, Soybean, Rice	Till et al. (2006); Till et al. (2004)
	Az-MNU	Rice	Till et al. (2007)
	Az	Barley	Talame et al. (2008)
	DES	Groundnut	Knoll et al. (2011)
Li-Cor, dHPLC	EMS	*Arabidopsis*	Colbert et al. (2001); McCallum et al. (2000)
Phenotypic (stress)	DEB, GR, FN	Rice	Leung et al. (2001)
dHPLC	EMS	Barley	Caldwell et al. (2004)
CE	MNU	Rice	Suzuki et al. (2008)
AGE	EMS	Bread wheat	Dong et al. (2009)
CE, HRM	EMS	Tomato	Gady et al. (2009)
HRM	EMS	*Arabidopsis*	Bush and Krysan et al. (2010)
MALDI-TOF	EMS	Oat	Chawade et al. (2010)
PAGE	EMS	Bread wheat	Uauy et al. (2009)

1.5.1 Discovery of Mutant in TILLING Populations

Uncommon mutation discovery is significant for fundamental research as well as biomedical and biotechnological advances. TILLING is a functional genomics technique used to identify unusual mutations in populations. Mutagenesis, DNA isolation and pooling, and high-throughput mutation identification in target genes are all part of it.

1.5.1.1 Direct Sequencing

Sanger-based technology is used in this strategy to screen a TILLING population; however, it is quite expensive. DNA sequencing might be regarded as the finest caliber of in-screening because it makes it simple to identify all mutations. While screening often concentrates on a limited number of genes, the existence of a reference genome potentially allows for the construction and analysis of full mutant genomes (Mishra et al., 2023). When a phenotype is clearly present but candidate gene is absent, this is extremely helpful. The price and speed requirements for sequencing technology are raised as a result, though.

1.5.1.2 Li-Cor

In a TILLING population the most popular technique for finding mutations is Li-Cor. It hinges on the precise mismatch base cleavage caused by repeated melting and reannealing of a PCR product generated from the target region. The occurrence of a hybrid DNA molecule with a single mismatch indicates the presence of mutation in it. A selective endonuclease, usually Cel-1 or Endo-1, is used to split the fragment, producing two shorter fragments that can be distinguished using polyacrylamide gel electrophoresis (Colbert et al., 2001). The forward and reverse PCR primers can be modified to include fluorescent dye-tags of different colors, which the Li-Cor device uses to identify the amplified fragments. In diploid organisms, a single Li-Cor can run a 96-lane gel and its sensitivity is high enough to allow up to 16-fold sample pooling, for a total of 768 samples per run. Large hexaploid genomes have a higher level of complexity; hence during screening, this number is drastically decreased. The Li-Cor technique also has a number of intrinsic disadvantages that need to be taken into account. The ratio between the amounts of the cleavage enzyme and the PCR product, as well as the concentrations of fluorescent dye-primer and DNA, all affect the outcomes and must be optimized. Moreover, a specialized tool is needed for effective fluorescent fragment identification and respectable throughput. In comparison, amplicons produced by a Li-Cor system can be as long as 1.5 kb, making them the longest of all techniques. Endo-1 and Cel-1 are both rather pricey, but there is a methodology that explains how to directly separate Cel-1 from celery stalks (Till et al., 2006). The resulting enzyme extract, CJE (celery juice extract), can take the place of pure enzyme in a number of applications, resulting in a large cost savings per reaction. CODDLE, the most well-liked bioinformatics tool for creating primers for Li-Cor, combines primer functional analysis with an algorithm that determines the gene locations where harmful mutations are most likely to occur based on the selected mutagen and gene structure. While ParseSNP can forecast the anticipated impact of an inserted SNP on protein function, GelBuddy automates band detection in electrophoretic gels for postrun gel analysis.

1.5.1.3 High-Performance Liquid Chromatography

A high-performance liquid chromatography (HPLC)–based method can be considered a sensitive screening choice and was used in initial TILLING study (Caldwell et al., 2004). Like the Li-Cor procedure, HPLC is used to separate samples after Cel-1 mismatch cleave enzyme has been added to them. Two additional elution peaks with a combined size equal to the initial PCR product would be the sign of a heterozygous mutation (McCallum et al., 2000b). Although 32-fold pools are conceivable, an 8-fold pool of samples for a diploid organism is advised, allowing 8 samples to be analyzed at once (Caldwell et al., 2004). Its potential as a high-throughput screening tool would be limited by the requirement for several HPLCs to process numerous samples at once.

1.5.1.4 Electrophoresis

Agarose or polyacrylamide (PAGE) gel electrophoresis has been suggested as a low-cost substitute for Li-Cor devices for high-throughput screening. Ethidium bromide (EtBr) is utilized to visualize the fragments after agarose gel separation rather than fluorescent dyes. The authors claim that it is possible to create an eightfold pool with a maximum amplicon length of 3 kb (Raghavan et al., 2007). This technique has been used successfully to search a wheat population for waxy and hard grain mutants using a fourfold pool on thin (4 mm) gels (Dong et al., 2009). Agarose gel electrophoresis may be the best approach for TILLING on a budget because it doesn't need specialized equipment and Cel-1 can be replaced with celery juice extract (CJE) (Till et al., 2006). However, because the approach is less sensitive than Li-Cor, more Cel-1 must be used per sample, underscoring the importance of making your own CJE.

1.5.1.5 Capillary Electrophoresis

TILLING populations can also be screened using CE, or electrophoresis in capillaries (Suzuki et al., 2008). The material is mixed with EtBr, placed into glass capillaries, and electrophoretically separated after being split with Cel-1/Endo-1. By using UV light to excite DNA-bound EtBr at the capillary's end, it is possible to detect the presence of DNA and produce a digital absorption spectrum over time. A mutant strand will cause new peaks to appear on the graph. With a maximal fragment size of roughly 1.5 Kb, Li-Cor is comparable. Moreover, an eightfold pool can be distinguished using the detection limit (Suzuki et al., 2008). In contrast to conventional electrophoresis, conformation-sensitive capillary electrophoresis (CSCE) does not require enzymatic degradation (Gady et al., 2009). This approach excludes Cel-1/Endo-1 while doing PCR and melt-annealing. Instead, a semidenaturing gel (CAP) is inserted into the capillary, allowing homoduplexes and heteroduplexes to be distinguished by the impact of mismatch on migration rate. Although the authors advise a fourfold pooling, it is possible to construct an eightfold pool of diploid DNA using this technique (Gady et al., 2009). When intercalating dyes are employed in place of fluorescent primers, the sensitivity of all capillary electrophoresis techniques is slightly decreased. The device can be updated to simultaneously control 96 lanes, and analysis takes only 5–10 minutes per run. The high instrument cost associated with CE needs a large initial investment.

1.5.1.6 Melt at High Resolution (HRM)

When the intercalating dyes are attached to DNA, it gives a fluorescence glow and this technique is used in HRM. DNA strands gradually separate as the temperature rises, releasing the dye and reducing the overall fluorescence. Thermal and luminescence graphs are used to depict the outcomes. Due to the mismatched base, there will be an alteration in the melting temperature, and this mutation will cause an alteration in the graph. By comparing heterozygotes' standardized melting curves to those of homozygotes or wildtype samples, heterozygotes are quickly identified (Gundry et al., 2003; Wittwer et al., 2003). Although its sensitivity, HRM is constrained by amplicon GC length and content, with a usual read-only covering 150–500 bp, which is substantially shorter than Li-Cor and CE. HRM is extremely useful when the target is a specific region that has been shown to impact protein structure or when the gene of interest has multiple short exons, making a short read length feasible. The need for specialized software to interpret the various melt-curves is a drawback. Standard qPCR equipment can perform HRM with a straightforward software modification, giving them an appropriate platform for first TILLING screens. Wheat (Dong et al., 2009), tomato (Gady et al., 2009), medaka (Ishikawa et al., 2010), and *Arabidopsis* (Bush and Krysan, 2010) mutations have all been effectively identified using HRM (Dong et al., 2009; Ishikawa et al., 2010; Gady et al., 2009).

1.5.1.7 Matrix-Assisted Laser Desorption Ionization Time-of-Flight

Matrix-assisted laser desorption ionization time-of-flight (MALDI-TOF) spectroscopy was developed in 1985 and has since grown to be an essential analytical tool in the study of polymer chemistry and proteomics. MALDITOF has also been employed in the high-throughput SNP detection sector. There is now only one standardized, high-throughput method available; it is known as Mass Cleave and was developed by SEQUENOM. Presently, this has not been fully utilized in SNP identification (Van Den Boom and Ehrich, 2007). To produce detectable tiny RNA fragments, this technique uses a T7-R&DNA polymerase synthesis step followed by an RNAse degradation phase. Once fragments have been located, it is possible to put them back together in silico to reconstruct the screening PCR result and locate mutations. A novel diamino benzophenone (DABP) matrix for the study of nucleotides was introduced recently. DABP has the same resolution and sensitivity as standard 3-HPA (3-hydroxypicolinic acid) but has a 100-fold higher salt tolerance (Fu et al., 2006). This matrix provides a straightforward and elegant substitute for 3-HPA for SNP analysis because the presence of even minute amounts of K^+ and Na^+-ions in the sample solution substantially lowers the sensitivity of the assay. Compared to Li-Cor-based methods, MALDI-TOF is rather simple. There is no need for individual step optimization or enzyme titration during the enzymatic degradation phases because they are simple and reliable. Heterozygote mutations can be detected in a hexaploid organism in addition to its sensitivity by using this technique. Since the technique does not depend on heteroduplex formation, homozygous mutations can be accurately detected without combining samples. A homozygous mutation would be more noticeable because it would cause a mass peak on the MALDI graph to vanish. This suggests that in late-stage TILLING populations with a rising homozygous mutation rate, MALDI-TOF-based screens are more crucial. The initial MALDI-TOF based SNP finding procedure was used in concrete evidence screening that was published (Chawade et al., 2010). By lowering response quantity, moving to a DABP matrix that is more salt tolerant, and creating software for automated sample screening, we were able to modify and enhance the SEQUENOM MALDITOF procedure for TILLING applications. Without sacrificing sensitivity, our improved approach reduced the size of the reaction in half and used just one-eighth of the initial enzyme concentration. We also developed new tools to precisely locate novel SNPs. TILLING screening using MALDI-TOF devices might be a helpful addition to existing screening techniques while we wait for more alternative options, and it might even be a good replacement for significant expenditures in Li-Cor technology. This seems to be certainly relevant for laboratories, MALDI-TOF technology is currently a key component of the facility due to the wide range of uses it has.

1.5.1.8 Advancing Technologies

Next-generation sequencing (NGS) has highly increased the likelihood of detecting mutation in the whole-genome. Because sequencing costs have decreased due to greater technical precision, increased output, and expanded dimensions in recent years, NGS has a lot of potential for TILLING. The most prevalent NGS platforms are the Roche Applied Science 454 Genome Sequencer FLX Ti and the Illumina (Solexa)

Genome Analyzer. Illumina only offers up to 100 bases per read, compared to 454's typical read length of 750 bases, but it produces a much greater amount of sequencing data. Moreover, the data quality, read length, and quantity of generated sequences are all constantly changing with these technologies. With the GS-FLX+ system, for instance, Roche just implemented read durations of up to 1 KB. There are already several proof-of-concept techniques for integrating NGS into TILLING applications. One or more genes of interest can be tested with a single FLX-454 run by using three-dimensional pooling. A tomato TILLING population used in experiments shows that up to 12,000 samples can be analyzed at once on a single 454-picotiter plate (PTP) utilizing Key Point technology (Rigola et al., 2009). Moreover, high-throughput TILLING using Illumina sequencing has been modified to screen populations of rice and wheat (Tsai et al., 2011). The technique called CAMBa (Coverage Aware Mutation Calling using Bayesian analysis) verified previously discovered mutations with fewer false positives while also identifying additional mutations missing by CJE mismatch-cleave-based TILLING (Tsai et al., 2011). Because of the enormous amount of data produced by NGS, bioinformatics expertise and access to computing power are essential for processing. With more known methods, PacBio RS, a novel technology based on sequencing of single molecule, is now accessible. The read length for this instrument exceeds 1 kilobyte on average, and often more than 10% of reads are between 1.5 and 2.5 kilobytes in length, with some reads exceeding 4500 base pairs (Flusberg et al., 2010). With the help of recent technological developments, the sequencer can now produce about 35 MB of sequencing data every run. This method, which has not yet been modified for TILLING, will be especially helpful for nonsequenced genomes without a prior alignment scaffold. SNP finding has also been done using NGS in addition to direct screening. An average coverage of 5× and a depth of more than 90% were found in a recent NGS investigation of 17 wild and 14 farmed soybean genomes. For QTL mapping and association investigations, this study discovered 205,614 tag SNPs with a high allelic diversity (Lam et al., 2010). Six premium maize cultivars were subjected to NGS analysis, which revealed 100,000 SNPs, 30,000 insertion-deletion polymorphisms, and presence/absence variation in several genes (Lai et al., 2010). These papers show how high-throughput technologies are becoming more important in areas other than mutation detection.

1.6 FROM GENOTYPE TO PHENOTYPE

The prime objective of TILLING is to find mutations in crucial genes and then connect those changes to a particular feature, in contrast to conventional screening methods used by researchers. However, this approach can only be used if a gene linked to the desired attribute has been found and its sequence is available. It is now possible to determine that the mutations are most likely to change the structure of a protein or cause translation to fail, producing a nonfunctional product, using maps and software. Experiments in morphology, histology, physiology, or biochemistry can subsequently be used to confirm the indicated probable phenotypes. Though theoretically straightforward, the screening process and subsequent evaluation could lead to several difficulties. In most nonmodel systems, a whole genomic sequence is not available, making it difficult, if not impossible, to find enhancer and promotor alterations upstream of the gene of interest. The paradox that a single mutation may not always affect a cell's overall function, even if it is predicted to be damaging, adds another layer of complexity. A poor or nonexistent rate of mutation penetration may be caused by the expression of homologs of the target gene. This is particularly true for hexaploid plants, where there may be homologs of the important gene in each of the three genomes and two more alleles can make up for a defective allele. Hence, it is necessary to identify knockout mutations in all alleles by tedious screenings, followed by the time-consuming crossing of the various mutations within the same genome. The final trait's development may be greatly delayed as a result. Despite these limitations, many organizations have reported success in establishing a connection between genotypic variation and novel phenotypes in several crop species, especially in wheat, where attributes associated with the waxy phenotype and grain hardness are being developed (Slade et al., 2005; Dong et al., 2009; Sestili et al., 2010); in soybean, where TILLING has ascertained to be useful in identifying genetic changes in the FAD1, 2, and 3 genes to boost the oleic acid content (Dierking and Bilyeu, 2009); and in sorghum, where lignin content has been decreased by mutation of COMT (Xin et al., 2008).

1.6.1 Identification of New Traits in Mutated Populations

Two major steps are essential in detecting and selecting mutant plants with improved traits: mutant screening and validation.

1.6.2 Screening at Biochemical Level

Identification of genetic mutations is the main goal of TILLING. This does not, however, rule out the use of TILLING populations and other mutant populations for phenotypic screening. The main distinction between genotypic and phenotypic screening is exemplified in Figure 1.5. It is impossible to gauge macromolecular proportion and the quantity of bioactive substances like lignin and other lipids, fibers, and starch in the field. Secondary plant cell walls contain lignin, which adds stiffness. Foragers avoid lignin because it prevents microbial enzymes from breaking down cell-wall polysaccharides and is indigestible on its own. Because they consume less energy, crop cultivars with reduced lignin levels in their cell walls are recommended for animal feed. The Wiesner test, also known as the phloroglucinol-HCl assay, is a quick and affordable technique for identifying changed lignin levels in seeds. In total, 17 lines from an oat TILLING population were found after 1824 lines were screened. These lines had seeds with less lignin stain intensity. For additional evidence, precise assessment of lignin levels in mutant seeds was carried out using the acetyl-bromide technique (Chawade et al., 2010, Iiyama and Wallis, 1988; Vivekanand et al., 2014).

Another important breeding goal is to increase the number of dietary elements that directly reduce plasma cholesterol levels by preventing cholesterol efflux or absorption. For instance, cereals often include mixed-linkage (1 3), (1 4)-D-glucan soluble fiber, with oat and barley having the largest quantities. The ß-glucan concentration of 1500 randomly selected lines from an oat TILLING population was assessed with the

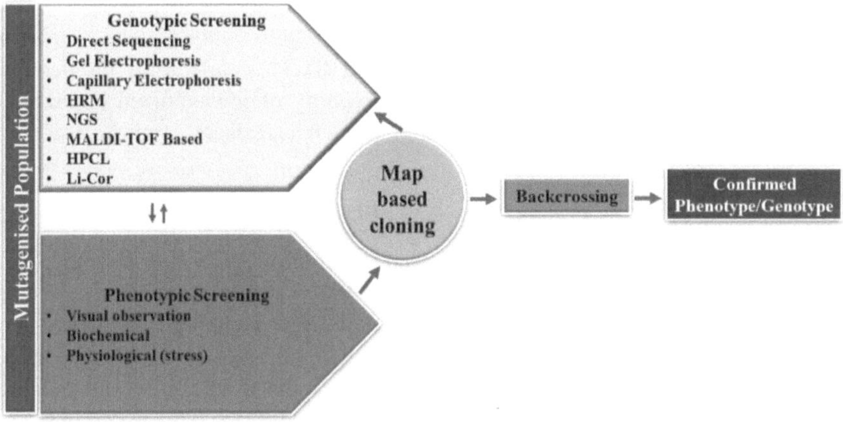

FIGURE 1.5 Outline of many techniques for screening mutated people and creating new stable characters.

help of Megazyme assay kit (McCleary and Codd, 1991; Sikora et al., 2013). There were lines with higher concentrations of glucan. Also few lines showed less than half of the original cultivar (Belinda). With the rise in TILLING-populations, we predict that these populations will undergo advanced biochemical assays in addition to TILLING, or genetic screening, to identify key characteristics. Phenotypic screening has the advantage of allowing for the easy identification of the desired attribute. In contrast to genotypic screening, it is impossible to pinpoint the precise mutation(s) that caused the trait. There are countless such instances that show the effectiveness of biochemical screens in the literature (Barkawi et al., 2010; Reiter et al., 1997; Park et al., 2007).

1.6.3 Screening at Physiological Level

Around the world, fungi pose a serious threat to agriculture. The problem is expected to get worse due to global climate change, which is predicted to cause milder winters and increased humidity levels. A pathogen that is particularly problematic and very relevant on a global scale is fusarium. Fusarium, which has over a thousand different species, infects important agricultural crops and causes disease. In addition, a variety of mycotoxins produced by *Fusarium* spp. build in grain, make their way into the food chain, and cause serious health concerns to both people and animals. *Fusarium* head blight (FHB) disease, for which there are now no effective control techniques, is especially problematic. Fungicide treatments for FHB result in variable and erratic results, occasionally significantly deteriorating mycotoxin contamination (McMullen et al., 1997). Unfortunately, it does not seem like breeding populations vary enough to find and create disease-resistant lines. On the contrary, mutagenized populations can be used to discover resistant lines with a high genetic component, even for features that vary greatly with extrinsic influences, such as disease resistance. Designing an in vitro assay with such strict selection to enable the identification of specific, rare lines with potent disease resistance is the problem. By creating a Petri dish assay to recognize *Fusarium*-tolerant oat from a mutant population with a more genetic variation, this notion was put to the test (Chawade et al., 2010). This early study implies that, with proper technique design, morphologic screening of mutagenized populations may be utilized to identify complicated features, such as resistant strains.

1.7 FROM PHENOTYPE TO GENOTYPE

To find out the relevant genetic trait in a mutagenized population to be completely meaningful, phenotypic screening must be succeeded by a description of the molecular event underlying the altered feature. In plants with sequenced genomes, new phenotypes can be obtained using a mix of whole-genome resequencing, microarrays, and linkage maps. In comparison to wildtype samples, this offers a thorough perspective of changes in gene expression and newly discovered SNPs. This is demonstrated by the discovery that the semidwarf phenotype present in many marketed varieties of rice is caused by a mutation in the GA20 oxidase. Through genetic mapping it was evaluated that the concerned character was linked to a part of chromosome 1. The rice reference genome was used as a starting point for the identification and sequencing of a potential GA gene in that location because it was known that the dwarf phenotype had lower amounts of GA. The lower GA levels are explained by a 280-bp deletion that was found in the sequence and rendered the protein inactive (Spielmeyer et al., 2002). Microarray technology has also been successfully applied to link individual traits to genome-wide changes in rice and *Arabidopsis* (Singer et al., 2006; Edwards et al., 2008). However, for SNP detection, next-generation methods like Illumina sequencing currently surpass more established microarray approaches (Huang et al., 2009). However, one method entailed looking for disease-causing mutations in EMS-induced *Arabidopsis* Col-0 mutants with sluggish growth and pale green leaves. Prior to getting crossed with the recessive mutants, the recessive mutants were bred with the *Landsberg erecta* ecotype. The DNA of 500 F2 individuals was then combined and sequenced by Illumina to achieve up to 22-fold genomic coverage. For the purpose of locating mutations in the segregating population, the SHORE map software was created. The *AT4G35090* gene had a mutation that caused a nonsynonymous codon shift from serine to asparagine (Schneeberger et al., 2009). Using other techniques, Austin et al. (2011) discovered three genes essential for cell wall production. To begin with, flupoxam sensitivity was tested in the *Arabidopsis* EMS-treated Col-0 mutants because it is known to interfere with cell wall synthesis or integrity. The mutants were subsequently bred with a *L. erecta* ecotype. Illumina GA sequencing was used to extract and screen the F2 population's genomic DNA. They were capable of correctly determining the basic mutations and, consequently, the genes in charge of the phenotype by applying an internal statistical technique (Austin et al., 2011). In the absence of a reference genome, it could be challenging to detect a mutation because it is not necessary for it to occur in an exon of the candidate gene. The downstream effect can be caused by a variety of mutations, including promotor mutations, mutations that alter genome structure, mutations that occur upstream in the regulatory pathway, and different mutations of micro-RNA. It can be incredibly challenging and time-consuming to analyze these aspects thoroughly in the absence of a reference genome. In such circumstances, a first step would be to collect as many mutants as possible and assess each one separately, resequencing all relevant genes and carrying out qPCR assays to find any potential changes in candidate gene expression. It is challenging, but not impossible, to establish a genotype-phenotype link using this technique. Feiz et al. (2009) connected Puroindoline a and b mutations in an EMS-mutagenized population to wheat grain hardness using EST libraries rather than a fully sequenced genome. Genetic maps sometimes lack known gene interactions, denying researchers a helpful tool for narrowing down the list of possible genes.

1.8 STABLE MARKERS INTROGRESSION INTO BREEDING POPULATIONS

The phenotypic variances between individual plants in the field always exist due to variable environmental conditions, though the present exclusive varieties are genetically quite uniform. In addition to showing variances in overall plant design, cultivars cultivated at various locations under various fertilization, insect, and weed control, and climatic circumstances also differ in the quantity of specific organic substances and metabolic processes. However, depending on the way each mutation mediates the phenotypic, the impact of the environmental element differs. Therefore, there will be less variety in the expression of a trait if the genetic component is high. Leaf form, leaf color, and pubescence on the leaves are examples of observable features that are genetically fixed and do not alter noticeably in response to the environment. As a result, these traits serve as identifiers to distinguish market types. In a perfect world, a steady, obvious property can also be linked to a more precise, but concealed, character trait. The discovery of phenotypes that are phenotypically stable and are associated with a certain genotype is therefore essential for a successful selection strategy. However, for crucial quality traits like high starch, fats, protein, and fiber composition; decreased amounts of hazardous components; and superior postharvest handling features, such relationships are not always obvious. More focused testing must be performed to identify these traits. Such tests often take a lot of time and money, and they cannot be performed on a lot of samples.

In contrast, using a mutagenized population with a high amount of variety increases the likelihood of finding a particular character and reduces the number of analyses necessary to find some quality feature. Additionally, there is a higher chance of finding uncommon mutations that disable transcription factors or other pleiotropic genes. Such mutations will penetrate more effectively, and environmental variables will have less of an impact on their phenotypes. The genetic component of this particular mutation is visible with the unaided eye during the entire growth season. Similar to obvious abnormalities, robust genetic alterations that can only be identified biochemically can also exist. The trait can be introduced into breeding lines deficient in the trait after being found in a mutant population and tested for genetic stability in the field. The use of a marker during introgression minimizes the number of essential crosses and removes as many random mutations from the mutagenized lines as possible. This indicator may be observable, biological, or molecular. A molecular marker, such as a mutation with a desired quality feature, is favored and has several benefits over traditional phenotypic selection. "Marker-assisted molecular selection" (MMAS) refers to this process. MMAS is neutral and unaffected by environmental variables because it is based on DNA. The characteristic can frequently be quantified as early as the seedling stage, and material for the assay can be gathered from any plant developmental stage. This reduces the need for labor, time, and field space. If the marker's correlation to the trait is sufficiently strong, molecular markers can also be employed to select for complicated traits. If there is a correlation between a molecular marker and disease resistance, disease resistance can be assessed without subjecting the plant to the pathogen. An understanding of how a particular mutation directly upregulates, downregulates, or removes a gene can serve as the foundation for MMAS. It will be tightly related to a certain trait in this instance. MMAS may also be indirect and based on a phenotypic correlation that is statistically significant. QTLs are SNP microsatellite markers that can be discovered via hybridization methods such as DNA sequencing, Southern blotting, PCR, MALDI-TOF, and others. Semagn et al. (2006) conducted a detailed examination of numerous marker types. Perhaps the most important feature of MMAS is its automation and high-throughput screening capability. Robots, fluorescence detection methods, automatic scripts, and other automated tools can be used to automate DNA isolation, pipetting, separation, and evaluation, which greatly speeds up the screening process.

1.9 CONCLUSION

Growing population, decreasing cultivatable land, and new and regionally shifting abiotic and biotic pressures are only a few of the rapidly rising problems for agricultural productivity that demand thoughtful analysis and creative solutions. Crop genetic improvement requires both recent developments and translational science to be successful in the long run. Induced mutagenesis is expected to remain a crucial technique for breeders since it is a quick and reasonably priced way to produce novel alleles and phenotypes. New technologies will also make it possible to identify the mutant alleles that were employed to produce successful mutant types, shedding light on gene function and crop output. Mutagenesis in breeding has once again matured in the last ten years. Breeders can now select for characters that were extremely challenging to breed for just a few years back thanks to plant mutagenesis, which tends to increase the genetic diversity of crop plants that have been inbred for millennia. This genetic diversity can then be used in combination with high-resolution phenotypic and genotypic screening techniques. Elite inbred cultivars with new genetic diversity introduced offer a rare chance to discover fresh features while maintaining the outstanding agricultural performance of the lines. We have reached a point where integrating this toolbox with much more traditional breeding techniques will save us both money and time thanks to the quick collection of genetic database from a variety of crop plants. While markers are produced during the process, this approach also enables stacking of the advantageous features, opening the door for the emergence of polygenic traits like tolerance to abiotic stress. Recombinant DNA technologies and GMOs will be well complemented by mutagenesis and high-resolution screening (GMOs) in the development of more climate- and population-adaptive food plants, despite still being constrained by the capabilities of the endogenous genome.

REFERENCES

Abe A, Kosugi S, Yoshida K, Natsume S, Takagi H, Kanzaki H, Matsumura H, Mitsuoka C, Tamiru M, Innan H, Cano L, Kamoun S, Terauchi R (2012) Genome sequencing reveals agronomically important loci in rice using MutMap. *Nature Biotechnology* 30(2):174–178.

Ahloowalia BS, Maluszynski M, Nichterlein K (2004) Global impact of mutation-derived varieties. *Euphytica* 135(2):187–204.

Annicchiarico P (2002) *Genotype X Environment interaction: challenges and opportunities for plant breeding and cultivar recommendations*. Food Agriculture Organization of the United Nations, FAO, Rome, Italy.

Auerbach C (1949) Chemical mutagenesis. *Biological reviews of the Cambridge Philosophical Society* 24(3): 355–391.

Auerbach C and Robson JM (1946) Chemical production of mutations. *Nature* 157(3984): 302–302.

Austin RS, Vidaurre D, Stamatiou G et al. (2011) Next generation mapping of *Arabidopsis* genes. *Plant Journal* 67(4): 715–725.

Bado S, Forster BP, Till BJ, Nielen S, Ghanim AMA, Lagoda PJL (2015) Plant mutation breeding: current progress and future assessment. *Plant Breeding Reviews.*

Balick MJ (1997) *Plants, People, and Culture: The Science of Ethnobotany*. Vol. 9, Scientific American Library, New York, NY, USA.

Barkawi LS, Tam YY, Tillman JA, Normanly J, and Cohen JD (2010) A high-throughput method for the quantitative analysis of auxins. *Nature Protocols* 5(10): 1609–1618.

Bush SM, and Krysan PJ (2010) iTILLING: a personalized approach to the identification of induced mutations in *Arabidopsis*. *Plant Physiology* 154(1): 25–35.

Caldwell DG, McCallum N, Shaw P, Muehlbauer GJ, Marshall DF, and Waugh R (2004) A structured mutant population for forward and reverse genetics in Barley (*Hordeum vulgare* L.). *Plant Journal* 40(1): 143–150.

Chawade A, Sikora P, Brautigamet M et al. (2010) Development and characterization of an oat TILLING-population and identification of mutations in lignin and beta-glucan biosynthesis genes. *BMC Plant Biology* 10: 86.

Chrispeels MJ and Sadava DE (2003) *Plants, Genes, and Crop Biotechnology*. Vol. 24, Jones and Bartlett, Boston, MA, USA, 2nd edition.

Cobb JN, Declerck G, Greenberg A, Clark R, McCouch S (2013) Next-generation phenotyping: requirements and strategies for enhancing our understanding of genotype-phenotype relationships and its relevance to crop improvement. *Theor Appl Genet (Theoretische und angewandte Genetik)* 126(4):867–887.

Colbert T, Till BJ, Tompa R et al. (2001) High-throughput screening for induced point mutations. *Plant Physiology* 126(2): 480–484.

Conrad LJ, Kikuchi K, Brutnell TP (2008) Transposon tagging in cereal crops. In: *The Handbook of Plant Functional Genomics*. Wiley-VCH Verlag GmbH & Co. KGaA, Weinheim, pp. 267–290.

Coolhaas C (1952) Large-scale use of F1 hybrids in "Vorstenlanden" tobacco. *Euphytica* 1(1):3–9.

Cuperus JT, Montgomery TA, Fahlgren N, Burke RT, Townsend T, Sullivan CM, Carrington JC (2010) Identification of MIR390a precursor processing-defective mutants in *Arabidopsis* by direct genome sequencing. *Proc Natl Acad Sci USA* 107(1): 466–471.

Darwin C (1859) *On the Origin of Species by Means of Natural Selection.* Vol. 9, J. Murray, London, UK, 1st edition.

Dierking EC and Bilyeu KD (2009) New sources of soybean seed meal and oil composition traits identified through TILLING. *BMC Plant Biology* 9: article 89.

Dong C, Dalton-Morgan C, Vincent K, and Sharp P (2009) A modified TILLING method for wheat breeding. *Plant Genetic* 2(1): 39–47.

Dong C, Vincent K, and Sharp P (2009) Simultaneous mutation detection of three homoeologous genes in wheat by high resolution melting analysis and mutation Surveyor. *BMC Plant Biology* 9: article 143.

Edwards JD, Janda J, Sweeney MT et al. (2008) Development and evaluation of a high-throughput, low-cost genotyping platform based on oligonucleotide microarrays in rice. *Plant Methods* 4(1): article 13.

Evans LT (1993) *Crop Evolution, Adaptation, and Yield.* Vol. 11, Cambridge University Press, New York, NY, USA.

FAO (2009) FAO's Director-General on how to feed the World in 2050. *Popul Dev Rev* 35 (4):837–839.

FAO-IAEA. 2011. *Mutant variety database.* http://mvgs.iaea.org/AboutMutantVarieties.aspx

Feiz L, Martin JM, and Giroux MJ (2009) Creation and functional analysis of new Puroindoline alleles in *Triticum aestivum. Theoretical and Applied Genetics* 118(2):247–257.

Fekih R, Takagi H, Tamiru M, Abe A, Natsume S, Yaegashi H, Sharma S, Sharma S, Kanzaki H, Matsumura H, Saitoh H, Mitsuoka C, Utsushi H, Uemura A, Kanzaki E, Kosugi S, Yoshida K, Cano L, Kamoun S, Terauchi R (2013) MutMap+: genetic mapping and mutant identification without crossing in rice. *PLoS One* 8(7):e68529.

Fiorani F, Schurr U (2013) Future scenarios for plant phenotyping. *Annu Rev. Plant Biol* 64:267–291.

Flusberg BA, Webster DR, Lee JH et al. (2010) Direct detection of DNA methylation during single-molecule, realtime sequencing. *Nature Methods* 7(6): 461–465.

Fu Y, Xu S, Pan C, Ye M, Zou H, and Guo B (2006) A matrix of 3,4-diaminobenzophenone for the analysis of oligonucleotides by matrix-assisted laser desorption/ionization time-of-flight mass spectrometry. *Nucleic Acids Research* 34(13): article e94.

Gady ALF, Hermans FWK, Van De Wal MHBJ, Van Loo EN, Visser RGF, and Bachem CWB (2009) Implementation of two high through-put techniques in a novel application: detecting point mutations in large EMS mutated plant populations. *Plant Methods* 5(1): article no. 13.

Gepts P (2002) A comparison between crop domestication, classical plant breeding, and genetic engineering. *Crop Science* 42 (6): 1780–1790.

Greene EA, Codomo CA, Taylor NE, Henikoff JG, Till BJ, Reynolds SH, Enns LC, Burtner C, Johnson JE, Odden AR, Comai L, Henikoff S (2003) Spectrum of chemically induced mutations from a large-scale reverse-genetic screen in *Arabidopsis. Genetics* 164(2):731–740.

Gundry CN, Vandersteen JG, Reed GH, Pryor RJ, Chen J, and Wittwer CT (2003) Amplicon melting analysis with labelled primers: a closed-tube method for differentiating homozygotes and heterozygotes. *Clinical Chemistry* 49(3): 396–406.

Henry IM, Nagalakshmi U, Lieberman MC, Ngo KJ, Krasileva KV, Vasquez-Gross H, Akhunova A, Akhunov E, Dubcovsky J, Tai TH, Comai L (2014) Efficient genome-wide detection and cataloging of EMS-induced mutations using exome capture and next-generation sequencing. *Plant Cell* 26:1382–1397.

Henry IM, Zinkgraf MS, Groover AT, Comai L (2015) A system for dosage-based functional genomics in poplar. *Plant Cell* 27(9):2370–2383.

Hillman GC and Davies MS (1990) Domestication rates in wildtype wheats and barley under primitive cultivation. *Biological Journal of the Linnean Society* 39(1): 39–78.

Hirochika H (2001) Contribution of the Tos17 retrotransposon to rice functional genomics. *Current Opinion in Plant Biology* 4(2):118–122.

Huang X, Feng Q, Qian Q et al. (2009) High-throughput genotyping by whole-genome resequencing. *Genome Research* 19(6): 1068–1076.

Hunter CT, Suzuki M, Saunders J, Wu S, Tasi A, McCarty DR, Koch KE (2014) Phenotype to genotype using forward-genetic Mu-seq for identification and functional classification of maize mutants. *Frontiers in Plant Science* 4:545.

Iiyama K and Wallis AFA (1988) An improved acetyl bromide procedure for determining lignin in woods and wood pulps. *Wood Science and Technology* 22(3): 271–280.

Ishikawa T, Kamei Y, Otozai S et al. (2010) High-resolution melting curve analysis for rapid detection of mutations in a Medaka TILLING library. *BMC Molecular Biology* 11: article 70.

Jankowicz-Cieslak J, Huynh OA, Bado S, Matijevic M, Till BJ (2011) Reverse-genetics by TILLING expands through the plant kingdom. *Emirates Journal of Food and Agriculture* 23(4): 290–300.

Jankowicz-Cieslak J, Till B (2015) Forward and Reverse Genetics in Crop Breeding. In: Jameel M. Al-Khayri, S. Mohan Jain, Dennis V. Johnson (eds) *Advances in plant breeding strategies: breeding, biotechnology and molecular tools.* vol 1, 1st edition. Springer, Switzerland, pp. 215–240.

Kharkwal MC, Shu QY (2009) *The role of induced mutations in World Food Security. Induced plant mutations in the genomics era.* Food and Agriculture Organization of the United Nations, Rome, Italy.

Knoll JE, Ramos ML, Zeng Y et al. (2011) TILLING for allergen reduction and improvement of quality traits in peanut (*Arachis hypogaea L.*). *BMC Plant Biology* 11: article 81.

Konzak CF (1957) Genetic effects of radiation on higher plants. *Quarterly Review of Biology* 32(1):27–45.

Kunter B, Bas M, Kantoglu Y, Burak M (2012) Mutation breeding of sweet cherry (*Prunus avium* L.) var. 0900 Ziraat. In: Shu Q, BP F, Nakagawa H (eds) *Plant Mutation Breeding and Biotechnology*. CAB International, Oxfordshire, pp. 453–463.

Kurowska M, Daszkowska-Golec A, Gruszka D, Marzec M, Szurman M, Szarejko I, Maluszynski M (2011) TILLING – a shortcut in functional genomics. *Journal of Applied Genetics* 52(4):371–390.

Lai J, Li R, Xu X et al. (2010) Genome-wide patterns of genetic variation among elite maize inbred lines. *Nature Genetics* 42(11): 1027–1030.

Lam HM, Xu X, Liu X et al. (2010) Resequencing of 31 wild and cultivated soybean genomes identifies patterns of genetic diversity and selection. *Nature Genetics* 42(12): 1053–1059.

Lee LS, Till B, Hill H, Huynh O, Jankowicz-Cieslak J (2014) Mutation and mutation screening. In: RJ H, Furtado A (eds) *Cereal Genomics, Methods in Molecular Biology, vol 1099.* Humana Press, Totowa, NJ, USA, pp. 77–95.

Leung H, Wu C, Baraoidan M et al. (2001) Deletion mutants for functional genomics: progress in phenotyping, sequence assignment, and database development. In Rice Genetics. Brar D, Hardy B and Khush G (eds). International Rice Research Institute 4: 239–251.

Li G, Chern M, Jain R, Martin JA, Schackwitz WS, Jiang L, Vega-Sanchez ME, Lipzen AM, Barry KW, Schmutz J, Ronald PC (2016) Genome-wide sequencing of 41 rice (*Oryza sativa L.*) mutated lines reveals diverse mutations induced by fast-neutron irradiation. *Molecular Plant* 9:1078–1081.

Mba C (2013) Induced mutations unleash the potentials of plant genetic resources for food and agriculture. *Agronomy* 3:200–231.

Mba C, Afza R, Bado S, Jain SH (2010) Induced mutagenesis in plants using physical and chemical agents. In: Davey MR, Anthony P (eds) *Plant cell culture: essential methods*. Wiley, Chichester, UK, pp. 111–130.

Mba CM, Afza R, Jankowcz-Cieslak J, Bado S, Matijevic M, Huynh O, Till BJ (2009) Enhancing genetic diversity through induced mutagenesis in vegetatively propagated plants. In: Shu QY (ed) *Induced plant mutations in the genomics era*. Food and Agriculture Organization of the United Nations, Rome, Italy, pp. 293–296.

McCallum CM, Comai L, Greene EA, and Henikoff S (2000a) Targeted screening for induced mutations. *Nature Biotechnology* 18(4): 455–457.

McCallum CM, Comai L, Greene EA and Henikoff S (2000b) Targeting induced local lesions IN genomes (TILLING) for plant functional genomics. *Plant Physiology* 123(2): 439–442.

McCleary BV and Codd R (1991) Measurement of (1 → 3), (1 →4)-β-D-glucan in barley and oats: a streamlined enzymic procedure. *Journal of the Science of Food and Agriculture* 55(2): 303–312.

McMullen M, Jones R, and Gallenberg D (1997) Scab of wheat and barley: a re-emerging disease of devastating impact. *Plant Disease* 81(12): 1340–1348.

Meeley RB, Briggs SP (1995) Reverse genetics for maize. *Maize Genetics Cooperation Newsletter* 69:67–82.

Mishra P, Padhan SR, Sahoo JP, et al. (2023) Molecular understanding of signaling compounds for optimizing cell signal transduction mechanism under abiotic stresses in crop plants. In: N.K. Gupta, Yuri Shavrukov, R.K. Singhal, Nikolai Borisjuk (eds). *Multiple Abiotic Stress Tolerances in Higher Plants*. CRC Press, Boca Raton, pp. 163–181.

Muller HJ (1927) Artificial transmutation of the gene. *Science* 66(1699): 84–87.

MVD (2016) Mutant variety database. http://mvd.iaea.org/.

Newhouse KE, Smith WA, Starrett MA, Schaefer TJ, Singh BK (1992) Tolerance to imidazolinone herbicides in wheat. *Plant Physiology* 100:882–886.

Panigrahi KK, Mohanty A, Padhan SR, Guru RKS (2021) Genotoxicity and DNA damage induced by herbicides and toxins in plants. In: Zeba Khan, Mohd Yunus Khalil Ansari, Durre Shahwar (eds). *Induced Genotoxicity and Oxidative Stress in Plants*. Springer Singapore, Singapore, pp. 29–63.

Panigrahi KK, Mohanty A, Padhan SR, et al. (2023) Recent advances in genome editing towards sustainable agriculture. In: Zeba Khan, Durre Shahwar, Yasmin Heikal (eds). *Genome Editing and Global Food Security*. Routledge, London, pp. 155–193.

Park SY, Chi MH, Junhyun J et al. (2007) High throughput phenotype screening pipeline for functional genomics in *Magnaporthe oryzae*. PROTOCOL (Version 1). Available at Protocol Exchange https://doi.org/10.1038/nprot.2007.168

Preuss SB, Britt AB (2003) A DNA-damage-induced cell cycle checkpoint in Arabidopsis. *Genetics* 164:323–334. https://doi.org/10.1093/genetics/164.1.323

Raghavan C, Naredo MEB, Wang H, et al. (2007) Rapid method for detecting SNPs on agarose gels and its application in candidate gene mapping. *Molecular Breeding* 19(2): 87–101.

Ray DK, Mueller ND, West PC, Foley JA (2013) Yield trends are insufficient to double global crop production by 2050. *PLoS One* 8:e66428. https://doi.org/10.1371/journal.pone.0066428

Reiter WD, Chapple C, and Somerville CR (1997) Mutants of *Arabidopsis thaliana* with altered cell wall polysaccharide composition. *Plant Journal* 12(2): 335–345.

Rigola D, van Oeveren J, Janssen A et al. (2009) High-throughput detection of induced mutations and natural variation using Key Point technology. *PLoS ONE* 4(3): article se4761.

Ronald P (2011) Plant genetics, sustainable agriculture and global food security. *Genetics* 188:11–20. https://doi.org/10.1534/genetics.111.128553

Schneeberger K, Ossowski S, Lanz C et al. (2009) SHOREmap: simultaneous mapping and mutation identification by deep sequencing. *Nature Methods* 6(8): 550–551.

Semagn K, Bjornstad A, and Ndjiondjop MN (2006) An overview of molecular marker methods for plants. *African Journal of Biotechnology* 5(25): 2540–2568.

Sestili F, Botticella E, Bedo Z, Phillips A, and Lafiandra D (2010) Production of novel allelic variation for genes involved in starch biosynthesis through mutagenesis. *Molecular Breeding* 25(1): 145–154.

Sigurbjoernsson B, Micke A (1974) Philosophy and accomplishments of mutation breeding. In: Polyploidy and induced mutations in plant breeding (Proceedings of two Meetings Joint 1 Mutagenesis for Crop Breeding and Functional Genomics 17 FAO/IAEA Division and EUCARPIA, Bari, Italy 1972). *International Atomic Energy Agency*, Vienna, pp. 303–343.

Sikora P, Tosh SM, Brummer Y, and Olsson O (2013) Identification of high β-glucan oat lines and localization and chemical characterization of their seed kernel β-glucans. *Food chemistry* 137(1–4): 83–91. https://doi.org/10.1016/j.foodchem.2012.10.007

Singer T, Fan Y, Chang HS, Zhu T, Hazen SP, and Briggs SP (2006) A high-resolution map of *Arabidopsis* recombinant inbred lines by whole-genome exon array hybridization. *PLoS Genetics* 2(9): article e144.

Slade AJ, Fuerstenberg SI, Loeffler D, Steine MN, and Facciotti D (2005) A reverse genetic, non-transgenic approach to wheat crop improvement by TILLING. *Nature Biotechnology* 23(1): 75–81.

Smartt J and Simmonds NW (1995) *Evolution of Crop Plants*. Longman Scientific and Technical, Harlow, UK; Wiley, New York, NY, USA, 2nd edition.

Spielmeyer W, Ellis MH, and Chandler PM (2002) Semidwarf (sd-1), "green revolution" rice, contains a defective gibberellin 20-oxidase gene. *Proceedings of the National Academy of Sciences of the United States of America* 99(13): 9043–9048.

Stadler LJ (1928a) Mutations in barley induced by X-rays and radium. *Science* 68(1756): 186–187.

Stadler LJ (1928b) Genetic effects of X-Rays in Maize. *Proceedings of the National Academy of Sciences of the United States of America* 14(1): 69–75.

Suzuki T, Eiguchi M, Kumamaru T, et al. (2008) MNU-induced mutant pools and high-performance TILLING enable finding of any gene mutation in rice. *Molecular Genetics and Genomics* 279(3): 213–223.

Talame V, Bovina R, Sanguineti MC, Tuberosa R, Lundqvist U, and Salvi S (2008) TILLMore, a resource for the discovery of chemically induced mutants in barley. *Plant Biotechnology Journal* 6(5): 477–485.

Tester M, Langridge P (2010) Breeding technologies to increase crop production in a changing world. Science 327:818–822. https://doi.org/10.1126/science.1183700

Till BJ, Cooper J, Tai TH et al. (2007). Discovery of chemically induced mutations in rice by TILLING. *BMC Plant Biology* 7: 19.

Till BJ, Reynolds SH, Weil C et al. (2004) Discovery of induced point mutations in maize genes by TILLING. *BMC Plant Biology* 4: 12.

Till BJ, Zerr T, Comai L, and Henikoff S (2006) A protocol for TILLING and Eco tilling in plants and animals. *Nature Protocols* 1(5): 2465–2477.

Tollenaar D (1934) Untersuchungen Über Mutation Bei Tabak. *Genetica* 16:111–152. https://doi.org/10.1007/bf01837860

Tsai H, Howell T, Nitcher R et al. (2011) Discovery of rare mutations in populations: tilling by sequencing. *Plant Physiology* 156(3): 1257–1268.

Uauy C, Paraiso F, Colasuonno P et al. (2009) A modified TILLING approach to detect induced mutations in tetraploid and hexaploid wheat. *BMC Plant Biology* 9: article 115.

Van Den Boom D, and Ehrich M (2007) Discovery and identification of sequence polymorphisms and mutations with MALDITOFMS. *Methods in Molecular Biology* 366: 287–306.

Van Harten AM (1998) *Mutation breeding: Theory and practical applications.* Cambridge University Press, Cambridge, England.

Vivekanand V, Chawade A, Larsson M, Larsson A, and Olsson O (2014) Identification and qualitative characterisation of high and low lignin lines from an oat TILLING population. *Industrial Crops and Products* 59:1–8. https://doi.org/10.1016/j.indcrop.2014.04.019

Wang T, Uauy C, Till B, and Liu CM (2010) TILLING and associated technologies. *Journal of Integrative Plant Biology* 52(11): 1027–1030.

Weigel D, Mott R (2009) The 1001 genomes project for Arabidopsis thaliana. *Genome Biology* 10:107. https://doi.org/10.1186/gb-2009-10-5-107

Westergaard M (1957) Chemical mutagenesis in relation to the concept of the gene. *Experientia* 13(6): 224–234.

Winkler S, Schwabedissen A, Backasch D, et al. (2005) Target selected mutant screen by TILLING in *Drosophila. Genome Research* 15(5): 718–723.

Wittwer CT, Reed GH, Gundry CN, Vandersteen JG, and Pryor RJ (2003) High-resolution genotyping by amplicon melting analysis using LCGreen. *Clinical Chemistry* 49(6–1): 853–860.

Wu JL, Wu C, Lei C, et al. (2005) Chemical- and irradiation induced mutants of indica rice IR64 for forward and reverse genetics. *Plant Molecular Biology* 59(1): 85–97.

Xin Z, Li Wang M, Barkley NA et al. (2008) Applying genotyping (TILLING) and phenotyping analyses to elucidate gene function in a chemically induced sorghum mutant population. *BMC Plant Biology* 8: article 103.

Yuan Y, Van Allen EM, Omberg L, et al (2014) Assessing the clinical utility of cancer genomic and proteomic data across tumor types. *Nature Biotechnology* 32:644–652. https://doi.org/10.1038/nbt.2940

2 Epigenetic Architecture of Complex Traits in Plants

Gurbachan S. Miglani, Maninder Kaur, Geetika Bhandari, and Gaganpreet Kaur*
School of Agricultural Biotechnology, Punjab Agricultural University, Ludhiana, Punjab, India
*Corresponding Email: miglanigs1945@gmail.com

2.1 INTRODUCTION

The journey from domestication to the modern era of epigenetics has tremendously revolutionized crop research. In the past, the dissection of polygenic phenotypic traits of agronomical, ecological, and economical interest employed principles of classical genetics (Cubillos et al., 2011). Its main purpose initially was to produce enough food for the increasing human population and to overcome the effects of industrialization as it caused reduction of crop land area (Kim et al., 2018). Although Green Revolution in the late 1960s brought tremendous change in crop yield, now, the science of "epigenetics", a term coined by Waddington (1942) and later broadened by several researchers, including Weinhold (2006), holds promise in the present era. Various epigenetic mechanisms include DNA modifications (mostly of cytosine methylation type), histone posttranslational modifications (PTLMs, e.g., acetylation, methylation, phosphorylation, ubiquitylation, ADP ribosylation SUMOlyation), and biogenesis of small noncoding RNAs (sncRNAs). The epigenetic mechanisms owing to regulatory flexibility lead to changes in phenotype without any change in the genotype (Weinhold, 2006; Allis and Jenuwein, 2016; Kumar et al., 2017; Li et al., 2017). The Greek prefix "epi" stands for "over", "above", "outside of", or "around" the classical genetic basis of inheritance (Kumar, 2017). The modification in the gene activity without alterations in the DNA sequence causes modifications that has the capability to transmit the epigenetic information to the daughter cells over many sexual generations (Weinhold, 2006; Johannes et al., 2008; Weigel and Colot, 2012).

Further, somatic epigenetics, referred to as the nonheritable epigenetic alteration in a gene, also does not rely on alteration in the sequence of DNA yet alters the gene expression levels without changing the amino acid sequence of the expressed protein. These somatic changes comprise DNA methylation, oxidative damage, and structural changes in chromatin DNA (Lahiri and Maloney, 2006; Mathers, 2006). Although we know the iceberg tip is genomics and single-nucleotide polymorphisms, its bottom is epigenetic mechanisms (Weinhold, 2006). Epigenetics for Darwinian

DOI: 10.1201/9781003392897-2

evolution revolves around the molecular processes that are independent of nucleotide sequence to control the genome activity and inherited to the next generation via mitosis or meiosis (Kumar and Singh, 2016; Cortijo et al., 2014a). Simultaneously, epigenetics is nondependent on *cis*- or *trans*-acting DNA sequence changes linked to heritable phenotypic variations (Johannes et al., 2008). The epigenetic signatures are the indication of the specific kind of regulation at the epigenetic locus (epilocus) (Schmitz et al., 2013).

Allis and Jenuwein (2016) have extensively reviewed the molecular basis of complex epigenetic processes. Kumar et al. (2017) and Springer and Schmitz (2017) have reviewed the history of epigenetic mechanisms and role of spontaneous and persuaded epigenetic variation and its regulation in crop improvement. Mounger et al. (2021) deciphered that the epigenetic processes form the basis of variation of trait within and among hostile populations. Processes like genomic shock, hybridization, and whole-genome duplication cause variation in reprogramming, gene expression, and epigenetic modifications contributing toward phenotypic variation. The epigenetics field has the potential to change the level of gene expression as it has also been reported that under nonstress conditions, epigenetic changes revert to the prestress state in short period of time (Kumar, 2017). Alterations in short interfering RNA and in levels of DNA methylation appear to be linked with heterosis and small noncoding RNAs (sncRNAs) that are specifically implicated in the establishment of DNA methylation by RNA-dependent DNA methylation (RdDM), conferring the role of epigenetics (Groszmann et al., 2011). Here we focus on the epigenetic architecture of complex traits of plants by dealing with the existing knowledge about epigenetic-linkage mapping and epigenetic quantitative trait loci (epi-QTL), founded on epigenetics and its role in crop improvement.

2.2 DNA METHYLATION AND HISTONE MODIFICATIONS IN REGULATION OF GENE EXPRESSION

The interplay between cytosine methylation and histone modifications marks the specificity of epigenetic phenomenon (Kumar et al., 2017; Springer and Schmitz, 2017). These epigenetic progressions play a part in controlling the gene expression in prokaryotes, single-cell eukaryotes, and multicell eukaryotes (humans, animals and plants).

2.2.1 DNA Methylation

The common epigenetic mechanism involves DNA methyltransferases (DNMTs) which selectively add methyl group to produce 5-methylcytosine ($5^{m}C$) (Kumar et al., 2017). The cytosine methylation follows a nonrandom pattern and it exists in the form of a bunch in particular sectors of the genome (Tran et al., 2005; Vaughn et al., 2007). Cytosine methylation has association with transposable elements and siRNAs (Becker et al., 2011). Mercé et al. (2020) have extensively reviewed the modulation of methylation patterns conferring epigenetics for its utilization in crop improvement.

DNA methylation has proven to be the best indicator contributing toward phenotypic variations (Lu et al., 2020). Plant has a potential mechanism of DNA methylation that allows them to survive with duplicated DNA (Schmitz et al., 2013). The entire epigenetic-based methylation mechanism depends upon the RdDM pathway recruiting small RNAs as a chaperon for the *de novo* DOMAINS REARRANGED METHYLTRANSFERASE 2 (DRM2) and maintenance DNA METHYLTRANSFERASE 1 (MET1) (Watson et al., 2014). The constancy of cytosine methylation with respect to environmental difference serves as a cue to explore role of epigenetics in crop improvement (Springer and Schmitz, 2017). The dense methylation at *met1* locus governs its stable transfer in the subsequent generations (Watson et al., 2014). DNA methylation of recently produced DNA strand occurs during DNA replication due to covalent modification of DNA using the methylation patterns of parental DNA strand as a template, in a faithful manner during cell division and its sites majorly include CG, CHG, and CHH (where H = A, C, or T) sequence settings throughout the plant genomes (Springer and Schmitz, 2017).

Cell cycle accounts for changes in the cytosine methylation dynamics during male gametogenesis that result in the induction of epigenetic variation in clonally propagated plants (Papareddy and Nodine, 2021). Along with methylation, demethylation in the promoter and coding area is also important for the activation of expression of particular genes under fluctuating environmental circumstances and plant developmental phases (Li et al., 2017). Thousands of hemimethylated and fully methylated positions are present in the same genotypes (Zou et al., 2017). The methylation pattern is heritable for about hundreds of generations and has a vital role in evolution and adaptation (Lu et al., 2020).

The differentially methylated regions (DMRs) undergo cosegregation with the genotype which proves it as an important avenue to pass the epigenetic marks through cell division (Schmitz et al., 2013). The detection of methylation "hotspots" linked with the trait of interest has the potential to provide deep insight into study of the purpose of plant genes and adaptation features of plant under diverse environments.

2.2.2 Histone Modifications

In mid-1960s, V.G. Allfrey and his group stated that the histone variants, histone modifications, chromatin remodeling, and regulatory RNAs (such as sncRNAs) are employed for epigenetic regulation (Allfrey et al., 1964). The histone proteins comprised numerous evolutionary conserved lysine (K), arginine ®, and serine (S) residues to carry out acetylation, methylation, and ubiquitylation processes (Millar and Grunstein, 2006). Histone PTLMs and chromatin architecture alterations affect cellular processes such as DNA replication, chromosome segregation, chromosome condensation, transcription, and DNA repair (Kumar et al., 2017). Ueda and Seki (2020) reveal that histone alterations are part of epigenetic regulatory networks for the regulation of abiotic stresses, viz., cold, heat, drought, dehydration, and salinity response.

2.2.3 Combined Effect of DNA Methylation and Histone Modifications

The histone modifications and cytosine methylation help in distinguishing euchromatin from heterochromatin on the basis of their patterns that confer gene activity or gene repression crucial for the regulation of gene expression activity (Allis and Jenuwein, 2016; Bird, 2002; Cedar and Bergman, 2009). Overall, the entire epigenetic mechanism revolves around cytosine methylation and histone modifications for controlling the phenotypic traits conferring gene expression in plant system. Zhao et al. (2021) observed that epigenetic regulation through cytosine methylation, histone modification, and chromatin remodeling along with epigenetic memory is involved in plant-heat interactions for the sustenance of growth and production of plants under harsh heat conditions.

2.3 VARIATIONS AMONG DNA METHYLATION PATTERNS

There are tremendous variations among the DNA methylation patterns within various populations. The DNA methylation-based epigenetic marks that undergo changes at several regions throughout the genome (Cortijo et al., 2014a) have a vital role in regulating gene expression (Suzuki and Bird, 2008), which affects phenotype, such as human disease (Watanabe and Maekawa, 2010) and plant development (King et al., 2010). The cytosine methylation variation has dynamic nature and there exists strong link between cytosine methylation and gene expression (Lu et al., 2020). The modification of methylation patterns has significant contribution toward variation in phenotypic traits through multiple gene expression. The variability in the methylation patterns in plant genomes serves as a supplement to account for the missing heritability.

DNA methylation exhibits varying patterns in retrotransposons (long terminal repeat and LINEs) and DNA transposons (terminal inverted repeats and helitrons) (Schmitz et al., 2013). The duplicated genes have ability to maintain methylation states in *Arabidopsis thaliana* (Widman et al., 2009; Schmitz et al., 2013). The cytosine methylation variation shows an association with the tissue and developmental stage-specific changes and transcriptional and posttranscriptional silencing of transgene in plants (Kumar, 2017). The first "epigenetic drug", 5-azacytidine, has the capability to stop cytosine methylation and thereby modify gene expression and phenotypes in fibroblast cell lines (Taylor and Jones, 1979; Allis and Jenuwein, 2016). In plants, John and Amasino (1989) provide the first evidence of DNA methylation observed from crown-gall tumor culture that showed the phenotypic differences in the regenerated plants. Methylation tends to block transcription factor binding to prevent transcription in *Arabidopsis* (Gehring and Henikoff, 2008). The pleiotropic effects contribute toward complexity of epigenetic variation based on molecular and morphological traits (Kooke et al., 2019). This epigenetic variation arises during developmental processes and stress exposures (Kumar, 2017). The epiallelic variation is responsible for the generation of metabolic difference in both quantitative and qualitative manner (Kooke et al., 2019).

2.3.1 Plasticity Behavior in Plants

The self-motivated changes in cytosine methylation, histone variants, small RNAs, chromatin, and transposable elements produce plasticity behavior in plants

(Bloomfield et al., 2014). Plasticity in plants is crucial for sustaining homeostasis under varying environmental conditions through epigenetic processes. The epigenetic regulation of developmental stages and changeovers plays an important role in crop establishment, growth, floral initiation, pollination, seed set, and ripening in plants through regulatory systems for conferring plasticity. These epigenetic fingerprints contribute toward the phenotypic variation through the control of gene transcription, developmental elasticity, and interactions with the environment (Long et al., 2011). Epigenetic variation causes phenotypic variation by virtue of phenotypic changes in plants (Boquete et al., 2021).

The epigenetic variation is responsible for phenotypic diversity in plants, followed by transmission to offspring for numerous generations, leading to epigenetic inheritance in the form of epialleles (Kooke et al., 2015; Kumar et al., 2017; Kooke et al., 2019). Environment-dependent chromatin modifications result in the development of epigenetic changes in gene function (El-Soda et al., 2014). The G × E interaction is essential in identifying the part of epigenetic differences in crop breeding through active selection and attention to maintain the germplasm to assure the epigenetic plasticity into new cultivars. The extensive G × E interactions linked with crop yield traits serve as a limiting factor in the identification of genes. The multienvironment trials for more than one year are required to explore the G × E interaction in QTL mapping (Kang, 1998; Yan and Kang, 2003; Jiang et al., 2020).

2.3.2 Stability of the Epigenetic Marks

For transgenerational stability, the epigenetic marks pass through germline for their transmission to the subsequent generations without being erased by any inherent mechanism, which is thus efficiently maintained (Kumar, 2017). More than 99% of methylated regions in DNA undergo inheritance across each generation in *A. thaliana* (Lu et al., 2020). Transgenerational inheritance is mainly acquired through environmental changes that employ all non-sequence-based effects which get transferred to the subsequent generations (Heard and Martienssen, 2014). The exploration of natural populations of undomesticated plant species increases the basic understanding of population epigenetics (Richards, 2006; Lu et al., 2020). Cytosine DNA methylation generally transfers information through cell divisions. It has been observed that rate of spontaneous transgenerational epiallelic variation is 10^3 times greater than the genetic mutation rate in *Arabidopsis* (Becker et al., 2011), with hypermethylated alleles connected with siRNA manufacture and transcriptional gene silencing (TGS) (Schmitz et al., 2013).

2.4 EPIALLELES

2.4.1 Transmission of Epialleles

Although epigenetic variation is based on the stability in the next generation, the epialleles are not always as stable as the primary DNA sequence (Becker et al., 2011). The sites of epiallelic conditions occur locally and distantly to other chromosomes

(Rangwala et al., 2006; Schmitz et al., 2013). Epialleles confer flexibility in transcriptional behavior for the generation of novel phenotypic variation. RdDM leads to the formation of random epialleles that have higher implications than genetic variation (Matzke and Mosher, 2014). About 97% of epialleles were transmitted through five meiosis using methylation-sensitive amplified polymorphism (MSAP) and retrotransposon (RT) epimarkers (Long et al., 2011).

The molecular approaches, viz., whole-genome, bisulfite, Roche sequencing, Chop PCR, microarray and northern blot hybridization, thermal asymmetric interlaced PCR (TAIL-PCR), quantitative reverse transcription PCR (qRT-PCR), and chromatin immunoprecipitation (CHIP) assay were used to detect the level of methylation, i.e., hypomethylation, hypermethylation, and differential methylation and conferred the detection of epiallele *lois lane* (Bondada et al., 2020), At2g10410 (Rangwala et al., 2006), R and S epialleles (Mittelsten Scheid et al., 2003), and NMR19-4 (He et al., 2018), respectively, in *A. thaliana* (Table 2.1). The consideration of epialleles as generation-dependent molecular phenotypes shows genome-wide map DNA sequence variations that have the ability to modulate the dynamic behavior across generations.

2.4.2 Inheritance of DNA Methylation Variants in RILs

Schmitz et al. (2013) performed epigenomic analysis of soybean RILs and deciphered that DMRs mostly cosegregate with the genotype and solved the theory of uncoupling in genotype and epigenotype. The epiallelic transgressive segregation also forms the basis of variation between parental lines and epigenetic recombinant inbred line (epi-RIL) studies (Johannes and Colomé-Tatché, 2011).

2.4.3 Natural and Induced Epigenetic Variation

The epigenetic difference comprises two types of epialleles, namely, natural and induced, conferring phenotypic traits (Springer and Schmitz, 2017). The natural epigenetic variation that resides in cytosine methylation patterns has the potential to alter gene expression through natural epialleles which can be utilized for crop improvement. The natural sources of epigenetic variation include spontaneous modifications, *cis* genetic modifications including transposon additions, structural relocations, and *trans* genetic changes including siRNA signals. The naturally occurring variation linked with DNA methylation has the potential to generate epiallelic variation which could be used in plant research and used as an additional source of variation within a species for crop improvement in future (Springer and Schmitz, 2017). Induced epigenetic variation includes mutations in the epi-RILs, chemical treatment with inhibitors of cytosine methylation and chromatin changes, directed epigenome editing, and tissue culture-based treatments.

The spontaneous or induced epigenetic variation-based epialleles have the capability to be stably inherited through mitosis (Springer and Schmitz, 2017). The natural and induced epialleles whose variation is based on the methylation patterns and not due to structural variation have the ability to impact the expression of neighboring

TABLE 2.1
Physiological Functions of Some Epialleles in Plant Thale Cress, *Arabidopsis thaliana* (L.) Heynh (Brassicaceae)

Plant material	Epimutations in target (gene or retrotransposons)	Epiallele (or its locus)	Level of methylation	Detection method	Functions	Reference
Diploid natural accessions CS 78917(Sorbo), CS78782(Cnt-1), CS77387(IP-Trs-0),CS77108(IP-Mos-1), CS77040(LIN S-5), CS76884(Got-22), CS76876(Geg-14), CS76616(Tscha-1), CS76573(Pla-0), CS76549(Me-0) **Tetraploid Wa-1 accessions** CS22644, CS28805, CS39006, CS6885, CS1586) and superman mutants: sup-5(CS3882), flo10-1/sup-2 (CS6225), clk-3(CS69095), cmt3-7(CS6365), kyp-2(CS6367), WT Col-0 and Ler stocks	Cadastral gene, *SUPERMAN* (*SUP*)	*lois lane* (*lol*)	Hypermethylation	Whole-genome sequencing and bisulfite sequencing analysis, Chop PCR	Silent to strong superwoman phenotypes affecting only the carpel whorl	Bondada et al. (2020)

(*Continued*)

TABLE 2.1 (Continued)
Physiological Functions of Some Epialleles in Plant Thale Cress, *Arabidopsis thaliana* (L.) Heynh (Brassicaceae)

Plant material	Epimutations in target (gene or retrotransposons)	Epiallele (or its locus)	Level of methylation	Detection method	Functions	Reference
Col, Ler, and Cvi accessions and mutant plants	Sadhu, a Novel *Arabidopsis* Retroposon	*Epiallelele locus At2g10410*	Hypomethylation	RNA extraction and microarray hybridization	Detection of transcript level polymorphism between three accessions	Rangwala et al. (2006)
Hygromycin-resistant line C *of A. thaliana* and the tetraploid wildtype	Hygromycin phosphotransferase transgene locus	*R and S epialleles*	Hypomethylated in C4R and hypermethylated and C4S lines	Northern hybridization, TAIL-PCR and Roche sequencing	Conferring difference in diploid and tetraploid species	Mittelsten Scheid et al. (2003)
Mutant lines from C24 background, pph-1 (SALK_ 000095) from Col-0 background	Naturally occurring DNA methylation variation region 19 (NMR19)	NMR19-4	Differential methylation	qRT-PCR, northern blot hybridization, CHIP assay	Leaf senescence and local climate adaptation	He et al. (2018)

genes (Becker et al., 2011). Three major classes of epialleles on the basis of underlying genetic variants are obligate, pure, and facilitated (Richards, 2006; Schmitz et al., 2013). The obligate epialleles are completely dependent on a genetic variant and pure epialleles are maintained independently of genetic variants, while the facilitated epialleles dependence breaks down because the genetic variant is under the influence of the epiallelic state (Richards, 2006; Schmitz et al., 2013).

2.4.4 Exploring the Influence of Epigenetic Modifications on Phenotype

The epigenome-wide association study has proved to be useful in exploring the influence of epigenetic modifications on phenotype on the basis of identification of genome-wide epimarkers. The preliminary criterion is the selection of plants with desired trait of interest, controlling phenotype based on the adaptability under extreme environmental conditions. It comprises three major steps (Gahlaut et al., 2020) (Figure 2.1). (A) The foremost step is the phenotyping of a population with respect to the target gene of interest in diverse germplasm. (B) The precise phenotyping leads to the identification of genome-wide methylation that delineates the genetic basis underlying the phenotypic variation among the genotypes understudy. (C) The last step includes the identification of methylated and unmethylated cytosine loci among different genotypes which interprets the epigenetic polymorphism. The combined approach of identifying epipolymorphisms with phenotypic variation through statistical analysis confers epigenotyping that leads to the identification of genome-wide methylation among number of targeted genotypes. The characteristic feature of stable inheritance of epigenetic scars determines the epigenetic diversity among different genotypes of the same population. This leads to working out the association of epialleles with phenotypic traits in the population, which finally results in the identification of epi-QTLs.

2.4.5 Phenotypic Variation: A Complex Case

QTL analysis based on phenotypic variation rather than QTL mapping based on DNA sequence or chromatin information holds strength (Johannes et al., 2008). The initial QTL scan at time point (T = 0) instituted on DNA or chromatin markers does not detect any QTL at any of the two loci (L1 or L2). This is due to the occurrence of "suppresses" or "buffers" for differential gene expression at nonpolymorphic epigenomic loci (epi-loci) that halt the DNA sequence variation. Later on, the environmental trepidation at T = 0 affects L1 (T = 1) and a sequence-dependent (*cis*) chromatin modification occurs in the next generation due to protected sequence variation release. At T = 1, QTL analysis deciphers $phQTL^{dna}$ and $phQTL^{epi}$ linked with chromatin markers. The hidden sequence variants improve the gain in phenotypic differences among populations. At T = 2, the generation undergoes attenuation and locus L1 triggers induction of chromatin changes at locus L2 (*trans*). The combined approach of chromatin level data in quantitative genetic investigations has the potential to provide scrutiny to the complex phenotypes in populations over time.

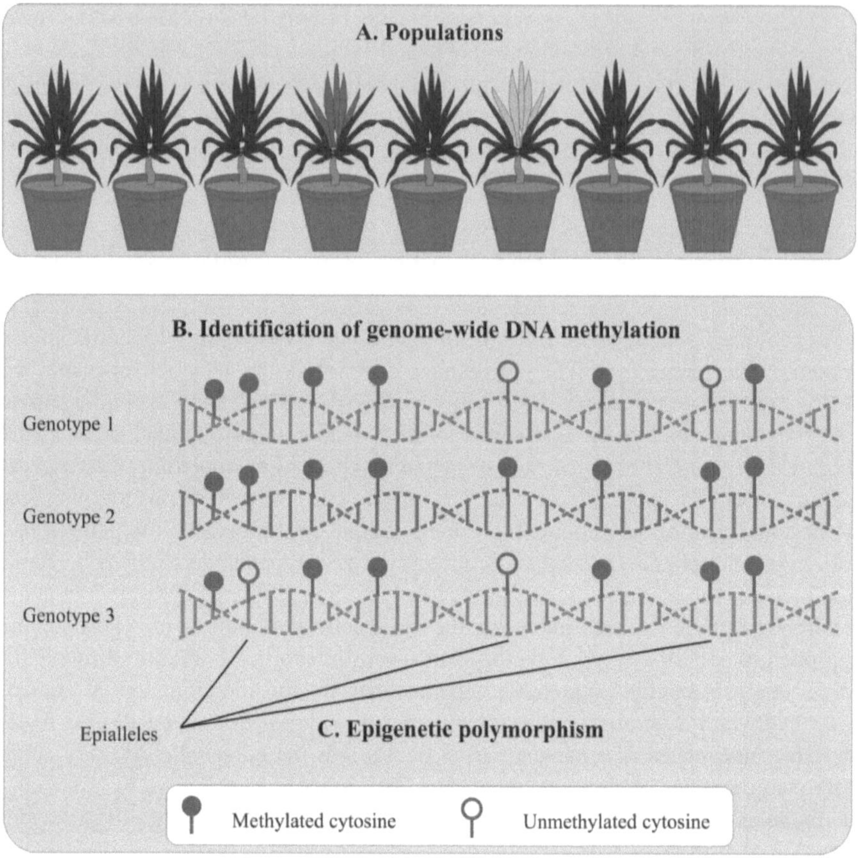

FIGURE 2.1 Identification of epialleles based on cytosine methylation represents epigenetic polymorphism among different genotypes of the population through epigenome-wide association study.

Source: Adapted from Gahlaut et al., 2020, published under Creative Commons Attribution License.

2.4.6 Understanding Contribution of Natural Epialleles toward Phenotypic Variation

The plants possess dozens of natural epialleles contributing toward phenotypic variation (Weigel and Colot, 2012). The self-reinforcing loop between CHROMOMETHYLASE 3 (CMT3) and KRYPTONITE involves cytosine methylation in the CHG context and histone H3 lysine 9 methylation (H3K9me) maintains the natural epiallele in *Arabidopsis* (Agorio et al., 2017) and it mainly depends upon the CG dinucleotide symmetry (Springer and Schmitz, 2017). The genome-wide cytosine methylation studies among ten *A. thaliana* lines suggest that normal epigenetic variation exists in *A. thaliana* methylome and report about 30,000 cytosines that undergo methylation (Becker et al., 2011).

2.4.7 Environmental Effect

The strong genetic gain in rice grain production under varying environmental conditions is achievable through the utilization of *Os*SPL14WFP epiallele (Kim et al., 2018). Sun et al. (2021) reported that chromatin, histone, and DNA changes based on epigenetic processes enable the plant to survive under stressful environments to cope up with the stress in future through the establishment of plant stress memory at the intergenerational and transgenerational levels. Epigenetics has the potential to delineate the G × E interactions that lead to the generation of different phenotypes (Liu et al., 2008). The impact of chromatin modification and DNA methylation depends upon the environmental changes that cause epigenetic modifications in gene function (El-Soda et al., 2014). The QTL analysis on epigenetic RILs performed under different environments reveals their power to give scrutiny to the underlying cytosine methylated loci linked with any detected variation and significant heritable variation.

2.4.8 Coevolution of Epigenetic and Genetic Variation

Natural epigenetic and genetic variations coevolve (Schmitz et al., 2013). The inheritance of widely held cytosine methylation variants is linked to Mendelian modes of transmission but some epigenetic variations do not obey the Mendelian principles of inheritance. The major potentials of epialleles include (a) determination of epigenetic variation; (b) effect of epimarks on phenotype; and (c) the inheritance pattern of epigenetic mark-linked superior phenotypes (Kumar, 2017).

2.5 CHANGES IN DNA METHYLATION STATUS MAY ACCOUNT FOR QTL

The majority of agronomically important plant characteristics display quantitative inheritance with low penetrance, exemplified through the research conducted in rapeseed under 10 natural environments, that have resulted in the documentation of hundreds of significant environment-specific QTL linked with yield and supplementary phenotypes in a "fixed" segregating population of homozygous lines (Shi et al., 2009). The epi-RILs are advantageous because cytosine methylation design variations in epi-RILs are constant across the generations (Johannes et al., 2009). However, the association of QTL with cytosine methylation status exhibits 60–90% heritability in terms of flowering time and primary root length in epi-RILs of *Arabidopsis* (Cortijo et al., 2014b). The whole-genome bisulfite sequencing and linkage mapping of methylation states uncovered local methyl-QTL.

2.5.1 Epigenetic Mapping versus Linkage Disequilibrium

Epigenetic mapping is better than linkage disequilibrium mapping (gap), conferring the causative genes contributing toward phenotypic variation. Epigenetics mapping has the potential to disentangle the intricate genetic and epigenetic regulatory system of contender genes governing multifaceted characters (Lu et al., 2020). The epi-QTL studies reveal that contender genes have association with multiple single-nucleotide polymorphisms (SNPs) which shows that the expression of these candidate genes is controlled by complex genetic mechanisms. The combinations of alleles for the production of desired traits form the basis of plant breeding. The knowledge of epialleles might prove to be helpful in plant breeding (Springer and Schmitz, 2017). Thus, stable and newly formed epigenetic variants over generations have significant utility for crop improvement.

2.5.1.1 Epigenetics Compared with Artificial Selection

Humans have exploited artificial selection in plants for a long time, which has had positive impact on various plant traits to shape the plants for better growth and performance under varying environments (Springer and Schmitz, 2017). Epigenetics is less time consuming as compared to artificial selection. The methylated regions have the potential to map genomic regions controlling the trait of interest. The epigenetics associated with the trait provides avenues for the identification of defective clones using DNA methylation profiling. The perturbation of DNA methylomes of crop plants results in the creation of fresh epiallelic differences in plant species (Springer and Schmitz, 2017). The first site-specific epigenome editing in plants was performed in *A. thaliana* by means of a zinc-finger nuclease (ZFN) fused to SUVH9 protein, which is integral to RdDM (Johnson et al., 2014). Different approaches employed for epigenome editing have been discussed (Miglani and Singh, 2020). Epigenome editing modifications induced in different plant species have also been reviewed (Miglani et al., 2020). Due to the stability of epigenetic variants in epigenetics, it would be readily acceptable for the stable improvement of agronomic traits.

2.5.1.2 Linkage Maps Based on Cytosine Methylation

First step to explore the epigenetic variation is the determination of the chromosomal distribution of $5^{m}C$ which is possible through the building of methylation-based linkage maps (Long et al., 2011). The cytosine methylation depicting linkage maps show polymorphism on the basis of DNA methylation patterns rather than variation among DNA sequences, conferring epigenetics phenomenon.

2.5.2 E-maps and epi-QTL

With the prevailing biotechnology-based advanced tools, researchers nowadays construct the epigenetic maps (E-maps) and detection of epi-QTL to target the trait of interest that confer scrutiny in the study of complex traits. The E-maps have been prepared through SSR and MSAP that have the potential to efficiently detect the DNA methylation polymorphisms to construct the methylation-based linkage maps in *Sorghum bicolor* (Duan et al., 2009) and *Brassica napus* (Long et al., 2011), respectively (Table 2.2). Zou et al. (2017) employed the amplified fragment single-nucleotide polymorphism and methylation (AFSM) marker to generate the highly dense cassava genetic linkage map with 18 linkage groups. Lu et al. (2020) constructed E-maps with a total of 19 major linkage groups (>50 cM) and 2463 MSAP markers that constitute the overall framework for the linkage E-map in *Populus* species.

TABLE 2.2
Construction of Methylation-based Linkage Maps in Agricultural Crops

Scientific name (common name)	Family	Population	Mapping Method	Linkage groups	Total map coverage (cM)	Markers	Reference
Brassica napus L. [Rapeseed]	Brassicaceae	BnTNDH mapping population	JoinMap 4.0	19	296.0	MSAP, retrotransposon-specific, and centromeric-specific markers	Long et al. (2011)
Manihot esculenta Crantz [Cassava]	Euphorbiaceae	KS population derived from crossing two noninbred lines with female parent KU50 and male parent SC124	JoinMap 4.1	18	2190.3	AFSM markers	Zou et al. (2017)
Populus alba, *P. glandulosa*, *P. tomentosa* [Poplar]	Salicaceae	Full-sib family population of interspecific hybridization between (*Populus alba* × *P. glandulosa*) with YX01 as the female and *P. tomentosa* LM50 as the male parent	JoinMap 4.1	19	2426.7	MSAP markers	Lu et al. (2019)
Sorghum bicolor (L.) [Indian millet]	Poaceae	F_2 population derived from a cross of B_2V_4 × 1383-2	Map maker/ EXP 3.0	11	483.6	MSAP and SSR	Duan et al. (2009)

Abbreviations: KS, two noninbred lines comprising female parent KU50 and male parent SC124; MSAP, methylation-sensitive amplified polymorphism; AFSM, amplified fragment single-nucleotide polymorphism and methylation; SSR, simple sequence repeats.

Methylome maps can delineate the methylation status in the coding and non-coding sections of genes, intergenic areas, transposable elements (TEs), and repetitive sequences to decipher the epigenetic mechanism for improvement in the crop species (Kumar, 2017). To date, a number of plant researchers have mapped QTL in the nuclear genome of crop plants. In contrast, investigating and mapping the epigenetic basis of composite traits is a tedious task (Cortijo et al., 2014a). The epi-QTL has a strong association with the phenotypic traits in *A. thaliana*, which has the potential to dissect complex traits (Lu et al., 2020). Therefore, the epi-QTL mapping methods hold promise to offer a better resolution to the genetic alternates conferring broad diversity in plant systems (Zou et al., 2017). The major target of epi-QTL studies is to map the DNA methylation sites within chromosomes for the detection of trait of interest (Long et al., 2011).

2.5.3 Epi-QTL Mapping in Crops

The epi-QTL mapping has the potential to detect the effect of epigenetic loci linked with the phenotypic variations to target trait of interest (Lu et al., 2020). It has vital role in unraveling the composite genetic and epigenetic regulatory system of contender genes that control the intricate characteristics. Various studies that are based on the implementation of epi-QTL mapping in crop species are compiled in Table 2.3. Cortijo et al. (2014b) identified DMRs that function as epi-QTL and are also responsible for 60–90% of the heritability for flowering time and primary root length traits in the epi-RIL populations of isogenic *Arabidopsis* lines. Moreover, these epi-QTLs are reproducible in nature that segregate according to differences in DNA methylation and are autonomous of changes in the DNA sequences. Long et al. (2011) employed MSAP and retrotransposon epimarkers and identified epi-QTLs linked with various agronomic traits, viz., yield, seed quality, and plant development, in *B. napus*.

Zou et al. (2017) detected candidate epi-QTL responsible for cold stress and storage root quality and yield in the cassava population. Around 318 methylated QTL genes showed significant association with cold tolerance in cassava crop species. Kooke et al. (2019) worked on *A. thaliana* epi-RILs to identify the effect of epigenetic variation due to the action of small RNAs. They identified 30 epi-QTL associated with plant growth and morphology. Lu et al. (2020) utilized linkage-linkage disequilibrium approach and identified epi-QTLs with MSAP markers that provide explanation for 1.7–44.5% of phenotypic differences along with 23 causal genes which participated in growth regulation and wood formation in *Populus* species. The epi-RILs are identical with respect to the DNA sequences and exhibit differences in methylation patterns due to the recombination-based difference(s) in the epigenetic mark(s) (Kumar et al., 2017). The variation in the epi-QTL interval has the potential to affect small RNAs followed by the alteration in the gene regulation mechanism (Kooke et al., 2019). Johannes et al. (2008) showed QTL mapping framework as the most useful approach to explain the epigenetic phenomenon over multigenerational data among the population.

TABLE 2.3
Epigenetic QTL Mapping in Crop Species

Plant	Family	Plant material	QTL method	Markers	Association of epi-QTL with trait of interest	Reference
Arabidopsis thaliana (L.) Heynh [Thale cress]	Brassicaceae	Epi-RILs	R/QTL	Methylation markers	Flowering time and primary root length	Cortijo et al. (2014b)
					Plant growth and morphology	Kooke et al. (2019)
Brassica napus [Rapeseed]	Brassicaceae	BnTNDH mapping population	WinQTLCart 2.5	Methylation-sensitive AFLP and retrotransposon epimarkers	Agronomic traits associated with yield, seed quality, and plant development	Long et al. (2011)
Manihot esculenta (Crantz) [Cassava]	Euphorbiaceae	KS population derived from crossing two noninbred lines with female parent KU50 and male parent SC124	MapQTL 5.0	Amplified fragment single nucleotide polymorphism and methylation markers	Cold stress, storage root quality and yield	Zou et al. (2017)
Populus alba, P. glandulosa, P. tomentosa [Poplar]	Salicaceae	Full-sib family population of interspecific hybridization between (*Populus alba* × *P. glandulosa*) with YX01 as the female and *P. tomentosa* LM50 as the male parent	MAPQTL 6.0	Methylation-sensitive amplification polymorphism markers	Plant growth and wood properties	Lu et al. (2019)

2.6 EPIGENETIC VARIATION CONTRIBUTES TO COMPLEX TRAITS

Epigenetic variation donates to intricate traits through the establishment of firm inheritance of compound traits such as flowering time as investigated in epi-RIL populations. The presence of two differing genomes and epigenomes in the same nucleus has the perspective toward alteration in gene expression and the trait of an organism (Greaves et al., 2012). Epigenome is a conceivable donor to heterosis, majorly in intraspecies hybrids with increased vigor and limited genetic diversity. The interaction between the epigenomes aided by the two parental sets of chromosomes leads to modified gene expression in F_1 hybrids (Greaves et al., 2012). This method has the capability to discover the epigenetic features of the chromatin inheritance system to study the epigenome associated with sequence-dependent and sequence-independent areas of difference and constancy. The epigenetic marks, such as cytosine methylation, have the potential to regulate gene transcription, developmental plasticity, and environment interactions contributing toward phenotypic variation. Moreover, the plant genomes exhibit wide variety of target sites and cytosine methylation regions rather than the animal genomes.

2.7 CHROMATIN THREE-DIMENSIONAL INTERACTIONS IN GENE EXPRESSION

Classically, DNA sequence variations hold the source of heritable traits and epigenetic variation in usual and investigational populations. Johannes et al. (2008) showed an association between epigenetics and complex traits which demonstrated the dynamic relationship between chromatin and DNA sequence features among complex traits linked with epigenetics. The natural genetic variation controls numerous molecular phenotypes related to trait of interest among individuals. The expression-based QTLs (eQTLs) for cell types, tissues, and conditions interpret noncoding genetic variations delineating complex organismal traits and diseases. Delaneau et al. (2019) have explained the genomic differences in proximal (*cis*) or distal (*trans*) gene regulation based on the study of gene expression, chromatin, and the three-dimensional genome conformation analysis. The grouping and activity of regulatory elements across individuals have revealed two genomic structures, viz., *cis*-regulatory domains (CRDs) and *trans*-regulatory hubs (TRHs), responsible for triggering gene expression. This has the potential to link the noncoding genetic difference and gene expression within and across the chromosomes.

Delaneau et al. (2019) have also characterized relationship between genetic variation, regulatory elements (REs), and gene expression to discover *cis*- and *trans*-regulatory harmonization. The population-scale dataset combines the activity of REs for monomethylated histone 3 at lysine 4 (H3K4me1), trimethylated histone 3 at lysine 27 (H3K4me3), and acetylated H3 at lysine 27 (H3K27ac). Following are the salient features of this study: (1) the well-delimited CRDs are involved in converting local chromatin structure into topologically associating domains (TADs) with higher organization. The consistency with a higher-order chromatin organization has been categorized into A and B nuclear compartments. The allelic coordination signal makes intricate and higher-order 3D nucleus organization interpreting that *trans* associations are not transcriptionally mediated. (2) CRDs and TRHs based on active REs offer a dense genome-wide map that shows association between REs and genes. The varying links are based on *cis* and *trans* coexpression of genes across different cell types. (3) The chromatin peaks discovery influenced by nearby genetic alternates and affects the activity of CRDs (aCRD-QTLs) and QTLs that ultimately impact the correlation within CRDs (sCRD-QTLs). The QTLs present near the genomic targets are mostly enriched within functional sections of the genome and show the overlap of genetic alternates and intricate traits which clearly display that CRDs are under strong genetic control. (4) The capture methodology of CRDs and TRHs affecting eQTLs affect gene expression. Overall, about three-fourths of the eQTLs influence the CRD activity and pronounce four specific types of genetic effects, viz., *cis*-eQTLs influencing distal genes, multiple *cis*-eQTLs linked with independent properties, multiple sporadic alternatives associated with cumulative effect, and *trans*-eQTLs. Delaneau et al. (2019) thus constructed the genome-wide map of the coordination between REs and noncoding genetic influences in *cis* and *trans* coexpression states that explained the complexity and specificity in terms of gene expression.

2.8 QUANTITATIVE MODELS TO PREDICT THE GENE EXPRESSION LEVEL

Karlić et al. (2010) derived quantitative models and recognized significant correlation between histone modification and gene expression levels. The histone modifications forecast gene expression motivated by high CG content promoters (HCPs) or low CG content promoters (LCPs). Likewise, the quantitative models linking H3K4me3 and H3K79me1 determine the gene expression levels in LCPs, whereas HCPs necessitate H3K27ac and H4K20me1 to trigger the gene expression. Overall, the quantitative models have determined the association of histone modifications and gene expression to predict the gene expression levels of a cell type.

The sessile autotrophs, i.e., monocotyledonous and dicotyledonous angiosperm species, are devoid of the internal thermoregulation of hot-blooded animals. The current breeding programs significantly enhance the yield and relative climate stability. The high global temperatures with increased amplitude and temporal change of climate parameters are responsible for global crop yield variability (Porter and Semenov, 2005; Ray et al., 2015). The crop plants overpower the thermal environments on diurnal and seasonal basis through plasticity behavior to regulate the epigenetic phenomenon and intricate genomic regulatory systems. Moreover, the plant cell compartmentalization shields the genomic central processing unit within the interphase nucleus in the plant system.

G • E interactions determine the phenotypic plasticity to maintain the crop performance, yield, and quality under variable environments (Bloomfield, 2014). The micronutrient potassium (K) triggers the metabolic adjustment to attain better yield with respect to salinity and cold and drought stresses. Generally, plant productivity is based on numerous factors including trivial oscillations in ambient temperature, as low as ±1°C (Porter and Semenov, 2005; Hüve et al., 2006; Blanchard and Runkle,

2011), and flexible retorts to maintain the plant-specific developmental stages (Muchow et al., 1990; McMaster and Wilhelm, 1997). Thus thermal environment fluctuations influence genetic processes that ultimately tend to affect crop yield and superiority (Craufurd and Wheeler, 2009; Bloomfield et al., 2014; Ray et al., 2015).

Contrastingly, thermal and ionic indications and conservative genetic models mediate crop traits and developmental phase fluctuations to discover the signal transduction paths and behavior (Hammond et al., 2011; McClung and Davis, 2010). The chromatin dynamics and epigenetic marks mediate plant thermosensory signaling (King, 2015). DNA, histones, nucleosomes, and chromatin changes depend upon the epigenetic marks and localized nuclear ionic situation. The exterior temperature and salt depend upon cytosine methylation and H2A standing. Likewise, the internal cation foci affect chromatin compression with intricate electrostatic contacts like histone alterations and divalent cations. The involvement of histone acetylases (HATs) and histone deacetylases (HDACs) in H4 and H3 acetylation form the complex chromatin code. The DNA methylation reversible changes influence nuclear chromatin and manage the difference in external ionic and temperature environments. The in-depth knowledge of epigenetic mechanism through the association of molecular mechanisms and G × E interactions trigger long-term improvement of crop performance.

2.9 HERITABLE BASIS OF COMPLEX TRAITS

Nowadays, epigenetic differences in natural and experimental populations display the dynamic relationship between chromatin and DNA sequence influences in composite characters (Kakutani et al., 1999; Riddle and Richards, 2002, 2005; Vaughn et al., 2007; Johannes et al., 2008; Zhang et al., 2008). The chromatin difference in experimental and natural populations serves as an added basis for phenotypic differences. Indeed, chromatin differences independent of DNA sequence polymorphisms have the potential to transmit information crosswise mitosis and meiosis in both mammals and plants at the cell, tissue, and whole organism level (Martienssen and Colot, 2001; Peaston and Whitelaw, 2006; Henderson and Jacobsen, 2007).

The stable alternative chromatin forms available (epialleles) crosswise generations are functionally inseparable from DNA sequence alleles at the population level. The stably transmitted epialleles are widely distributed among the plant system (Das and Messing, 1994; Soppe et al., 2000; Stam et al., 2002). Moreover, the meiotic transmission of epialleles shows intriguing patterns of instability (Rakyan et al., 2002), induced by genomic and environmental perturbations (Richards et al., 2006). The newly acquired epiallelic forms undergo stability and reversion across one or several generations that are dependent or independent of the nucleotide sequences (Richards et al., 2006; Penterman et al., 2007; Mathieu et al., 2007). The dynamic features are responsible for the creation and sustenance of composite traits in populations. The genome-wide portrayal of the temporal properties shows that the transgenerational association between DNA sequence alleles and epialleles leads to the generation of heritable phenotypes. Both transgenerational contexts (meiotic experiments) and intragenerational contexts (mitotic experiments) show that chromatin and DNA sequence data delineate the dynamic architecture of complex traits.

2.10 EXPERIMENTAL STRATEGIES FOR ANALYSIS OF COMPLEX TRAITS

The in-depth knowledge of multiple epigenomic mechanisms conferring expression variation might be useful for future studies in plant research (Schmitz et al., 2013). Moreover, plant epigenomes are highly diverse and dynamic responding to numerous biotic and abiotic stresses that initiate adaptive evolution among plant species. The epigenomics approach suggests complex genetic population patterns and heritability of methylation variants over multiple generations efficient to understand the methylation variants contributing toward phenotypic variation.

2.10.1 Single-Gene Perturbations

High-resolution genome-wide technologies have been widely employed to prepare genome-wide chromatin maps (Zilberman and Henikoff, 2007; Schones and Zhao, 2008; Suzuki and Bird, 2008). The epigenomic landscapes have been explored in several species, viz., *A. thaliana*, *Zea mays*, and human cells (Weber et al., 2005; Zhang et al., 2006; Penterman et al., 2007; Cokus et al., 2008). Most common assays include chromatin marks, including DNA methylation and numerous PTLMs of histone proteins. The forward genetic strategies attribute the chromatin difference to DNA sequence difference in investigational inhabitants. The role of a single gene or a small set of genes displays variability to reveal the epigenome consequences. The comparative analysis of genome-wide cytosine methylation contours of wild-type *A. thaliana* plants with mutants lacking genes shows the importance of cytosine methylation at several hundred genes positioned in *trans* to the acclimatizing loci (Zhang et al., 2006; Penterman et al., 2007; Cokus et al., 2008). Moreover, the impartial search for genetic loci impacting cytosine methylation and chromatin alterations has certain limitations, including requirement of the sequential trepidation of all genes and combinations trailed by a global epigenomic evaluation of each agitation (Mager and Bartolomei, 2005).

2.10.2 Multigenic Perturbations

The investigational populations resulting from crosses of diverse inbred parents are a powerful alternative to provide in-depth knowledge of epigenetic phenomenon. Zhang et al. (2008) revealed the transgenerational inheritance of cytosine methylation statuses through genome-wide cytosine methylation contours of two *A. thaliana* normal accessions to their first filial (F_1) descendants. The epialleles undergo stable transmission among plant populations (Riddle and Richards, 2002, 2005; Vaughn et al., 2007). The combinatorial approach of creation of more advanced crosses (e.g., an F_2 generation or RILs) joined with QTL mapping procedures has revealed the stability factor through sequence-independent chromatin inheritance (epigenetics) and *cis*- or *trans*-acting DNA-based features. Riddle and Richards (2002), working on RILs produced from two *A. thaliana* parents, showed cytosine methylation difference in nucleolus organizer regions (NORs) positioned at the tips of chromosomes 2 and 4. The interindividual variances in cytosine methylation in NORs have shown that one-half of the differences in the population were associated with QTLs connected to indicators in *cis*, whereas one-fifth show the same by *trans*-acting loci. The portion of the *cis* consequence has been attributed to the inheritance of parental DNA methylation frameworks that show the regional chromatin variation to resolve the self-governing as well as *cis* or *trans* DNA sequence-dependent mechanisms.

2.10.3 A Universal and Transgenerational Approach

Global and transgenerational approaches have been used to perform analysis at three levels: genome-wide sketching of DNA sequence, chromatin difference, and higher-level phenotypic evidence (Johannes et al., 2008). The connection of all these levels using genetically informative pedigrees has explained the connection between DNA sequence alleles and epialleles in a locus-by-locus way (both *cis* and *trans*) that shows its effect on transmitted phenotypes. Samples from investigational populations (such as RILs or F_2 offspring), natural ones (for example, humans), and pedigrees show diverse inheritance patterns. The RIL populations serve as remarkable experimental starting material to explain the epigenetic phenomenon. The cross between two diploid parents with different DNA sequence and chromatin contours lead to the generation of F_1 population. Likewise, brother-sister mating (selfing) produces F_2 population. Each F_2 individual is the consequence of an inbreeding process for multiple generations to generate RILs. After many such generations, the fully homozygous lines show only one haplotype. Among two epi-loci, one (L1) obeys while the other (L2) disobeys Mendelian inheritance due to metastability factor. The genome-wide profiling of each RIL by means of tiling array-based skill through chromatin and phenotypic deviations shows either meiotic or mitotic propagation. The nucleotide sequence has shown identical pattern and also the presence of both constant and metastable epialleles in both mitotic and meiotic cases. The RIL population explains the epiallelic changes over generational time under different DNA backgrounds. The DNA sequence stability and delicate chromatin subtleties across generations in a meiotic experiment is evaluated against the fixed genotypic background in a systematic manner, and across a range of diverse DNA backgrounds at the population level. This study thus has wide application in cell differentiation and cancer research. Thus, it has been evident that the time-dependent landscape of chromatin constancy is vital for mitotic rather than meiotic transitions.

2.10.4 From a Static to a Dynamic View

2.10.4.1 A Static View

Johannes et al. (2008) have deciphered the heritable source of phenotypic difference through two theoretical populations 1 and 2, conferring DNA sequence variation and epigenetic variation, respectively. The $phQTL^{dna}$ has been identified through sequence-based QTL analysis among population 1 at locus L1. Three DNA sequence variants exhibit higher phenotypic values in comparison with other three

DNA variants. Likewise, $phQTL^{epi}$ has also been identified from the chromatin-based QTL analysis of population 2 at locus L1, thereby displaying epigenetic variants with higher phenotypic values among other variants. Although both approaches hold true, incomplete propositions about the manner underlying the phenotypic variation in the single population show the cooccurrence of two sources of variation. Moreover, the biological model also displays the link between phenotypic variation and heritable basis.

Classically, quantitative genetics does not take into account epigenetic variation and traditional models establish the association of phenotypic variation to genotypic variation. The current QTL approaches, including linkage and association mapping, lead to the identification of stable DNA sequence variants displaying phenotypic variation. Alternative relationship of phenotypic variation to epigenetic variation generated through quantitative genetic models has shown that experimental populations are isogenic with indistinguishable DNA sequences but even so segregating epigenetic alternates (Kakutani et al., 1999). The considerable molecular markers in a genome-wide exploration are the chromatin states for epiallelic determining factor of phenotypic variation ($phQTL^{epi}$) in these populations. The $phQTL^{dna}$ shows alteration of gene products as well as gene expression, while $phQTL^{epi}$ influences gene expression levels or patterns. Either DNA sequence or epigenetic variation has been employed in mapping population consequent from diverse natural or experimental lines.

2.10.4.2 A Dynamic View

Richards (2006) has explained chromatin-level variation at a single locus to explain the extent of constancy across mitotic and meiotic transitions in addition to DNA sequence difference level reliance at the locus or anywhere in the genome. The relationship of genotype and the epigenotype has been classified into three categories: obligatory, facilitated, and pure (Johannes et al., 2008). An obligatory relationship shows deterministic overtone between genotype and epigenotype that leads to the inheritance of epialleles in a steady and strictly sequence-dependent style crossways meiosis and mitosis. Moreover, the DNA sequence difference occurs either in *cis* or in *trans* fashion. Such forced connection undergoes relaxation in the enabled grouping that induces changes in epiallelic states that can be transmitted to succeeding generations. Lastly, pure epigenetic difference classified into stable or metastable is based on the presence of epigenotypes that are completely independent of genotypes. The six RILs have been studied at two meiotic or mitotic generations ($T = 0$ and $T = 1$) individually for the four classes – sequence-dependent stable and sequence-dependent metastable; sequence-independent stable; and sequence-independent metastable – of epialleles at a single epilocus (Johannes et al., 2008). The organic models show the assumed associations from QTL analysis between DNA sequence, chromatin, and phenotypic difference. The epialleles donate to diverse degrees toward the phenotypic heritability in the first three states, but not at all in the last one, due to extreme uncertainty.

The traditional QTL analysis depends upon "sequence-based linkage" and "association mapping techniques" with constraint of incomplete or complete disengagement of DNA and chromatin difference (facilitated or pure epialleles). Although relationship of phenotype with epigenetic difference has been observed, this relationship is so only partly with DNA sequence difference, and sequence-based QTL estimations do not imprison the cause of heritable difference (Johannes et al., 2008). This limitation shows that heritability estimations do not delineate the molecular determining factors of phenotypes. The prevalence of low statistical power to sense $phQTL^{dna}$ with small effect sizes and epistasis also is a challenging situation. Additional tedious factor is the chromatin variation in *cis* or in *trans* position through obligate epiallele-based DNA sequence difference. The occurrence of number of QTLs noticed by using sequence-based methods and the difference that does not impact genes encoding harvests, DNA, or histone methyltransferases involved in chromatin control are nonetheless attributable to chromatin effects (Petronis, 2006).

2.10.5 Comprehensive Inventory of Different Types of Epialleles

The genome-wide sequence and chromatin statistics from each individual epiallele and probability model highlight that epialleles undergo classification with the epigenome according to DNA difference dependence in *cis* or in *trans* chromosomal positions, and degree of constancy occurs through variations in QTLs and epiallelic covariance evidence (Johannes et al., 2008). The epigenetic features have been revealed through the correlation between chromatin statistics at diverse time points. The multiple trait probability model tends to explain the covariation statistics with the detection of an epi-QTL DNA for the epiallele classification. The presence of "high" covariation in chromatin states at a given locus shows stable segregation of epialleles. The detection of an epi-QTL DNA explains the epiallelic statuses at the covariation structure in *cis* or in *trans* position which shows that steady transmission of epialleles is sequence dependent. The outcomes of QTL analysis can be envisioned through *cis-trans* plot that shows the transmission of epialleles in a sequence-dependent fashion. The transition is "constrained" by the genotypic context close by (*cis*) or far away (*trans*). Contrastingly, the statistics points that are sequence-independent epialleles fall outside of the plot (along the y-axis), representing epialleles that are either transmitted or not transmitted.

Overall, the analysis presents a complete inventory of the likely pervasiveness of different types of epialleles as well as their genomic dispersal. The epigenomic reference map for a specific population shows annotation according to its sequence difference dependency. The QTL method has the power to identify novel loci that are involved in chromatin control. In local cytosine methylation, none of the known genes related with *de novo* establishment and maintenance of cytosine methylation mapped to the region of one of the identified QTL intervals (Riddle and Richards, 2002). Hence, fine-mapping and sequencing of the QTL interval has the ability to yield the causal DNA alternates. Thus, epigenome QTL analysis results can be deciphered through "*cis-trans* plots" to separate categories belonging to the epi-loci that would display a heritable consequence on the phenotype.

2.10.6 Limitations of Epi-QTL Mapping

During introgression of epialleles (conferring trait of interest) there are possible chances of inheritance of undesirable traits including low fertility and reduced tiller number as detected in the recurrent parent backgrounds. Likewise, Kim et al. (2018) have demonstrated the same. So, the scrutiny parameters need to be taken into account during selection of parents utilized for crossing experiment in the breeding program and this may be due to the undesirable genome reconstitution of the parents. The significant limitation in epigenetics-based QTL mapping is that the number and strength of epi-QTL was significantly lower than the number and strength of QTL obtained during genetic investigations on classical RILs (Kooke et al., 2019). The larger source for total phenotypic variation is epigenetic, rather than genetic, in *Arabidopsis* metabolic analysis. Epigenetic variation shows large consequence on morphological difference, just as the genetic difference (Johannes et al., 2009; Kooke et al., 2015).

2.11 CONCLUSIONS

Epigenetics plays a vital role in the control of plant secondary metabolism in both leaves and flowers of *Arabidopsis*. Owing to its reversible nature, it is important in fluctuating environment. Moreover, it is necessary for phenotypic plasticity which results in the enhancement of adaptation under adverse environments. The epigenetic variation has the ability to transmit variation in the gene expression revealed through qRT-PCR studies. The construction of methylation-based linkage map deciphers the epigenetic phenomenon and potential to explore the machinery of target gene silencing in genetic transformation studies through the knowledge of high-density methylation positions. These methylation-based genetic maps have the potential to link the information between crop diversity, QTL, and physical map that will accelerate gene discovery and analysis in future. The detected QTL hotspot regions have the potential for introgression purpose in molecular marker-assisted selection (MAS) breeding. The epi-QTL mapping has the potential to perform the systematic mapping of complex traits. The detection of epialleles governing phenotypic variation eases the transfer of trait of interest in the elite genotype. The epi-QTL has an important part in the regulation of the tissue-specific accumulation of secondary metabolites in *A. thaliana*. Epigenetic transgene silencing has the potential to solve this ethical concern in near future. The transgenerational inheritance of epigenetic marks plays a crucial part in the improvement of stress tolerance among crop plant species. Epigenetics has the ability to remove the genetic erosion effect which is the decisive genetic bottleneck in many crop breeding programs. Although both epi-QTL and DNA-QTL have equal and strong environmental interactions, epi-QTL mapping has the potential to ease the modulation of complex traits in complex plant genome species.

The heritable loss of cytosine methylation process is the major cause of epi-RIL QTLs. Epi-QTL have all the essential properties to become targets of natural or artificial selection. The dissection of yield-related genes is time-consuming and laborious exercise but epigenetics-based detection of epi-QTL eases the detection of yield-related genes. The detection of alleles eases the gene pyramiding, thereby incorporating desirable alleles into the elite cultivar which is based on the performance of trait using an additive effect of different kinds of genes. The development of high-yielding lines on the basis of epigenetic alleles depends on high grain yield which is a major advantage. This makes it easy to conduct the promoter study that is directly linked with trait of interest and the epi-QTLs are also associated with promoter regions at the DNA methylation sites. The identification of functional epialleles conferring trait of interest like yield-related genes has the possibility to extend the breeder's choice to achieve high yield in rice. Utilization of epialleles depicting polymorphism and high reversion rate of epiallele are both effective for use in breeding programs. This will help the epigeneticists carry out large-scale epigenomic projects among plant populations in future.

REFERENCES

Agorio, A., Durand S., Fiume E., et al., 2017. An Arabidopsis natural epiallele maintained by a feed-forward silencing loop between histone and DNA. *PLoS Genet.* 13, e1006551. doi:10.1371/journal.pgen.1006551

Allfrey, V.G., Faulkner R., Mirsky A.E., 1964. Acetylation and methylation of histones and their possible role in the regulation of RNA synthesis. *Proc. Natl. Acad. Sci. U.S.A.* 51, 786–794. doi: 10.1073/pnas.51.5.786

Allis, C., Jenuwein, T., 2016. The molecular hallmarks of epigenetic control. *Nat. Rev. Genet.* 17, 487–500. doi:10.1038/nrg.2016.59

Becker, C., Hagmann, J., Muller J., Koenig D., Stegle O., Borgwardt K., Weigel. D., 2011. Spontaneous epigenetic variation in the *Arabidopsis thaliana* methylome. *Nature.* 480, 245–249. doi:10.1038/nature10555

Bird, A., 2002. DNA methylation patterns and epigenetic memory. *Genes Dev.*16, 6–21. www.genesdev.org/cgi/doi/10.1101/gad.947102

Blanchard, M. G., Runkle, E.S., 2011. The influence of day and night temperature fluctuations on growth and flowering of annual bedding plants and greenhouse heating cost predictions. *Hortscience.* 46, 599–603. doi:10.21273/HORTSCI 46.4. 599.

Bloomfield, J., Rose, T.J., King, G.J., 2014. Sustainable Harvest: Managing plasticity for resilient crops. *Plant Biotechnol J.* 12, 517–533. doi:10.1111/pbi.12198

Bondada, R., Somasundaram, S., Marimuthu, M.P., Badarudeen, M.A., Puthiyaveedu, V.K., Maruthachalam, R., 2020. Natural epialleles of Arabidopsis SUPERMAN display superwoman phenotypes. *Commun. Biol.* 3, 772. doi:10.1038/ s42003-020-01525-9

Boquete, M.T., Muyle, A., Alonso, C., 2021. Plant epigenetics: Phenotypic and functional diversity beyond the DNA sequence. *Am. J. Bot.* 108(4), 553–558. doi:10.1002/ajb2.1645

Cedar, H., Bergman, Y., 2009. Linking DNA methylation and histone modification: patterns and paradigms. *Nat Rev Genet.* 10, 295–304. doi:10.1038/nrg2540

Chan, S.W., Henderson I.R., Jacobsen S.E., 2005. Gardening the Genome: DNA methylation in *Arabidopsis thaliana. Nat. Rev. Genet.* 6, 351–360. doi:10.1038/ nrg1601

Cokus, S.J., Feng, S., Zhang, X., et al., 2008. Shotgun bisulphite sequencing of the Arabidopsis genome reveals DNA methylation patterning. *Nature.* 452, 215–219. doi:10.1038/ nature 06745

Cortijo, S., Wardenaar R., Tatché, M.C., et al., 2014a. Mapping the epigenetic basis of complex traits. *Science.* 343, 1145–1148. doi:10.1126/science.1248127

Cortijo, S., Wardenaar, R., Colomé-Tatché, M., Johannes, F., Colot, V., 2014b. Genome-wide analysis of DNA methylation in *Arabidopsis* using MeDIP-Chip. *Methods Mol. Biol.* 1112, 125–149. doi:10.1007/978-1-62703-773-0_9

Craufurd, P.Q., Wheeler, T.R., 2009. Climate change and the flowering time of annual crops. *J. Exp. Bot.* 60, 2529–2539. doi:10.1093/jxb/ erp196

Cubillos, F.A., Billi, E., Zörgö, E., Parts, L., Fargier, P., Omholt, S., Blomberg, A., Warringer, J., Louis, E.J., Liti, G., 2011. Assessing the complex architecture of polygenic traits in diverged yeast populations. *Mol. Ecol.* 20(7), 1401–1413. doi:10.1111/j.1365-294X.2011.05005.x

Das, O.P., Messing, J., 1994. Variegated phenotype and developmental methylation changes of a maize allele originating from epimutation. *Genetics.* 136, 1121–1141. doi:10.1093/genetics/136.3.1121

Delaneau, O., Zazhytska, M., Borel, C., et al., 2019. Chromatin three-dimensional interactions mediate genetic effects on gene expression. *Science.* 364(6439), eaat8266. doi:10.1126/science.aat8266

Duan, Y., Qian, J., Sun, Y., Yi, Z., Yan, M., 2009. Construction of methylation linkage map based on MSAP and SSR markers in *Sorghum bicolor* (L.). IUBMB Life. 61, e663–e669. doi:10.1002/iub.213

El-Soda, M., Boer, M. P., Bagheri, H., Hanhart, C. J., Koornneef, M., Aarts MGM., 2014. Genotype-environment interactions affecting preflowering physiological and morphological traits of *Brassica rapa* grown in two watering regimes. *J. Exp. Bot.* 65, 697–708. doi:10.1093/jxb/ert434

Gahlaut, V., Zinta, G., Jaiswal, V., Kumar, S., 2020. Quantitative epigenetics: a new avenue for crop improvement. *Epigenomes.* 4(25). doi:10.3390/epigenomes4040025

Gehring, M., Henikoff, S., 2008. DNA methylation and demethylation in Arabidopsis. *Arabidopsis Book.* 6, e0102. doi:10.1199/tab.0102

Greaves, I.K., Groszmann, M., Ying, H., Taylor, J.M., Peacock, W.J., Dennis, E.S., 2012. Trans chromosomal methylation in Arabidopsis hybrids. *Proc. Natl. Acad. Sci. U.S.A.* 109(9), 3570–3575. doi:10.1073/pnas.1201043109

Groszmann, M., Greaves, I.K., Albert, N., Fujimoto, R., Helliwell, C.A., Dennis, E.S., Peacock W.J., 2011. Epigenetics in plants – Vernalisation and Hybrid Vigour. *Biochim. Biophysica. Acta* 1809, 427–437. doi:10.1016/j.bbagrm.2011.03.006

Hammond, J.P., Mayes, S., Bowen, H.C. et al., 2011. Regulatory hotspots are associated with plant gene expression under varying soil phosphorus (P). supply in *Brassica rapa. Plant Physiol.* 156, 1230–1241. doi:10.1104/pp.111.175612

He, L., Wu W., Zinta G., et al., 2018. A naturally occurring epiallele associates with leaf senescence and local climate adaptation in *Arabidopsis* accessions. *Nat. Commun.* 9, 460. doi:10.1038/ s41467-018-02839-3

Heard, E., Martienssen, R.A., 2014. Transgenerational epigenetic inheritance: myths and mechanisms. *Cell.* 157(1), 95–109.

Henderson, I.R., Jacobsen S.E., 2007. Epigenetic inheritance in plants. *Nature.* 447, 418–424. doi:10.1038/nature05917

Hüve, K., Bichele I., Tobias M., Niinemets U., 2006. Heat sensitivity of photosynthetic electron transport varies during the day due to changes in sugars and osmotic potential. *Plant Cell Environ.* 29, 212–228. doi:10.1111/j.1365-3040.2005.01414.x

Jiang, F., Li, Z.W., Yin, X.F., Zhang, Y., Guo, R., Kang, M.S., Fan, X., 2020. Impact of donor QTL on grain yield and gray leaf spot of four recombinant inbred lines of maize. *Crop Sci.* 60, 841–851. doi:10.1002/csc2.20111

Johannes, F., Colomé-Tatché, M., 2011. Quantitative epigenetics through epigenomic perturbation of isogenic lines. *Genetics.* 188(1), 215–227. doi:10.1534/genetics.111.127118

Johannes, F., Colot, V. Jansen, R., 2008. Epigenome dynamics: a quantitative genetics perspective. *Nat. Rev. Genet.* 9, 883–889. doi:10.1038/nrg2467

Johannes, F., Porcher E., Teixeira F.K., et al., 2009. Assessing the impact of transgenerational epigenetic variation on complex traits. *PLoS Genet.* 5(6), e1000530. doi:10.1371/journal.pgen.1000530

John, M.C., Amasino, R.M., 1989. Extensive changes in DNA methylation patterns accompany activation of a silent T-DNA *ipt* gene in *Agrobacterium tumefaciens*-transformed plant cells. *Mol Cell Biol* 9, 4298–4303. doi:10.1128/ mcb.9.10.4298-4303.1989

Johnson, L.M., Du J., Hale C.J., et al., 2014. SRA- and SET-domain-containing proteins link RNA Polymerase V occupancy to DNA methylation. *Nature*. 507, 124–128. doi:10.1038/ nature12931

Kakutani, T., Munakata, K., Richards, E.J., Hirochika, H., 1999. Meiotically and mitotically stable inheritance of DNA hypomethylation induced by *ddm1* mutation of *Arabidopsis thaliana*. *Genetics*. 151, 831–838. doi:10.1093/genetics/151.2.831

Kang, M.S., 1998. Using genotype-by-environment interaction for crop cultivar development. *Adv. Agron*. 62, 199–252. doi:10.1016/S0065-2113(08)60569-6

Karlić, R., Chung, H.-R., Lasserre, J., Vlahoviček, K., Vingron, M., 2010. Histone modification levels are predictive for gene expression. *Proc. Natl. Acad. Sci. U.S.A.* 107(7), 2926–2293. doi:10.1073/ pnas.0909344107

Kim, S.R., Ramos J.M., Hizon R.J.M., et al., 2018. Introgression of a functional epigenetic Osspl14wfp allele into elite indica rice genomes greatly improved panicle traits and grain yield. *Sci. Rep.* 8, 3833. doi:10.1038/s41598-018-21355-4

King, G.J. 2015. Crop epigenetics and the molecular hardware of genotype × environment interactions. *Front. Plant Sci.* 6, 968. doi:10.3389/fpls.2015.0096

King, G.J., Amoah, S., Kurup, S., 2010. Exploring and exploiting epigenetic variation in crops. *Genome*. 53, 856–868. doi:10.1139/G10-059

Kooke, R., Johannes F., Wardenaar R., et al., 2015. Epigenetic basis of morphological variation and phenotypic plasticity in *Arabidopsis thaliana*. *Plant Cell* 27(2), 337–348. doi:10.1105/tpc. 114.133025

Kooke, R., Morgado, L., Becker, F., et al., 2019. Epigenetic mapping of the Arabidopsis metabolome reveals mediators of the epigenotype-phenotype map. *Genome Res*. 29, 96–106. doi:10.1101/gr. 232371.117

Kumar, S., 2017. Epigenetic control of apomixis: a new perspective of an old enigma. *Adv. Plants Agric. Res.* 7, 00243. doi:10.15406/apar.2017.07.00243

Kumar, S., Singh, A., 2016. Epigenetic regulation of abiotic stress tolerance in plants. *Adv. Plants Agric. Res.* 5, 00179. doi:10.15406/apar.2016.05.00179

Kumar, S., Singh, A.K., Mohapatra, T., 2017. Epigenetics: history, present status and future perspective. *Indian J. Genet Plant Breed.* 77, 445–463. doi:10.5958/0975-6906. 2017.00061.X

Lahiri, D.K., Maloney, B., 2006. Genes are not our destiny: the somatic epitype bridges between the genotype and the phenotype. *Nat. Rev. Neurosci.* 7(12), 976. doi:10.1038/ nrn2022-c1

Li, Y., Kumar, S., Qian, W., 2017. Active DNA demethylation: mechanism and role in plant development. *Plant Cell Rep.* 37,1–9. doi:10.1007/s00299-017-2215-z

Liu, L., Li Y., Tollefsbol T.O., 2008. Gene-environment interactions and epigenetic basis of human diseases. *Curr. Issues Mol. Biol.* 10(1–2), 25–36.

Long, Y., Xia, W., Li, R., et al., 2011. Epigenetic QTL mapping in *Brassica napus*. *Genetics*. 189(3), 1093–1102. doi:10.1534/genetics.111.131615

Lu, Q., Hong, Y., Li, S. *et al.*, 2019. Genome-wide identification of microsatellite markers from cultivated peanut (*Arachis hypogaea* L.). *BMC Genomics*. 20(799). https://doi.org/ 10.1186/s12864-019-6148-5

Lu, W., Xiao, L., Quan, M., Wang, Q., El-Kassaby, Y.A., Du, Q., Zhang, D., 2020. Linkage-linkage disequilibrium dissection of the epigenetic quantitative trait loci (epiQTLs) underlying growth and wood properties in Populus. *New Phytol.* 225(3), 1218–1233.

Mager, J., Bartolomei, M., 2005. Strategies for dissecting epigenetic mechanisms in the mouse. *Nat Genet.* 37, 1194–1200. doi:10.1038/ng1664

Martienssen, R.A., Colot, V., 2001. DNA methylation and epigenetic inheritance in plants and filamentous fungi. *Science.* 293, 1070–1074. doi:10.1126/science.293.5532.1070

Mathers, J.C., 2006. Nutritional modulation of ageing: genomic and epigenetic approaches. *Mech. Ageing Dev.* 127(6), 584–589. doi:10.1016/j.mad.2006.01.018

Mathieu, O., Reinders, J., Caikovski, M., Smathajitt, C., Paszkowski, J., 2007. Transgenerational stability of the *Arabidopsis* epigenome is coordinated by CG methylation. *Cell.* 130, 851–862. doi:10.1016/j.cell.2007.07.007

Matzke, A.M., Mosher, R.A., 2014. RNA-directed DNA methylation, An epigenetic pathway of increasing complexity. *Nat. Rev.* 15, 394–408. doi:10.1038/nrg3683

McClung, C.R., Davis, S.J., 2010. Ambient thermometers in plants: from physiological outputs towards mechanisms of thermal sensing. *Curr. Biol.* 20, R1086–R1092. doi:10.1016/j.cub.2010.10.035

McMaster, G.S., Wilhelm, W.W., 1997. Growing Degree-Days: one equation, two interpretations. *Agric. For. Meteorol.* 87, 291–300. doi:10.1016/S0168-1923(97)00027-0

Mercé, C., Bayer, P.E., Fernandez, C.T., Batley, J., Edwards, D., 2020. Induced methylation in plants as a crop improvement tool: progress and perspectives. *Agronomy* 10, 1484. doi:10.3390/agronomy10101484

Miglani, G.S., Kaur, A., Kaur, L., 2020. Plant gene expression control using genome- and epigenome-editing technologies. *J. Crop Impr.* 34(1), 1–63.

Miglani, G.S., Singh, R., 2020. Epigenome editing in crop improvement. Chapter 5. In: M.S. Kang (Ed.) *Quantitative Genetics, Genomics and Plant Breeding.* CABi Publishing, Oxon, U.K., pp. 44–70.

Millar, C., Grunstein, M., 2006. Genome-wide patterns of histone modifications in yeast. *Nat. Rev. Mol. Cell Biol.* 7, 657–666. doi:10.1038/nrm1986

Mittelsten Scheid, O., Afsar, K., Paszkowski, J., 2003. Formation of stable epialleles and their paramutation-like interaction in tetraploid *Arabidopsis thaliana. Nat Genet.* 34(4), 450–454. doi:10.1038/ng1210

Mounger, J., Ainouche M.L., Bossdorf O., et al., 2021. Epigenetics and the success of invasive plants. *Philos. Trans. R. Soc.* doi:10.1098/rstb.2020.0117

Muchow, R.C., Sinclair, T.R., Bennett, J.M., 1990. Temperature and solar radiation effects on potential maize yield across locations. *Agronomy J.* 82, 338–343. https://doi.org/10.2134/agronj1990.00021962008200020033x

Papareddy, R.K., Nodine, M.D., 2021. Plant Epigenetics: Propelling DNA methylation variation across the cell cycle. *Curr. Biol.* 31(3), R129–R131. doi:10.1016/j.cub. 2020.11.049

Peaston, A.E., Whitelaw, E., 2006. Epigenetics and phenotypic variation in mammals. *Mamm. Genome.* 17, 365–374. doi:10.1007/s00335-005-0180-2

Penterman, J., Huh J.H., Hsieh T.F., Fischer, R.L., 2007. Genomic imprinting in *Arabidopsis thaliana* and *Zea mays*. In: Olsen O. L., editor. *Endosperm – Development and Molecular Biology*. 1. Vol. 6. Heidelberg: Springer Berlin; pp. 219–239. doi:10.1007/7089_2007_112

Petronis, A., 2006. Epigenetics and twins: Three variations on the theme. *Trends Genet.* 22, 347–350. doi:10.1016/j.tig.2006.04.010

Porter, J.R., Semenov, M.A., 2005. Crop responses to climatic variation. *Philos. Transactions Royal Soc.* London. Series. B, Biological Sciences. 360, 2021–2035. doi:10.1098/rstb.2005.1752

Rakyan, V.K., Blewitt, M.E., Druker, R., Preis, J.I., Whitelaw, E., 2002. Metastable epialleles in mammals. *Trends Genet.* 18(7), 348–351. doi:10.1016/s0168-9525(02)02709-9

Rangwala, S.H., Elumalai R., Vanier C., Ozkan H., Galbraith D.W., Richards. E.J., 2006. Meiotically stable natural epialleles of sadhu, a novel Arabidopsis retroposon. *PLoS Genet.* 2(3), e36. doi:10.1371/journal.pgen.0020036

Ray, D.K., Gerber, J.S., MacDonald, G.K., West, P.C., 2015. Climate variation explains a third of global crop yield variability. *Nat. Commun.* 22, 5989. doi:10.1038/ ncomms6989

Richards, E.J., 2006. Inherited epigenetic variation – Revisiting soft inheritance. *Nat. Rev. Genet.* 7, 395–401. doi:10.1038/nrg1834

Riddle, N.C., Richards, E.J., 2002. The control of natural variation in cytosine methylation in *Arabidopsis. Genetics.* 162, 355–363.

Riddle, N.C., Richards, E.J., 2005. Genetic variation in epigenetic inheritance of ribosomal RNA gene methylation in Arabidopsis. *Plant J.* 41, 524–532. doi:10.1111/ j.1365-313X.2004.02317.x

Schmitz, R.J., He Y., Valdés-López, O., et al., 2013. Epigenome-wide inheritance of cytosine methylation variants in a recombinant inbred population. *Genome Res.* 23(10), 1663–1674. doi:10.1101/gr.152538.112

Schones, D., Zhao, K., 2008. Genome-wide approaches to studying chromatin modifications. *Nat. Rev. Genet.* 9, 179–191. doi:10.1038/nrg2270

Shi, J., Li, R., Qiu, D., et al., 2009. Unraveling the complex trait of crop yield with quantitative trait loci mapping in *Brassica napus. Genetics.* 182, 851–861. doi:10.1534/ genetics.109.101642

Soppe, W.J., Jacobsen, S.E., Alonso-Blanco, C., Jackson, J.P., Kakutani, T., Koornneef, M., Peeters, A., 2000. The late flowering phenotype of *Fwa* mutants is caused by gain-of-function epigenetic alleles of a homeodomain gene. *Mol. Cell.* 66(1), 791–802. doi:10.1016/s1097-2765(05)00090-0

Springer, N., Schmitz, R., 2017. Exploiting induced and natural epigenetic variation for crop improvement. *Nat. Rev. Genet.* 18, 563–575. doi:10.1038/ nrg.2017.45

Stam, M., Belele C., Dorweiler J.E., Chandler V.L., 2002. Differential chromatin structure within a tandem array 100 kb upstream of the maize *b1* locus is associated with paramutation. *Genes Dev.* 16, 1906–1918. doi:10.1101/gad.1006702

Sun, C., Ali, K., Yan, K., Fiaz, S., Dormatey, R., Bi, Z., Bai, J., 2021. Exploration of epigenetics for improvement of drought and other stress resistance in crops: a review. *Plants.* 10(6), 1226. doi:10.3390/plants10061226

Suzuki, M.M., Bird, A., 2008. DNA methylation landscapes: Provocative insights from epigenomics. *Nat. Rev. Genet.* 9, 465–476. doi:10.1038/nrg2341

Taylor, S.M., Jones, P.A., 1979. Multiple new phenotypes induced in 10T1/2 and 3T3 cells treated with 5-azacytidine. *Cell.* 17, 771–779. doi:10.1016/0092-8674(79)90317-9

Tran, R.K., Henikoff, J.G., Zilberman, D., Ditt, R.F., Jacobsen, S.E., Henikoff, S., 2005. DNA methylation profiling identifies CG methylation clusters in Arabidopsis genes. *Curr. Biol.* 15, 154–159. doi:10.1016/j.cub.2005.01.008

Ueda, M., Seki, M., 2020. Histone modifications form epigenetic regulatory networks to regulate abiotic stress response. *Plant Physiol.* 182(1), 15–26. doi:10.1104/pp.19.00988

Vaughn, M.W., Tanurdzic M., Lippman Z., et al., 2007. Epigenetic natural variation in *Arabidopsis thaliana. PLoS Biol.* 5, 1617–1629. doi:10.1371/journal.pbio.0050174

Waddington, C.H., 1942. The Epigenotype. *Endeavour* 1, 18–20. doi:10.1093/ije/dyr184

Watanabe, Y., Maekawa, M., 2010. Methylation of DNA in cancer. *Adv. Clin. Chem.* 52, 145–167. doi:10.1016/s0065-2423(10)52006-7

Watson, C.J., Collier, P., Tea, I., et al., 2014. Hypoxia-Induced epigenetic modifications are associated with cardiac tissue fibrosis and the development of a myofibroblast-like phenotype. *Hum. Mol. Genet.* 23(8), 2176–2188. doi:10.1093/hmg/ddt614

Weber, M., Davies, J.J., Wittig, D., Oakeley, E.J., Haase, M., Lam, W.L., Schübeler, D., 2005. Chromosome-wide and promoter-specific analyses identify sites of differential DNA methylation in normal and transformed human cells. *Nat. Genet.* 37(8), 853–862. doi:10.1016/s0065-2423(10)52006-7

Weigel, D., Colot, V., 2012. Epialleles in plant evolution. *Genome Biol.* 13(10), 249. doi:10.1186/gb-2012-13-10-249

Weinhold, B., 2006. Epigenetics: The science of change. Environ. Health Perspect. 114(3), A160–A167. doi:10.1289/ehp.114-a160

Widman, N., Jacobsen S.E., Pellegrini, M., 2009. Determining the conservation of DNA methylation in *Arabidopsis. Epigenetics.* 4, 119–124. doi:10.4161/epi.4.2.8214

Yan, W., Kang, M.S., 2003. *GGE Biplot Analysis: A Graphical Tool for Breeders, Geneticists, and Agronomists.* Boca Raton, FL: CRC Press.

Zhang, K., Mosch, K., Fischle, W., Grewal, S.I., 2008. Roles of the Clr4 methyltransferase complex in nucleation, spreading and maintenance of heterochromatin. *Nat. Struct. Mol. Biol.* 15, 381–388. doi:10.1038/nsmb.1406

Zhang, X., Yazaki J., Sundaresan A., et al., 2006. Genome-wide high-resolution mapping and functional analysis of DNA methylation in Arabidopsis. *Cell.* 126(6), 1189–1201. doi:10.1016/j.cell.2006.08.003

Zhao, J., Lu, Z., Wang, L. Jin B., 2021. Plant responses to heat stress: physiology, transcription, noncoding RNAs, and epigenetics. *Int. J. Mol. Sci.* 22(1), 117. doi:10.3390/ijms22010117

Zilberman, D., Henikoff, S., 2007. Genome-wide analysis of DNA methylation patterns. *Dev.* 1346(1), 3959–3965. doi:10.1242/dev.001131

Zou, M., Lu, C., Zhang, S., et al., 2017. Epigenetic map and genetic map basis of complex traits in cassava population. *Sci. Rep.* 7, 41232. doi:10.1038/srep41232

3 Induced Mutagenesis for Gene Pool Expansion with Special Attention on Pulse Crops

*Biswajit Pramanik and Sandip Debnath**
Department of Genetics and Plant Breeding, Palli Siksha Bhavana (Institute of Agriculture), Visva-Bharati University, Sriniketan, West Bengal, India
[1]biswajit1996pramanik@gmail.com,
[2]*sandip.debnath@visva-bharati.ac.in,

3.1 INTRODUCTION

The rapidly overexpanding human population, with an estimation of crossing 9 billion by 2050, will create food scarcity all over the globe (Konuma, 2016; Sar et al. 2022). Hence, to meet this rising demand for proper nutrition through higher production of food can be considered the need of the hour in recent times (Das et al. 2023). Enhancement in the field of crop genetic resources is a significant component of attempts for securing global food supply as well as nutrition issues (Nelson, 2020; Pramanik et al. 2023b). Alongside, genetic diversity with the desired characteristics is the utmost requirement for plant breeding in the agriculture sector (Hawkesford and Griffiths, 2019;). Mutation, or a sudden heritable change in the genetic architecture of an organism, is a natural process to reproduce novel gene variants (alleles). Notably, most of the genetic changes occurring to any organism, including plants, are the unidentified effects of mutation (San Martin, 2021). Several mutation-causing agents, like various radiation (both particulate and nonparticulate) and certain chemicals, can be implemented to bring about mutations and produce genetic variants from which selection process of desired mutants can be conducted (Van Eenennaam, 2019). Basically, the variation, produced by mutation and used in the whole process of natural selection, is the key driving force in evolution (Nei, 2007).

Mutation breeding, the innovative branch of plant breeding, has now merged with the molecular era with the advancement of molecular genetics and several DNA technologies (Raina et al. 2016). Mutations are known to be the heritable alterations in any organism's phenotype through the changes in its genotypic level. Generally, chemical alterations at the gene level are the primary causes of these modifications, which possess the potential to result in new and heritable characteristics in plants. Therefore, crop plants can be exposed to these chemical agents for mutations to occur,

 DOI: 10.1201/9781003392897-3

from which desirable variants can be selected [21]. Mutation breeding is actually a new generation practice of plant breeding in which new varieties of crops with desirable characteristics can be originated utilizing various protocols of induced mutations (Mir et al. 2020).

'Mutation breeding' was termed by Freisleben and Lein in order to develop mutant lines by induced mutagenesis in agriculture (Lamo et al. 2017). Mutagenesis is actually an inventive procedure of stable alteration of the genetic information of an organism, which generally takes place due to several occurrences of misprinting in DNA repairing process.

Mutation can be described as a natural process through which several novel gene variants or alleles can be originated. Allele recombination on homologous chromosomes and their independent assortment at meiosis-I have a significant effect on the generation of various varieties acquired through this method (Bulankova et al. 2021). In the past, numerous food crops have undergone mutation breeding to generate varieties according to the desired characteristics. Induced mutagenesis is one of the most efficient tools for the identification of several critical regulatory genes and their molecular and biochemical regulating pathways (Chaudhary et al. 2019).

In this ongoing era of plant breeding, due to immense advancement in plant molecular biology as well as biotechnology and plant genomics, new impulses in plant mutation research, ranging from fundamental theories of mutagenesis to involvement of reverse genetics, are being experienced. The more knowledge about the huge potentialities and extensive outcomes of mutagenic techniques is discovered, the more engagement of breeders is now occurring with greater efficiency and superior sophistication. Therefore, this chapter predominantly aims to evaluate mutation breeding, its historical success, applicability, benefits, and drawbacks in plant breeding.

3.2 HISTORY OF PLANT BREEDING TO THE ONSET OF MUTAGENESIS

Plant breeding may be regarded to have progressed through several distinct phases like domestication, selection, hybridization, novel genetic variation, and genetic engineering. The first phase consisted of domestication, during which the prehistoric humans brought wild plant species under human management. Domestication began almost 11,000 years ago, when humans are believed to have started in agriculture. Domestication is the first and the most critical step as it makes a plant species accessible to crop improvement efforts. There must have been considerable selection, both by natural and ethnological efforts, during that particular phase. Therefore, domestication may be regarded as the most basic method of plant breeding as all the other protocols become applicable to a plant species only after their successful domestication. This method continues till today and is likely to continue for some time in future. This is particularly true in the case of timber trees, medicinal plants, and plants satisfying some special requirements. During the long period of prehistoric and historic cultivation, natural selection has certainly acted on the domesticated species. It is very likely that humans also exerted some selection, either knowingly or unknowingly. Movement of humans from one area to another brought about the movement of the cultivated plant species as well. Deliberate and planned introduction into an area of new plant species or varieties from other parts of the globe is an integral part of plant breeding in recent times.

In the era of second phase, conscious and planned selection was practiced to utilize the natural genetic variation existing in the crop species. This resulted in a number of improved varieties, especially of self-pollinated crops. Several researchers used selection methods, including plant selection and progeny test, to develop some useful cereal varieties. According to Patrick Shireff, only the variation of heritable nature responded to selection, and this variation arose through natural mutation and natural hybridization. Vilmorin further developed the progeny test and implemented this method successfully in the improvement of sugar beet (*Beta vulgaris*). The individual plant selection method was developed in detail by Nilsson-Ehle and his associates in 1900. In 1903, Johannsen proposed the pure line theory that provided the genetic basis for individual plant selection. These and other studies resulted in two basic selection schemes, namely, mass and pureline selections, for self-pollinated crop. The schemes, especially pureline selection, gave rise to several improved varieties of different crops. However, once the natural genetic variability was exhausted, these selection schemes were no longer attractive breeding procedures, although mass selection is still used for purification and maintenance of pureline varieties. In contrast, selection continues to be an attractive breeding approach for cross-pollinated crops, especially in those where heterosis has not become economically feasible.

Even before selection was refined as a breeding tool, hybridization began to be employed for crop improvement. The main feature of this phase, phase three, of plant breeding is the planned and controlled hybridization between diverse, selected genotypes or varieties to create new and more desirable combinations among the existing genes. Babylonians and Assyrians pollinated date palm artificially as early as 700 B.C. In the 17th century, several varieties of heading lettuce were developed in France, some of which were still in cultivation during the 1990s. In 1717, Thomas Fairchild produced the first artificial hybrid, popularly known as "Fairchild's mule", by crossing carnation with sweet Williams. Around 1800, Knight developed several varieties of apples, pears, peaches, grapes, etc. by hybridization. The breeding methods based on hybridization principle acquired a scientific basis with the elucidation of various genetic and cytoplasmic principles, which followed the rediscovery, in 1900, of Mendel's work published originally in 1866. Subsequently, the detailed studies of East and Shull on inbreeding in maize paved the way for the development of hybrid varieties, first in maize and later in various crop species. Inter-varietal hybridization continues and is the most likely to continue, to be the major approach for the improvement of self- and cross-pollinated as well as asexually propagated crop species.

Distant hybridization, i.e., hybridization between two distinct species belonging to the same genus or different genera, has been employed for the transfer of new genes, especially for resistance to biotic and abiotic stresses, from related wild species into crop species and of alien cytoplasm with a view to generate cytoplasmic male sterility. Somatic hybridization (fusion between protoplasts) extends the scope of distant hybridization to sexually incompatible species combinations. Notable contributions have been made, particularly in the development of disease-resistant varieties. Distant hybridization dates back to 1717 and is credited to Thomas Fairchild, but its agricultural application is much more recent, e.g., production of *Raphanobrassica* by Kerpenchenko in 1928. Somatic hybridization began in 1972 when Carlson and coworkers successfully obtained an interspecific hybrid by fusing the protoplasts of *Nicotiana glauca* with those of *Nicotiana langsdorffii*. However, it is just beginning to deliver potential commercial products.

The fourth phase of plant breeding began with the objective of creating novel genetic variations not available in a crop species. New alleles of existing genes were created by inducing mutations using radiation like X-rays, γ-rays, etc., or chemical mutagens such as ethylmethane sulfonate (EMS) and methylmethane sulfonate (MMS). Induction of mutations became feasible following the demonstration in 1927 by Muller of mutagenic action of X-rays. The first mutation breeding program was launched in 1929 in Sweden by Ake Gustafsson and coworkers. Most notably, the first variety that evolved through this protocol was released in 1950.

3.3 MUTATION AND MUTATION BREEDING

Mutation may be defined as the permanent and relatively rare change in the number or sequence of nucleotides in any organism. The genetic improvement of crop plants for various economic characteristics through the use of induced mutations is referred to as mutation breeding. It is commonly used in both self-pollinated and asexually propagated species. It may be natural or induced. The major advantage of the employment of induced mutations has been the possibility to correct one or few negative characters or to get new gene combination, which is desirable without changing the major part of the organism's total genetic architecture (Debnath et al. 2022).

3.3.1 The Foundation Concept of Mutagenesis

The term mutation was introduced by Hugo de Vries in 1900, although mutations were known to occur in animals and plants before this. For example, a short-legged sheep was actually discovered by an English farmer in the 18th century. Later, mutagenic action of X-rays was discovered by Muller in 1927 while experimenting on *Drosophila*. Later, in 1928, Stadler introduced the mutagenic effects of both γ- and X-rays in barley (*Hordeum vulgare*) and maize (*Zea mays*), respectively. Muller was awarded the Nobel Prize in 1946 in recognition for this work. In 1946, Auerbach and Robson reported that nitrogen mustards produced mutation in *Drosophila*. Subsequently, an array of chemicals with mutagenic action were described later on. Immediately after Muller's discovery in 1927, mutation breeding programs were initiated in Sweden, USSR, and Germany. The Swedish mutation breeding program was started in 1929 by Nilson-Ehle and accordingly continued by Ake Gustafsson and coworkers. Mutation breeding attracted considerable attention during 1950s and 1960s, and several countries further initiated mutation breeding. Although initial concept regarding mutagenesis was to create new genes, later it became clear with further advancement that this protocol possessed the ability to create novel genetic variation, not novel genes.

3.3.2 Types of Mutations

The term "mutation" is now used in a relatively strict manner to describe those few alternations which affect the molecular makeup of a gene. Generally, these alterations are called gene or point mutations. Besides, various structural changes in the chromosomes such as deletions, duplication, inversions, and translocations also bring about phenotypic changes in organisms, which are referred to as chromosomal mutations (Figure 3.1a).

Therefore, based on the extent of phenotypic effect on plants, mutations are categorized into two distinct classes (Gaul, 1964), i.e., macromutations and micromutations (Figure 3.1a).

i. *Macromutations*: These generate a large identifiable phenotype effect on individual plants. These are generally oligogenic in nature, due to which the mutants can easily be selected in the M_2 generation.
ii. *Micromutations*: These generate a little phenotypic effect that can only be recognized on the basis of the size of the population. On the contrary, these mutations are polygenic in nature, due to which selection is effective only in or after M_3 generation.

The method of identification of mutations has also been categorized into several classes (Figure 3.1a).

i. *Morphological mutations*, which include alterations in the external form including shape, color, and size, e.g., dwarfism in pea, kernel color in corn, etc.
ii. *Lethal mutations*, which refer to such changes in the genotype, that causes death of the individual. These are among the easiest to score mutations. For example, the chlorophyll mutation named albino (caused by deficiency of chlorophyll) is lethal.
iii. *Biochemical mutations*, which are generally detected through the surfacing of any deficiency. Supplementing the mutant with that specific nutrient or the chemical compound for which it is deficient can help it to overcome the same. Such mutagenic studies have been conducted on bacteria and fungi.
iv. *Resistant mutations*, which are detected by their capability to grow in the presence of any antibiotic or pathogenic agent to which the control is susceptible.
v. *Conditional mutations*, which allow the mutant phenotype to be expressed only under specific restrictive conditions such as high or low temperature. On the other hand, while exposed to normal or permissive conditions, normal phenotype is exhibited by the same mutant. These kinds of mutations are generally considered while studying cell cycle. Furthermore, both the morphological and lethal mutations are also considered under macromutations.

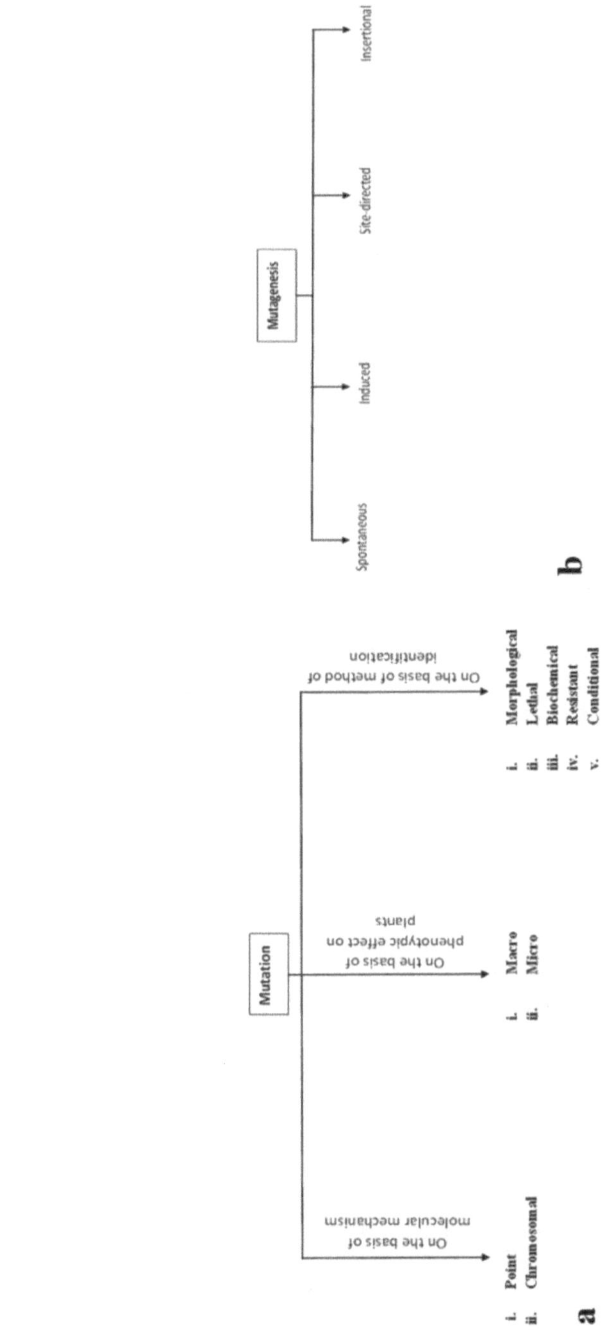

FIGURE 3.1 (a) Classification of mutation on various basis. (b) Types of mutagenesis practiced on various organisms.

3.3.3 Types of Mutagenesis and Mutation Breeding

Mutagenesis is the result of mutation which occurs through altering the genetic architecture of any organism (Figure 3.1b). It can either happen naturally or be created manually in a laboratory. On the other hand, gene transfer between or within species in order to create organism with desirable characteristics is generally called genetic modification (Oladosu et al. 2016). Mutations not only cause structural changes in an encoded protein but also have the potential of reduction or full loss of its expression. Notable to mention, mutations can be stated harmful to a cell or organism if any change in the sequence of DNA affects all the copies of the protein encoded by it (Drummond and Wilke, 2008).

3.3.3.1 Spontaneous Mutation

Mutations named spontaneous mutations are relatively infrequently found in natural populations. Naturally occurring mutations, as opposed to those brought on by human intervention, are known as spontaneous mutations (Yali and Mitiku, 2022). It is challenging to identify the source of these mutations since they appear at random. Spontaneous mutations have a nonzero likelihood of happening even in an uncontaminated healthy cell. Almost 10,000 times per cell per day in humans experience oxidative DNA damage, compared to 100,000 times per cell per day in rats. To locate spontaneous mutations, a particular change might be employed.

In the DNA double helix, cytosine is frequently deaminated to uracil, which is a major source of naturally occurring point mutations. After replication, a mutant daughter cell is produced when a T•A base pair substitutes the wildtype C•G base pair. Another reason for spontaneous mutations is errors in DNA replication. Even though replication is often done with great fidelity, mistakes sometimes happen (Ohno, 2019).

3.3.3.2 Induced Mutagenesis in Plants

The availability of significant genetic diversity, which a plant breeder may combine to produce new kinds, is essential to the accomplishment of a crop advancement effort. The result of both spontaneous and forced mutations is this diversity. The most popular technique for breeding plants through sexual reproduction is now induced mutations. Breeders have used both physical as well as chemical mutagens to induce mutations in order to create new crop types. Chemical mutagenesis, which is known to result in point mutations, is widely recognized to be less effective than induced mutagenesis caused by physical mutagens. There are several physical mutagens that may be used in mutagenesis, including UV rays, β-rays, γ-rays, X-rays, fast and thermal neutrons, and electron and ion beams. The effectiveness of the mutagen determines how frequently the intended mutants manifest. Application of stronger mutagens with various mutation spectra is therefore crucial for the induction of variability.

This breeding approach has been revitalized by the fast improvement of plant molecular genetics and genomics in areas crucial to mutation breeding. Mutation breeding is anticipated to directly benefit from rapid scientific and technological breakthroughs in molecular genetics and genomics (Shu, 2009). Generally, access to sufficient varieties for a single trait can provide the breeder the opportunity to improve plant breeding (Swarup et al. 2021). Applied genetics consists of both plant breeding and mutation breeding. One of the most important developments in genetics was the discovery of experimental mutagenesis around the turn of the 20th century, which eventually led to the breeding of plant mutants. Genetic variety in advantageous traits is required for plant breeding for agricultural improvement. Radiation and several chemicals are mutagenic agents that produce desirable mutation (Oladosu et al. 2016).

Worldwide, 2252 mutants of plant species have been produced. While 1585 were produced directly, 667 were released as a result of hybridization with mutants. Out of a total of 2252, 1700 mutant variations have been introduced in crops grown from seeds and 552, in species grown vegetatively. Among seed-propagated species, rice with 873 mutant varieties, barley with 307 mutant varieties, and wheat with 265 mutant varieties are the 3 species with the most mutant types produced.

Maximum variations have developed as a result of radiation. The most mutant variations were produced by EMS among chemical mutagens. The Mutant Variety Database has 3222 entries, 2456 seed propagated plants, and 367 vegetatively propagated plants. Based on the common name of the entry, the categorization above was made. Rice occupies the first position among all other crops in having the highest number of mutant varieties produced to date, followed by barley and chrysanthemum. The next positions are engaged by wheat, soybean, and maize, respectively (Jankowicz-Cieslak et al. 2017) (Figure 3.2a). Surprisingly, the majority (48%) of the mutant varieties included in the Mutant Variety Database have improved agronomic and botanical characteristics. This could be because botanic and agronomic characteristics are visible and screening often doesn't require specific equipment (Zong et al. 2019).

3.3.3.3 Insertional Mutagenesis

This kind of mutagenesis is caused by DNA insertions, either by transposon element activation or genetic transformation and insertion of T-DNA (T-DNA insertion mutagenesis) (transposon mutagenesis or transposition mutagenesis). The resultant mutants are transposon mutants or insertion mutants (Wilson et al. 1996).

3.3.3.4 Site-Directed Mutagenesis

This type of mutagenesis is the process of altering a DNA molecule at a particular place (Bezie et al. 2021). This is achieved in plants by homologous recombination of the T-DNA fragment with native DNA molecules and genetic transformation. The replacement of a foreign molecule with an indigenous DNA segment, which can be as small as only one nucleotide, characterizes the mutants (Miglani, 2017).

3.3.4 History of Plant Mutagenesis

The first accounts of mutant crops date back to 300 B.C. in China, marking the beginning of the history of mutation induction. However, from the start to the present, there have been major changes that have increased the frequency of mutations (Oladosu et al. 2016). Mutations have provided the genetic variations that drive the evolution of species. Changes in species have been important for agricultural goals including species domestication and crop improvement in addition to adaptability to the natural environment (Barghi et al. 2020).

Var. chlorina, a tobacco variety with a light green color that was created in Indonesia in 1936, was the first induced mutant variety. Over the following few decades, a number of further induced mutant variations, like the tulip var, were introduced in a range of plant species. An ancient Chinese text contains the earliest account of mutant selection in plant breeding for grain maturity and other features. The larger celandine "incisa" mutant, which was the first verified (spontaneous) plant mutation, was originally documented in 1590. Mutation induction activities peaked in the 1950s, 1960s, 1970s, and 1980s, with significant successes in terms of mutant variety releases. Early in the 21st century, mutation technologies made a resurgence as a consequence of quicker and better understanding of mutagenesis and related fields, which resulted in new applications (Nazarenko et al. 2018).

The use of molecular and genomic methods for mutant screening and characterization allowed mutation breeding to embrace and make use of the most recent findings and scientific advancements in plant genomes and molecular biology research (Koornneef and Meinke, 2010).

Several nations have used induced mutagenesis and mutation breeding to create superior mutant varieties of a variety of significant agricultural crop species, including cereals, pulses, oilseeds, vegetables, fruits, fibers, and ornamentals. These nations include China, India, Pakistan, Bangladesh, Vietnam, Thailand, Italy, Sweden, the United States, Canada, and Japan, with China consisting of the maximum number of mutant varieties, followed by Japan and India (Figure 3.2b). To date, a range of traits have been improved in mutant varieties, including yield, maturity, quality, and resistance to biotic and abiotic stresses (Pathirana et al. 2011).

3.3.5 Induction of Mutation

Cosmic and UV rays are the main agents that cause spontaneous mutation. Spontaneous or natural mutations occur at very low frequency, i.e., 10^{-6} per gene in an individual. After the study of H. J. Muller on *Drosophila* (1927) and Stadler (1928) on barley, it was discovered that this frequency could be increased by implementing external mutagenic agent such as X-ray. Later in an experiment, Nilson-Ehle produced two-rowed barley mutant with dense ears and very stiff straw. Further, Gustafsson named these mutants as erectoids. Apart from being higher yielder, these mutants could tolerate fertilizer toxicity. After the discovery of mustard gas as a potent chemical mutagen by Auer Bach and Robson (1946), several other agents have been uncovered that are effective enough to enhance the frequency of induction of mutations.

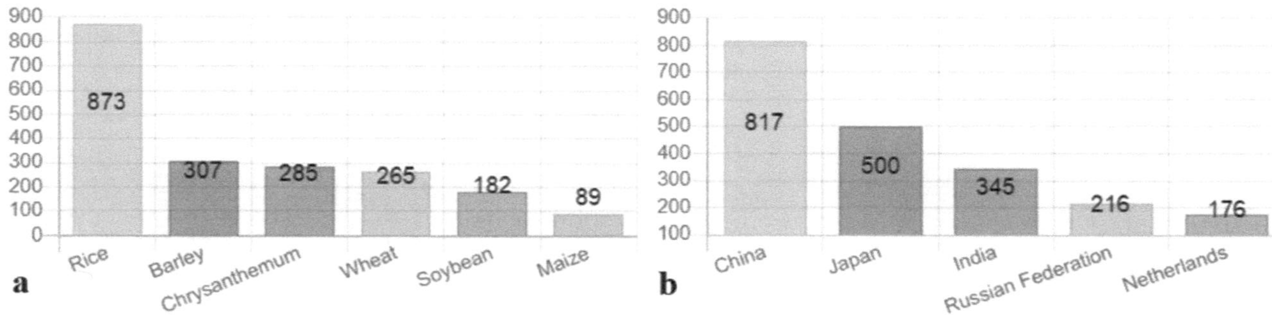

FIGURE 3.2 (a) Crops and (b) nations with highest numbers of mutant varieties all over the world.

Source: https://mvd.iaea.org

3.3.6 Types of Mutagens

The mutation-inducing agents are basically called mutagens. These mutagens are classified into following two categories according to their nature (Table 3.1):

a) *Physical*: X-rays, γ- rays, UV radiations, β-particles, fast and thermal neutrons, and accelerating particles.
b) *Chemical*: Base analogues, antibiotics, alkylating agents, acridines, azides, hydroxylamine (NH_2OH), and nitrous acid (HNO_2)

Despite a variety of mutagenic agents being readily available, it is still challenging to coordinate the induction of mutation with intended character expression. Ionizing radiations, such as X-rays and γ-rays, are typically selected because of their ease of use, excellent penetration capacity, repeatability, higher mutagenic frequency, and little disposal issues. The UV rays can only penetrate a small amount of tissue, which limits their application to pollen grains. However, UV radiation works quite well to irradiate protoplasts and cells of grown plants. The radiation dosage depends on the explosive's length and radiation intensity. It is measured in terms of the number of ionizations that take place and is given in Roentgen ® units. Irradiation dosage is often stated in terms of kR or Gray (Gy) in mutant breeding studies, where 1 Gy = 100 rad and 1 kR = 10 Gy. Radiation absorbed dosage is measured in units of rad, where 1 rad is equivalent to 100 erg/g or 10^{-2} J/kg and can be represented as rad per second, minute, or hour.

The most often utilized chemical mutagens include ethyleneimine (EI), diethyl sulfate (DES), methyl nitroso urea (MNH), ethyl nitroso urea (ENH), and ethyl methane sulphonate (EMS), which are all members of a particular group of alkylating agents. All of these substances interact with DNA by alkylating with the purine and pyrimidine bases as well as the phosphate groups. A chemical mutagen's dosage is primarily influenced by concentration, time, and temperature of treatment. Also, presoaking, postwashing, postdrying, and various conditions of the solution used such as pH, concentrations of present metallic ions, carrier agents, and seed storage are modifiable parameters of mutagenesis. EMS dosages for causing morphological mutations range from 0.01 to 0.8%. From species to species, there are very few variations in the dosage to be employed in the therapy (Siddiqui, 1999). Effective doses of γ-rays to cause chlorophyll as well as morphological changes in lentil were reported to range from 214 to 218 Gy (Pramanik et al. 2023a), while 250 Gy was established as the 50% mortality dosage (LD_{50}) in a previous study (Malik et al. 1998). On the other hand, Rajput et al. (2001) discovered the directly proportional relationship between the frequency of chlorophyll mutants and the dose of γ-rays. In that particular study, they found maximum and minimum number of mutants when exposed to 600 and 200 Gy, respectively. The largest number of mutations are produced with the least amount of risk at the mutagen's optimal dosage. A preliminary therapy can be used to identify an ideal dosage. Overdosing on mutagens will result in a large number of plant deaths, while underdosing will result in low mutation frequencies. Lower frequencies, however, could have the benefit of inducing fewer undesired background mutations in addition to the desired mutation. The result of induced mutations has

TABLE 3.1
Different Categories of Mutagens with Their Mode of Actions

Types of mutagens			Mode of action	Examples
Physical (Radiation)	Ionizing radiation	Particulate radiation	Breakage of H-bonds and sugar-phosphate moiety, cross-linkage between DNA strands.	α-Rays, β-rays, fast and thermal neutrons
		Nonparticulate radiation		X-rays, γ-rays
	Nonionizing radiation		Induction of purine and pyrimidine dimer formation.	UV rays
Chemical	Alkylating agents		Alkylate phosphate groups, purine and pyrimidine bases. Leads to mispairing or loss of bases.	EI, EO, EMS, MMS, EES, DES, NEU, NMU, MNNG, Sulfur and nitrogen mustards
	Acridines		Intercalates between bases disrupting their alignment and pairing. Results in deletion or addition of bases.	Acriflavine, Proflavine, Acridine orange, Acridine yellow, EtBr
	Base analogues		Base pair substitution.	5-Azocytidine, 5-bromo-deoxyuridine, 2-Aminopurine, Hypoxanthine, Maleic hydrazide, 6-Mercapto purine
	Others		Replacement of amino group with a hydroxyl (-OH) group, conversion of cytosine to a modified base.	HNO_2, NH_2OH, NaN_3

Abbreviations: DES, diethyl sulfonate; EES, ethyl ethane sulfonate; EI, ethylene amine; EMS, ethyl methane sulfonate; EO, ethylene oxide; EtBr, ethidium bromide; MMS, methyl methane sulfonate; MNNG, N′-methyl-N-nitro-N-nitroso-guanidine; NEU, N-nitroso-N-ethyl urea; NMU, N-nitroso-N-methyl urea.

been discovered to be influenced by a number of variables, including dosage rate, native plant species, type of target tissues, water content, oxygen level, and temperature of the target tissue, The water content of the seed has a significant impact on the frequency of mutations. Irradiating seeds that contain 12–14% water often result in a greater mutation frequency. In case of chemical mutagenesis, presoaking of the seeds in water for 8–12 hours and their posttreatment drying play a very crucial role.

3.3.7 Mutagenic Practices

3.3.7.1 Choice of Materials

To achieve clearly stated goals through mutant breeding, the selection of parent material is essential. Programs for breeding desirable mutations should be carefully thought out, well organized, and large enough to pick out the low frequencies that are most likely to occur. Improvement of regionally adapted variants or the creation of novel alleles not present in germplasm collections may be the goals of mutation breeding. While a locally adapted variety must be employed when the goal is to introduce a new variety through mutation breeding, any plant introduction can be used to develop new germplasm for use in cross-breeding. It is very important that the variety chosen for mutagenesis be among the finest recently released kinds. Two or more kinds must be used for mutagenesis because various genotypes respond to mutagens differently. Selecting a high-producing variety with good adaptation is usually useful for enhancing one or two particular features. All plant components can be altered in some way, particularly when using a certain kind of mutagen. Although treating seed is more usual, crops that are grown vegetatively can also benefit from treating entire plants, dormant cuttings, bulbs, tubers, and corns. Before treatment, soaking seeds in water produces favorable outcomes. Because it can withstand physical conditions that are often only tolerated by nonliving molecules, seed is the substance that is treated the most frequently. When treating seeds, the LD 50 dose—enough to prevent germination by around 50%—is typically employed to achieve positive outcomes. A mutagen's dosage and pace, or length of administration, differ depending on the type of plant and should be established through testing.

3.3.7.2 Selection of Mutants for Monogenic Traits

The type of population created through mutagenic treatment depends on the crop plant's reproductive strategy and the genetics of the desired improvement in character. The following plan can be used in a mutation breeding program to enhance a variety for a trait controlled by major genes (Figure 3.3):

i. About 500–600 seedlings are given the necessary mutagen treatment. The M_1 generation consists of the plants that are grown from treated seeds. The M_1 is heavily planted, and each plant's seeds are collected separately. It is important to carefully watch the M_1 plants to detect any dominant alterations.
ii. Seeds collected from M_1 single plants in plant to progeny rows are used to raise the M_2 generation. Plants are collected individually from progeny rows thought to contain mutant alleles. M_2 exhibits the recessive mutations. The

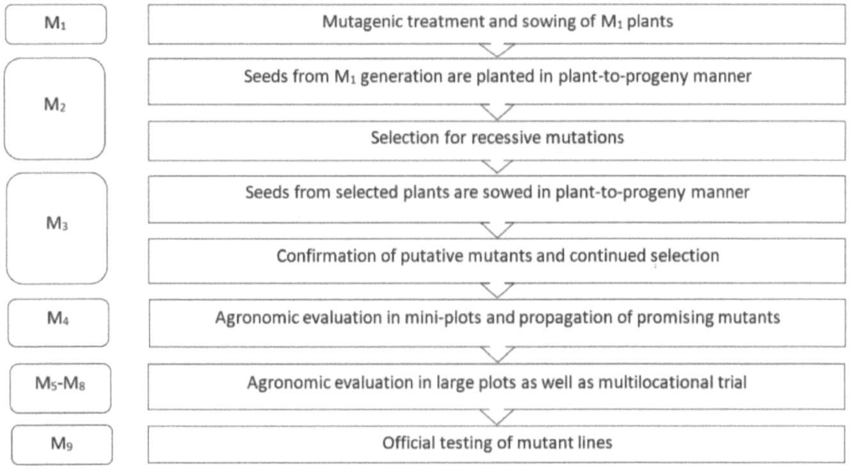

FIGURE 3.3 A representative plan for mutagenic treatment for monogenic traits.

size of the M_2 generation should be sufficient to meet the targeted goals of the mutant breeding program.

iii. To raise the M_3 generation, the chosen M_2 plants are cultivated in progeny rows. While the heterogeneous progenies are subsequently exposed to selection of mutant plants, if any, which can be cultivated to develop M_4 homogeneous progenies, the homogeneous mutant progenies are collected as bulks in the M_3 generation.

iv. The uniform mutant progenies from the M_4 generation are then assessed, ideally in duplicated trials. If there are any segregating descendants, they must be discarded.

v. In M_5, a preliminary yield experiment is conducted coupled with an appropriate check to assess the superior mutant progenies. Before being introduced as a new variety, the potential stable mutant progenies/lines undergo additional testing over the course of two to three years in many locations. Low-producing mutant lines may be kept and used in a hybridization procedure to produce variability.

3.3.7.3 Selection of Mutants for Polygenic Traits

Polygenic features experience genetic variation as a result of induced mutagenesis. This variance is often up to 50% more than that of the F_2 generation. Irradiation can occasionally cause the mean of the M_2 population to drop, although some M_3 families outperform the control or F_2 populations. Therefore, both positive and negative mutations are produced. Positive mutations offer opportunities for selection-based enhancement of polygenic features. M_1 and M_2 are grown in the same manner as for monogenic characteristics for polygenic traits. The robust plants in M_2 are chosen and harvested individually to produce distinct plant offspring in M_3. In the M_3 generation, few homogenous rows are harvested as bulks while inferior progeny rows are discarded. The discovery of mutants with changed quantitative features, such as partial or horizontal disease resistance, is aided by careful inspection of a large number (1000) of M_3 progeny rows. Bulked seeds from M_3 rows that are uniform can be evaluated in M_4 yield trials with a proper check. The superior offspring are next put to the test in repeated, multilocation yield experiments with an appropriate check. Again, M_4 may reject the segregating progenies. Outstanding homogenous progenies may be offered as new kinds following two to three years of testing at several locations.

3.3.8 Importance of Mutation in Plant Breeding

The development of molecular genetics and DNA technology has ushered in a molecular age for plant breeding, including mutant breeding (Chen et al. 2019). Molecular mutant breeding has the ability to increase the effectiveness and power of plant breeding when compared to conventional methods (Hasan et al. 2021). Crop improvement methods such as physical or chemical plant mutagenesis are well established. Numerous DNA technologies are now accessible for crop improvement thanks to quick advances in molecular genetics and genomics, which have also transformed breeding and developed molecular plant breeding approaches (Mussgnug et al. 2017). Another important role that mutant breeding has played and will continue to play is the creation of agricultural products with innovative traits. New mutant kinds of legumes and cereals with low phytate content or high resistant starch content are ideal examples, as was previously stated. Numerous food crop mutant breeding efforts are actively seeking novel variants that are antinutrient less or antinutrient free, or varieties with high concentrations of micronutrients, antioxidants, vitamins, and other nutrients (Zhao and Shewry, 2011).

As mutation research advanced, several essential genetic studies (dominance determination, mapping, allelism testing, etc.) were conducted. Later, the implementation of molecular biology provides the opportunity to explore the molecular basis of mutagenic events, which could be further exploited in gene discovery and breeding. Molecular genetics methods are developing swiftly, and it is anticipated that molecular mutation breeding will be significantly impacted by bioinformatics tools for processing massive amounts of data (Hallajian, 2016).

Long-term success depends on crop genetic improvement, which requires a combination of creative inventions and translational science (Hall and Richards, 2013). Because it is a rapid and inexpensive approach to creating new alleles and phenotypes, we think that breeders will continue to find use for induced mutagenesis as a helpful tool. Furthermore, new technologies will shed insight into gene function and agricultural production and enable the identification of the mutant alleles that were used to create successful mutant varieties (Jankowicz-Cieslak et al. 2017). Mutations are advantageous for both genetic diversity and gene functional studies. Examining mutants using forward genetics (from phenotype to gene) or reverse genetics (from gene to phenotype) can be used to assess gene function (Stainier et al. 2017).

Sweden was the first country to demonstrate the value of mutant breeding for crop improvement, starting with the practical plant breeding of agricultural plants using

X-ray and UV-generated mutations. They found mutants with extraordinarily stiff, taller straw, thick heads, and late maturity.

These mutants yielded more than the mother variety and generated more straw. The development of dwarf wheat and rice varieties during the Green Revolution is a prominent illustration of mutation breeding carried out through effective exploitation of mutant genes.

3.4 ACHIEVEMENTS THROUGH MUTATION BREEDING

Some advantages of mutant breeding include increased yield, earliness, stress resistance, salt tolerance, waterlogging tolerance, and bold seed size. Improved varieties of wheat, rice, and barley have been created through mutation breeding. Numerous crop types have been developed with high yield, higher quality, early maturity, dwarfness, disease resistance, and low toxin content. Inducing male sterility through mutation has also been used to reduce the cost of producing hybrid seeds, broaden genetic variation, and facilitate crop adaptability.

According to an earlier report, 2252 mutant versions of plant species have been produced worldwide. While 1585 were produced directly, 667 were released as a result of hybridization with mutants. Out of a total of 2252, 1700 mutant variations have been introduced in crops grown from seeds, whereas 552 were raised vegetatively. Among seed-propagated species, rice (434 mutant variations), barley (269), and wheat are the three species with the most mutant types produced (222). Maximum variations have developed as a result of radiation. Maximum mutant variants among chemical mutagens were produced as a result of EMS (Ahloowalia et al. 2004).

3.4.1 Genetic Constraints in Improving the Productivity in Pulses

Due to their extremely small genetic basis, pulses have limited potential for genetic enhancement through conventional methods. Further accelerating genetic degradation is the continuous use of restricted cultivars in hybridization to develop superior variety. The enormous responsibility of increasing genetic variety to widen the genetic basis and break the production plateau is given to breeders. One key strategy for creating genetic diversity that may be effectively used to the breeders' benefit is induced mutation. Through mutation breeding, a significant number of variations have been created around the world, many of which have caused an economic revolution among farmers. In addition, socioeconomic issues as well as biotic and abiotic pressures limit the production of key pulses. Because they are high in N and P, illnesses and insect pests like to target pulses. In India, the majority of pulses are cultivated in unreliable environments with little soil fertility. Nearly 90% of the land used for growing pulses receives rain, and as a result, their output is reduced by drought and heat stress. Additionally, alkaline and acidic soils are an issue in the country's semiarid and desert regions. In our nation, pulses are still mostly farmed on subpar soils with minimal inputs. The low output is made worse by the high rate of blossom drop and ovule abortion. The output of pulses is further hampered by the lack of high-quality seeds.

3.4.2 Effective Doses for Induction of Mutations

The kind of mutagen, the crop species, and the genotype within a crop species all affect the effective dosage for inducing mutations. In order to establish the most effective dosage for a certain crop and variety, a number of tests must be carried out. In most cases, it is discovered that a dosage that is only a little bit below the lethal dose 50 (LD_{50}) or growth reduction 50 (GR_{50}) is beneficial in causing important mutations. Dosages for gamma radiation have been standardized for mutant breeding in several pulse crops based on the results of past investigations. Mungbean and urdbean respond well to doses between 300 and 400 Gy, but pigeonpea responds well to doses between 100 and 200 Gy. It has been discovered that cowpea 200–300 Gy and chickpea 300–400 Gy are both effective in inducing beneficial mutations. By using the abovementioned effective dosages, a number of mutant varieties—either direct or mutant derivatives—have been issued in mungbean, urdbean, pigeonpea, and cowpea. In recent years, LD_{50} values for various crops and radiosensitive tests of various crops for electron beam have both been determined (mungbean: 500 Gy; urdbean: 400 Gy; chickpea: 300 Gy; and cowpea: 270 Gy). The efficiency of a mutagen is determined by its capacity to cause chlorophyll mutations in any mutation breeding scheme. In the M_2 generation, a wide variety of mutations that impact the development of chlorophyll are observed. Similar to other crops, these pulse crops have been extensively reported to contain albinos (totally devoid of chlorophyll and therefore dying within a few days), xanthas (yellow-colored), chlorinas (yellow with a tinge of greenness), and viridis (viruscent) (Bulankova et al. 2021). The most numerous species were discovered to be xanthas, followed by chlorina and albinos. Additionally, it has been discovered that populations treated with electron beams as opposed to populations treated with gamma rays had a greater overall incidence of chlorophyll mutations (Chaudhary et al. 2019; Chen et al. 2019). Variegated chlorophyll mutations that affect various plant sections have also been identified.

3.5 MUTATION BREEDING IN PULSES

The induced mutants that have the desired agronomic qualities can either be issued as variety right away or can be employed in hybridization with cultivars or other mutants to create mutant varieties (Table 3.2). Ionizing radiation in particular has been used as a physical mutagen in the development of the majority of these mutant species. Despite the fact that mutation is random, acquiring desirable mutant features depends on three things: the effectiveness of mutagenesis, the plant material used as a starting point, and an effective screening procedure.

3.5.1 Mutations for Morphological Changes

Over the years, a wide range of morphological macromutants in pulse crops have been identified. In various pulse crops, mutations have been found that affect various characteristics of stem (viz. plant height, internode length), leaf (shape and margin of leaf, number of leaflets), seed (size and shape of seeds, color of testa, pod wall

TABLE 3.2
Some Representative Mutant Varieties of Pulses with Their Improved Characteristics

Pulses	Varieties	Special features
Mungbean	TM-2000-2	Suitable for rice fallows and resistant to powdery mildew
	TM-96-2	Resistant to powdery mildew and *Corynespora* leaf spot
	TJM-3	Resistant to powdery mildew, yellow mosaic virus and *Rhizoctonia* root rot diseases
	TMB-37	Tolerant to yellow mosaic virus with early maturity
	Binsmung-5	Tolerant to yellow mosaic virus
	Binamash-1	
	MUM-2	Early maturing dwarf variety with significant tolerance against yellow mosaic virus
	TARM-18	Resistant to powdery mildew
	TARM-1	
	LGG-407	
	TARM-2	Resistant to powdery mildew with higher yield
	TAP-7	Tolerant to powdery mildew and high yielding
	CO-4	High yielder, 85 days maturity and drought tolerant
	Pant M-2	High yielder, resistant to yellow mosaic virus
	BM-4	Bushy erect type, tolerant to yellow mosaic mildew and powdery mildew
	OUM 11-5	Resistant to powdery mildew, yellow mosaic virus and *Corynespora* leaf spot
Pigeonpea	PKVTARA	Resistance to *Fusarium* wilt disease
	TJT-501	High yielding, Early maturing, Tolerant to *Phytophthora* blight
	TT-401	High yielding, tolerant to pod borer and pod fly damage
	TAT-10	Early maturing with medium-sized seed
	TT-6	Large seeded
	Trombay Vishakha-1	Increased seed size
	CO-3	High yielding, bold seeded
	CO-5	Early maturing type with photoinsensitive and drought tolerant
	TAT-5	Early type (140 days) with Increased seed size
Black gram	TU40	Powdery mildew resistance
	TU 94-2	Resistant to yellow mosaic virus with higher yield ability
	Sarla	Resistant to yellow mosaic virus
	TAU-2	High yielding
	TPU-4	Large seeded and high yielding
	TAU-1	Higher yield, large seeded and resistant against powdery mildew
	CO-4	Early maturing, erect and determinate type
	Vamban-2	Glabrous pods, tolerant to drought

(Continued)

TABLE 3.2 (Continued)
Some Representative Mutant Varieties of Pulses with Their Improved Characteristics

Pulses	Varieties	Special features
Cowpea	TRC-77-4	Suitable for rice-based cropping system
	TRC-901	Suitable for summer season and resistant to cowpea mosaic virus
Chickpea	CM-88 CM-98	Resistant against *Ascochyta* blight and Fusarium wilt
	RSG-2	High pod number, early, high yielding with salinity tolerance
	Pusa-408	High yielding with 140–150 days maturity
	Pusa-413	High yielding, Wilt resistant, 130–140 days maturity
	Pusa-417	High yielding, wilt resistant, semierect, dwarf, 110–130 days maturity
	BGM-547	High yielder, mature within 135 days and tolerant to wilt, root rot and stunt diseases
Lentil	Ranjan (S-256)	High yielding, spreading type 110 days of maturity
	PL 77-2	Spreading type, high yielding tolerant to wilt and *Ascochyta* blight
	Rajendra Masoor-1	Cold tolerant
	RH-44	Tolerant to herbicide viz. Imidazolinone
	M1-30 M1-596	Drought tolerant
	ML-9 ML-27	Erect and bushy type
	AEL 12/30/91	High yielding, early maturing type

proportion, etc.), plant itself (such as stature and type of plant growth), root length, time of flowering and maturity, pollen sterility, phyllotaxy, etc. For use in mutation breeding, several of these mutants are significant genetic resources. The mutants with pods above the canopy might be exploited to create types that are compatible with mechanized harvest. Similar to this, the cowpea determinate mutants might be utilized to reconstruct the plant ideotype for creating genotypes that mature at the same time (Souframanien et al. 2020). The early maturing mutants found in cowpea, mungbean, and chickpea are useful donors for shortening growth length and may allow plants to avoid reaching a critical water stress point. The long root mutants seen in mungbean, urdbean, cowpea, and chickpea may show promise in settings where there is less water stress. These pulse crop mutants for seed size and color might be utilized to create market ideotypes based on local customer preferences. Additionally, epistatic interactions in cowpea have been seen to cause many characters to change at the same time in response to a single mutation (Pandey and Dhanasekar, 2004).

3.5.2 Mutations for Disease Resistance

The first mutant mungbean variety with superior resistance against powdery mildew (TARM-2) in Asia was developed through induced mutagenesis, and eventually, a series of powdery mildew-resistant mungbean varieties were evolved by utilizing the previous mutant (Reddy et al. 2008). Development of yellow mosaic virus (YMV)–resistant varieties was also a successful outcome in this field, which have been successfully used in hybridization, leading to development of YMV-resistant varieties. Mutants with resistance to cowpea aphid-borne mosaic virus and leaf crinkle diseases have also been identified in cowpea (Souframanien et al. 2020) and are currently being utilized in the crossing program toward varietal development.

3.5.3 Mutations for Pest Resistance

It was discovered that the wild urdbean from the Trombay Hills (*Vigna mungo* var *silvestris*) has bruchid resistance, and it has been registered (INGR10133) as a possible genetic donor for bruchid resistance. These genotypes were used in the mutation breeding program, which resulted in the development of many mutant descendants (Gopala Swamys et al. 2016). In mungbean, bruchid resistance has been identified in a special genotype called "Thokalwadi wild" that belongs to the *Vigna radiata* var *sublobata* group (Dhole and Reddy, 2016). This genotype is currently being used in hybridization with our high-yielding mutants to create bruchid-resistant varieties.

3.5.4 Mutations for Biochemical Parameters

In chickpea, blackgram, cowpea, and mungbean, mutations altering phytic acid (PA) content have been found, and low PA mutants have been obtained (Souframanien et al. 2020). To lower the PA content to acceptable levels while maintaining the physiological balance, these mutants are being used in varietal development programs. According to Dhole and Reddy (2016), PA has also been shown to enhance the innate resistance of mungbean against biotic stressors including pests and diseases, and drought tolerance capacity in chickpea too (Joshi-Saha and Reddy, 2015).

The astringent, polyphenolic macromolecules known as tannins (or tannoids) attach to and precipitate proteins as well as a variety of other organic substances, such as amino acids and alkaloids. The breeding effort seeks to lessen these antinutritional elements without compromising their protective function. Cowpea mutants with reduced tannin content may be exploited to create pronutritional variants.

3.5.5 Mutations for Genetic Biofortification

The two common mineral micronutrient deficits in the world—iron (Fe) and zinc (Zn)—can be addressed by genetic biofortification. A promising candidate for biofortification is the chickpea, a food legume that is produced and consumed all over the world. It is especially beneficial for poor countries and people who are vegetarians out of choice or because animal protein is unaffordable. A preferable breeding goal for chickpea might be genetic biofortification for increased Fe and Zn with moderate

PA. Genotypes with consistently low/moderate PA and high Fe and/or Zn concentration were found (Dhole and Reddy, 2016).

3.5.6 Mutations for Varietal Development

Direct mutants with desirable agromorphological qualities or hybrids with mutants for the introduction of mutant traits into superior genetic backgrounds might be the products of mutation breeding. Positive mutations regarding the yield-attributing traits like plant height, number of branches per plant, clusters per plant, seeds per pod, pods per plant, etc. have been found in urdbean, mungbean, pigeonpea, cowpea, and chickpea. When elite cultivar T-9 was crossed with big seed mutants UM-196 (dark green leaf mutant) and UM-201 in urdbean, high-yielding cultivars TAU-1, TAU-2, and TPU-4 were created. Similar to this, three early maturing and high-producing varieties (TT-401, TJT-501, and PKVTARA) of pigeonpea have been created in addition to a rapid neutron-induced big seed size mutant variety TT-6 that was crossed with ICPL 84008. Kopergaon and TARM-2 were used in mungbean crosses to create the early maturing variety TMB-37 for the Northeast Plain zone. Due to its greater yield and earliest maturity, the State of Punjab has recently reintroduced this variety. A prime example of a dwarf determinate mutant being immediately issued as a variety in India is the cowpea variety TRC-77-4 (Souframanien et al. 2020). Some mutant cultivars, such as the mungbean variants TM-96-2, TM2000-2, TU-40, and TRC-77-4 (cowpea), are also appropriate for niches like rice fallows.

3.5.7 Novel Mutations

The globally conserved Arabidopsis Terminal Flower 1 (TFL1) gene homolog was found to contain a new SNP mutation in three gamma ray–induced determinate mutants in cowpea. The TFL protein's stability and function were adversely affected, resulting in loss of function, by the nonsynonymous exonic point mutation caused by the conversion of cytosine (C) to adenine (A) (Pro-136 to His) in determinate mutants (Dhanasekhar and Reddy, 2015). Additionally, in several cultivars of chickpea, agronomical mutants such as early maturity, tall and erect plant type, salt-tolerant mutant, and slow-transpiring mutant have been discovered. Mutants in mungbean were also shown to be drought resistant, have long roots and huge seeds, and mature early and synchronously. In comparison to the control (20.5 cm root length, 0.573 mm leaf thickness, and 40.9°C leaf canopy temperature), the drought-tolerant mutant had considerably longer roots (40 cm), thicker leaves (0.704–0.803 mm), and lower leaf canopy temperature (32.4°C) (Dhole and Reddy, 2018). In the population that had been exposed to an electron beam, a mutant with terminal flowers in cluster beans was found. The development of cluster bean cultivars of the determinate type will benefit from this mutant.

3.6 IMPACT OF MUTANT VARIETIES

Due to their improved production characteristics, several of the mutant pulse varieties created are highly well liked by the farming community. The 1985 introduction of the mutant urdbean variety TAU-1, which was created in conjunction with Dr. PDKV, Akola, is still highly well liked by Maharashtra farmers. In Maharashtra, this cultivar is widely farmed and takes up more than half of the space used for urdbean. Nearly 60% of the pigeonpea plantings in Madhya Pradesh are taken up by the variety TJT-501, which was created in partnership with JNKVV, Raipur. It is also one of the best varieties, garnering 12% of the nation's breeder seed credits [51]. Although it was initially produced for the North East Plain Zone, farmers all throughout the country like the mungbean variety TMB-37 for its early maturity and appropriateness for summer growing. In the state of Punjab, this variety has been readopted and rereleased. Similar to this, the TT-401 pigeonpea variety, which was originally intended for the central zone, has shown promise in southern states like Tamil Nadu, Andhra Pradesh, and Karnataka, where it has produced a record yield of 2500 kg/ha. In Andhra Pradesh and Chhattisgarh, respectively, the mungbean varieties TM96-2 (in collaboration with ANGRAU, Lam, AP) and TM2000-2 (in conjunction with IGKVV, Raipur) with resistance against powdery mildew and *Corynespora* leaf spot occupy sizable areas beneath rice fallows. The recently developed, high-yielding TU-40 type of urdbean is becoming more and more well liked in the southern regions. High yields are being produced by Maharashtrian farmers who grow PVK-TARA pigeonpea, particularly when drip irrigation is used.

3.7 ADVANTAGES OF MUTATION BREEDING

A crop species may be made very variable by mutation breeding, which can then be exploited to improve the germplasm and generate new cultivars. As opposed to hybridization, mutant breeding has a significant time benefit when creating new varieties. In general, it takes 11–12 years to generate a new variety through hybridization, but mutation breeding only needs 8–9 years (Toker et al. 2007). By changing alleles at known loci as well as at previously undiscovered loci, in addition to changing linkage groups, mutations may, in comparison, produce both qualitative and quantitative variation in a shorter amount of time. *Leguminosae* exhibit variety that is limited to a level that is present in the parents used in hybridization, although new variability may be created by mutations for many desirable features (Toker and Cagirgam, 2004). If the gene is present in the treated plant, mutagenesis can produce any form of variant seen in any plant family's germplasm.

3.8 DIFFICULTIES FACED IN MUTATION BREEDING

For various plant species, the mutation frequency varies, and even within a species, there are variations in the genotypic responses to various mutagens and their various treatments. The frequency of favorable mutations is extremely low, at only approximately 0.1%. The strategies utilized to manage various mutant generations, the efficacy of screening approaches, and the size of the population developed in M_1 and subsequent generations all have a role in how successful mutation breeding is. The bigger M_2 population should be cultivated and rigorously examined at various phases of growth and development to extract attractive mutants in order to achieve a greater success rate. Large-scale population screening is more time consuming, labor intensive, and expensive in terms of manpower and resources. While some mutations are undesirable since they cause chromosomal abnormalities and deletions, others have pleiotropic effects caused by related gene(s). Such mutants sometimes require backcrossing to parents or modified types, which is time consuming and labor intensive once again. Furthermore, it is difficult to break the unfavorable connection between a gene of interest and bad genes, which limits the application of mutation breeding for crop advancement on a regular basis.

3.9 CONCLUSION

With conventional breeding and mutational breeding working together, it is possible to create elite varieties of pulses suitable for various agroclimatic zones, increase our nation's production of pulses, and improve its nutritional security. Pulses have a narrow genetic base, which limits their genetic variability. The agricultural population has benefited immensely from the pulse crop types created by mutation breeding throughout the years by improving their standard of living. Induced mutations have also grown in significance in plant molecular biology during the past ten years as a strategy for locating and isolating genes and researching their structure and function. Plant breeders must have a thorough understanding of the genes governing significant agronomic and quality features in order to formulate effective plans and carry out breeding programs. In many instances, several mutant alleles provide the genetic diversity sources for crop breeding as well as functional research of the targeted gene. Therefore, by improving yield potential and stability, induced mutations might further increase global food production both directly and indirectly.

REFERENCES

Ahloowalia, B. S., Maluszynski, M., Nichterlein, K., 2004. Global impact of mutation-derived varieties. *Euphytica,* 135, 2, 187–204. https://doi.org/10.1023/B:EUPH.0000014914.85465.4f

Auerbach, C., Robson, J.M., 1946. Chemical production of mutations. *Nature.* 157, 302. https://doi.org/10.1038/157302a0

Barghi, N., Hermisson, J., Schlötterer, C., 2020. Polygenic adaptation: a unifying framework to understand positive selection. *Nat. Rev. Genet.* 21, 12, 769–781. https://doi.org/10.1038/s41576-020-0250-z

Bezie, Y., Tilahun, T., Atnaf, M., Taye, M., 2021. The potential applications of site-directed mutagenesis for crop improvement: a review. *J. Crop Sci. Biotechnol.* 24, 3, 229–244. https://doi.org/10.1007/s12892-020-00080-3

Bulankova, P., Sekulić, M., Jallet, D., Nef, C., Van Oosterhout, C., Delmont, T.O., Vercauteren, I., Osuna-Cruz, C.M., Vancaester, E., Mock, T., Sabbe, K., 2021. Mitotic recombination between homologous chromosomes drives genomic diversity in diatoms. *Curr. Biol.* 31, 15, 3221–3232. https://doi.org/10.1016/j.cub.2021.05.013

Chaudhary, J., Deshmukh, R., Sonah, H., 2019. Mutagenesis approaches and their role in crop improvement. *Plants,* 8, 11, 467. https://doi.org/10.3390/plants8110467

Chen, K., Wang, Y., Zhang, R., Zhang, H., Gao, C., 2019. CRISPR/Cas genome editing and precision plant breeding in agriculture. *Annu. Rev, Plant Biol.* 70, 667–697. https://doi.org/10.1146/annurev-arplant-050718-100049

Das, A., Mahanta, M., Pramanik, B., Purkayastha, S., 2023. Genetically modified crops and crop species adapted to global warming in dry regions. In *Enhancing Resilience of Dryland Agriculture Under Changing Climate: Interdisciplinary and Convergence Approaches* (eds) Anandkumar Naorem and Deepesh Machiwal (pp. 385–409). Singapore: Springer Nature Singapore. https://doi.org/10.1007/978-981-19-9159-2_19

Debnath, S., Sarkar, A., Perveen, K., Bukhari, N.A., Kesari, K.K., Verma, A., Chakraborty, N.R., Tesema, M., 2022. Principal component and path analysis for trait selection based on the assessment of diverse lentil populations developed by gamma-irradiated physical mutation. *BioMed Res. Int.* 2022, 1–14. https://doi.org/10.1155/2022/9679181

Dhanasekar, P., Reddy, K.S., 2015. A novel mutation in TFL1 homolog affecting determinacy in cowpea (*Vigna unguiculata*). *Mol. Genet. Genom.* 290, 55–65. https://doi.org/10.1007/s00438-014-0899-0

Dhole, V.J., Reddy, K.S., 2016. Association of phytic acid content with biotic stress tolerance in mungbean (*Vigna radiata* L. Wilczek). *Phytoparasitica,* 44, 261–267. https://doi.org/10.1007/s12600-016-0514-5

Dhole, V.J., Reddy, K.S., 2018. Genetic analysis and variability studies in mutants induced through electron beam and gamma rays in mungbean (*Vigna radiata* L. *Wilczek). Electron. J. Plant Breed.* 9, 1, 304–312.

Drummond, D.A., Wilke, C.O., 2008. Mistranslation induced protein misfolding as a dominant constraint on coding-sequence evolution. *Cell,* 134, 2, 341–352. https://doi.org/10.1016/j.cell.2008.05.042

Gaul, H., 1964. Mutations in plant breeding. *Rad. Bot.* 4, 3, 155–232. https://doi.org/10.1016/S0033-7560(64)80069-7

Hall, A.J., Richards, R.A., 2013. Prognosis for genetic improvement of yield potential and water-limited yield of major grain crops. *Field Crops Res.* 143, 18–33. https://doi.org/10.1016/j.fcr.2012.05.014

Hallajian, M.T., 2016. Mutation breeding and drought stress tolerance in plants. In: *Drought Stress Tolerance in Plants*, Vol 2 (eds) Hossain M., Wani S., Bhattacharjee S., Burritt D., Tran LS. Springer, Cham, pp. 359–383. https://doi.org/10.1007/978-3-319-32423-4_13

Hasan, N., Choudhary, S., Naaz, N., Sharma, N., Laskar, R. A., 2021. Recent advancements in molecular marker-assisted selection and applications in plant breeding programmes. *J. Genet. Eng. Biotechnol.* 19, 1, 1–26. https://doi.org/10.1186/s43141-021-00231-1

Hawkesford, M.J., Griffiths, S., 2019. Exploiting genetic variation in nitrogen use efficiency for cereal crop improvement. *Curr. Opin. Plant Biol.* 49, 35–42. https://doi.org/10.1016/j.pbi.2019.05.003

Jankowicz-Cieslak, J., Mba, C., Till, B.J., 2017. Mutagenesis for crop breeding and functional genomics. In: Biotechnologies for plant mutation breeding (eds) Jankowicz-Cieslak J., Tai T.H., Kumlehn J., Till B.J., Springer, Cham, pp. 3–18. https://doi.org/10.1007/978-3-319-45021-6_1

Joshi-Saha, A., Reddy, K.S., 2015. Repeat length variation in the 5′UTR of myo-inositol monophosphatase gene is related to phytic acid content and contributes to drought tolerance in chickpea (*Cicer arietinum* L.). *J. Exp. Bot.* 66, 19, 5683–5690. https://doi.org/10.1093/jxb/erv156

Konuma, H., 2016. Status of world food security and its future outlook, and role of agricultural research and education. *J. Sustain. Agric.* 10, 2, 69–75. https://doi.org/10.11178/jdsa.10.69

Koornneef, M., Meinke, D., 2010. The development of Arabidopsis as a model plant. *Plant J.* 61, 6, 909–921. https://doi.org/10.1111/j.1365-313X.2009.04086.x

Lamo, K., Bhat, D.J., Kour, K., Solanki, S.P.S., 2017. Mutation studies in fruit crops: a review. *Int. J. Curr. Microbiol. Appl. Sci.* 6, 12, 3620–3633. https://doi.org/10.20546/ijcmas.2017.612.418

Lynch, M., Blanchard, J., Houle, D., Kibota, T., Schultz, S., Vassilieva, L., Willis, J., 1999. Perspective: spontaneous deleterious mutation. *Evolution,* 53, 3, 645–663. https://doi.org/10.1111/j.1558-5646.1999.tb05361.x

Malik, I.A., Chaudhry, M.S., Ashraf, M., Erskine, W., 1998. Radio-sensitivity and mutability in lentil (*Lens culinaris* Medik.) as related to seed size. *J. Genet. Breed.* 52, 9–15

Miglani, G.S., 2017. Genome editing in crop improvement: present scenario and future prospects. *J. Crop Improv.* 31, 4, 453–559. https://doi.org/10.1080/15427528.2017.1333192

Mir, A.S., Maria, M., Muhammad, S., Ali, S.M., 2020. Potential of mutation breeding to sustain food security. *Genet. Var*. 1–15.

Mussgnug, J.H., 2017. Nuclear transformation and toolbox development. In: *Chlamydomonas: Molecular Genetics and Physiology* (ed) Hippler M., Springer, Cham, pp. 27–58. https://doi.org/10.1007/978-3-319-66365-4_2

Nazarenko, M., Lykholat, Y., Grygoryuk, I., Khromikh, N., 2018. Optimal doses and concentrations of mutagens for winter wheat breeding purposes. Part I. Grain productivity. *J. Cent. Eur. Agric.* 19, 1, 194–205. https://doi.org/10.5513/JCEA01/19.1.2037

Nei, M., 2007. The new mutation theory of phenotypic evolution. *PNAS,* 104, 30, 12235–12242. https://doi.org/10.1073/pnas.0703349104

Nelson, R., 2020. International plant pathology: past and future contributions to global food security. *Phytopathology,* 110, 2, 245–253. https://doi.org/10.1094/PHYTO-08-19-0300-IA

Ohno, M., 2019. Spontaneous de novo germline mutations in humans and mice: rates, spectra, causes and consequences. *Genes Genet. Syst.* 94, 1, 13–22. https://doi.org/10.1266/ggs.18-00015

Oladosu, Y., Rafii, M. Y., Abdullah, N., Hussin, G., Ramli, A., Rahim, H. A., Miah, G., Usman, M., 2016. Principle and application of plant mutagenesis in crop improvement: a review. *Biotechnol. Biotechnol. Equip.* 30, 1, 1–16. https://doi.org/10.1080/13102818.2015.1087333

Pandey, R.N., Dhanasekar, P., 2004. Radiation induced mutant in cowpea. *JRF.* 1, 1, 1–4.

Pathirana, R., 2011. Plant mutation breeding in agriculture. *Plant Sci. Rev.* 6, 032, 107–126. https://doi.org/10.1079/PAVSNNR20116032

Pramanik, B., Debnath, S., Rahimi, M., Helal, M.M.U., Hasan, R., 2023a. Morphometric frequency and spectrum of gamma-ray-induced chlorophyll mutants identified by phenotype and development of novel variants in lentil (*Lens culinaris* Medik.). *PLoS One*, 18, 6, e0286975. https://doi.org/10.1371/journal.pone.0286975

Pramanik, B., Sar, P., Bharti, R., Gupta, R., Purkayastha, S., Sinha, S., Chattaraj, S., Mitra, D., 2023b. Multifactorial role of nanoparticles in alleviating environmental stresses for sustainable crop production and protection. *Plant Physiol. Biochem*. 201, 107831. https://doi.org/10.1016/j.plaphy.2023.107831

Raina, A., Laskar, R. A., Khursheed, S., Amin, R., Tantray, Y. R., Parveen, K., Khan, S., 2016. Role of mutation breeding in crop improvement—past, present and future. *Asian Res. J. Agric.* 2, 2, 1–13.

Rajput, M.A., Sarwar, G., Siddiqui, K.A., 2001. Development of high yielding mutants in lentil. *Mutation Breeding Newsletter.* 45, 35–36.

Reddy, K.S., Dhanasekar, P., Dhole, V.J., 2008. A review on powdery mildew disease resistance in mungbean. *J. Food Leg,* 21, 151–155.

San Martin, W., 2021. Global nitrogen in sustainable development: four challenges at the Interface of science and policy. In *Life on Land. Encyclopedia of the UN Sustainable Development Goals.* (eds) Leal Filho W., Azul A.M., Brandli L., Lange Salvia A., Wall T., Springer, Cham. pp. 485–499. https://doi.org/10.1007/978-3-319-95981-8_114

Sar, P., Pramanik, B., Debnath, S., 2022. CRISPR/Cas: a new horizon in crop improvement. In: *Futuristic Trends in Biotechnology* (eds) Debnath S., Manjula I.K., Shah K.R., Bhattacharjee S., IIP International Publishers, New Delhi, India, pp. 9–27.

Shu, Q.Y., 2009. Turning plant mutation breeding into a new era: molecular mutation breeding. In: *Induced Plant Mutations in the Genomics Era* (ed) Shu Q.Y., Food and Agriculture Organization of the United Nations, Rome, pp. 425–427.

Siddiqui, B.A., 1999. Mutagenesis: Tools and techniques—A practical view. In: *Breeding in crop plants: Mutations and in vitro mutation Breeding* (eds) Siddiqui B.A., Khan S., Kalyani Publishers, Ludhiana, pp. 20–34.

Souframanien, J., Saha, A.J., Dhole, V.J., Dhanasekar, P. and Misra, G., 2020. Genetic improvement of pulse crops through induced mutation and biotechnological approaches. *IANCAS*, p.71.

Stainier, D.Y., Raz, E., Lawson, N.D., Ekker, S.C., Burdine, R.D., Eisen, J.S., Ingham, P.W., Schulte-Merker, S., Yelon, D., Weinstein, B.M., Mullins, M.C., 2017. Guidelines for morpholino use in zebrafish. *PLoS Genet.* 13, 10, e1007000. https://doi.org/10.1371/journal.pgen.1007000

Swamy, S.G., Mahalakshmi, M.S., Souframanien, J., 2016. Evaluation of certain blackgram varieties for resistance to pulse bruchid, *Callosobruchus maculatus* (F.). *J. Res. Angrau,* 44, 8–13.

Swarup, S., Cargill, E.J., Crosby, K., Flagel, L., Kniskern, J., Glenn, K.C., 2021. Genetic diversity is indispensable for plant breeding to improve crops. *Crop Sci.* 61, 2, 839–852. https://doi.org/10.1002/csc2.20377

Toker, C., Cagirgan, M.I., 2004. Spectrum and frequency of induced mutations in chickpea. *ICPN,* 11, 8–10.

Toker, C., Yadav, S.S., Solanki, I.S., 2007. Mutation breeding. Lentil: an ancient crop for modern times. pp.209–224. https://doi.org/10.1007/978-1-4020-6313-8_13

Van Eenennaam, A.L., Wells, K.D., Murray, J.D., 2019. Proposed US regulation of gene-edited food animals is not fit for purpose. *npj Sci, Food.* 3, 1, 1–7. https://doi.org/10.1038/s41538-019-0035-y

Wilson, K., Long, D., Swinburne, J., Coupland, G., 1996. A Dissociation insertion causes a semidominant mutation that increases expression of TINY, an Arabidopsis gene related to APETALA2. *The Plant Cell,* 8, 4, 659–671. https://doi.org/10.1105/tpc.8.4.659

Yali, W., Mitiku, T., 2022. Mutation breeding and its importance in modern plant breeding. *J. Plant Sci.* 10, 2, 64–70. https://doi.org/10.11648/j.jps.20221002.13

Zhao, F. J., Shewry, P. R., 2011. Recent developments in modifying crops and agronomic practice to improve human health. *Food Policy*, 36, 94–101. https://doi.org/10.1016/j.foodpol.2010.11.011

Zong, X., Yang, T., Liu, R., Zhu, Z., Zhang, H., Li, L., Zhang, X., He, Y., Sun, S., Liu, Q., Li, G., 2019. Genomic Designing for Climate-Smart Pea. In: *Genomic Designing of Climate-Smart Pulse Crops* (ed) Kole C., Springer, Cham, pp. 265–358. https://doi.org/10.1007/978-3-319-96932-9_6

4 Transposon Mutagenesis for Functional Genomics in Cereals

Shabneek Kaur, Jessica Kaur, Umesh Preethi Praba, Gurwinder Kaur, Gautam Chhabra, Rajveer Singh, and Yogesh Vikal
School of Agricultural Biotechnology, PAU, Ludhiana, Punjab, India
*Corresponding Email: yvikal-soab@pau.edu

4.1 INTRODUCTION

Genetic variation is a significant driver of evolutionary diversification and a source of phenotypic variability. Early in the 20th century, researchers demonstrated that exposing *Drosophila* and wheat to X-rays may significantly speed up the relatively slow pace of spontaneous mutation. These accomplishments are remarkable since they anticipate experimental proof that DNA is a heritable substance. This highlights a significant advantage in crop breeding regarding induced mutations: previous knowledge of genes and their function is not required for the development of new varieties. With the introduction of the first mutant variety of tobacco in the 1930s, mutation induction has thus been a crucial technique for crop breeding and functional genomics. Induced mutations, as a complement to plant mutation breeding, have been widely used for functional genomics in experimental organisms and crops. Despite its success, physical and chemical mutagenesis has led to issues like chimerism in cereal plants. Therefore, in the middle of the 20th century, the transposon mutagenesis method saved the day. Barbara McClintock, the first to study the activity of chromosomes in maize, discovered transposons, often known as jumping genes. This research resulted in the identification of the first transposable element, and transposon mutagenesis has since been used as a biological tool.

Transposable elements are a simple and adaptable way to disrupt plant genes and test their function. By engineering transposons to carry reporter genes and regulatory signals, the expression of target genes can be monitored and to some extent manipulated. Transposable elements disrupt gene function by inserting themselves into coding and regulatory areas. This can lead to either loss-of-function or gain-of-function, phenotypes that depict how the target gene performs in the context of an optimal chromosome. Loss-of-function mutations are frequently used to identify crucial genes, resulting in recessive lethality so that they can be preserved in

DOI: 10.1201/9781003392897-4

heterozygous diploids. This method has been used to identify genes in a variety of cereal crops, including rice, oats, maize, wheat, barley, sorghum, and pearl millet.

In order to precisely understand systems biology, functional genomics aims to develop and apply high-throughput genome-wide experimental techniques for the analysis of the function of genes identified and mapped by structural genomics (e.g., genome sequencing projects). Thus, reverse genetics is a strong method for identifying the biological purposes of gene sequences that are currently unknown. It is simple to analyze the transposon insertion mutations, but it is impossible to transform and regenerate independent transposon lines by positive/negative selection for transposition in order to obtain the thousands of insertion lines needed to saturate the entire genome (Krysan et al., 1999; Sundaresan et al., 1995; Tissier et al., 1999).

The following two benefits of transposon insertional mutagenesis are its main strengths: (a) first, it provides a direct means for determining the function of a gene product in planta revealing a direct relationship between the gene sequence and the phenotype (Krysan et al., 1999); and (b) second, it may lead to the isolation of tagged mutant genes through the recovery of the DNA sequences flanking the insertion sequence (Alonso et al., 2003; Azpiroz-Leehan and Feldmann, 1997; Krysan et al., 1999), which may, in turn, lead to the isolation of wild-type gene sequence.

In most of the important cereal species except maize, the creation of large transposon insertion populations is also specifically a challenging task due to either the lack of suitable transposons or the availability of useful methods for the activation of endogenous transposons. Nonetheless, in the recent past, increasingly major efforts have been undertaken toward the creation of large insertion mutagenesis populations for forward- and reverse-genetics analyses in at least three cereal species including rice, maize, and barley (H. S. Balyan et al., 2008). This book chapter aims at finding and analyzing the functions of various genes found through transposon mutagenesis, particularly in cereal crops, and the forthcoming future aspirations of the same.

4.2 BACKGROUND

Barbara McClintock initially identified transposons, also known as transposable elements, in maize (McClintock, 1950). McClintock was researching the progeny of self-pollinated maize plants produced by crosses between individuals who had a defective copy of chromosome 9 in the early 1940s. The telomeres on these plants were absent. Because of this study, the first transposable element was discovered, and transposon mutagenesis has since been utilized for biological studies. The first known active DNA transposon in vertebrates was *Tol2*. Because it induced albinism when it was inserted into the tyrosinase gene of Japanese medaka fish, it was first identified in those fish (Koga et al., 1996). These elements occur widely in plants including cereal species (Kumar and Bennetzen, 1999). One of the most widely used transposons is Tn5 which contains two IS50 elements as inverted terminal repeats.

The insertion and transposition abilities of the transposons have been widely harnessed in cereal species such as maize and rice. Most transposable elements have a tendency to preferentially transpose into gene-rich regions (Ito et al., 1999). The gene containing the transposon generally results in knockout phenotypes. These characteristics have made transposon tagging an effective method for functional analysis and gene isolation (Hamer et al., 2001).

Initially, only maize was used for the use of transposon mutagenesis and tagging. The maize transposable elements Ac/Ds, En/Spm, and MuDR/Mu proved useful for isolation of several maize genes (Fedoroff et al., 1983; Robertson, 1978; Sundaresan, 1996; Walbot, 1992, 2000). Later it was used extensively in isolation of genes from other crops like rice, petunia, and snapdragon (*Antirrhinum majus*). The maize Ac/Ds (Baker et al., 1986) and En/(I)Spm (Pereira and Saedler, 1989; Masson and Fedoroff, 1989) transposons have been shown to transpose in a number of heterologous hosts extending their utility for insertion mutagenesis in species that do not have active endogenous transposon (Ito et al., 2004; Shimamoto et al., 1993; Sundaresan, 1996; Upadhyaya et al., 2006). Additionally, the heterologous transposons have advantage over the endogenous transposons for insertional mutagenesis as these may be used after suitable in vitro modification (Pereira, 1998), and moreover the identification of tagged gene responsible for mutation is easier.

In comparison to rice and maize, a much smaller scale of transposon insertion mutagenesis has been effectively used in barley. The potential application of insertional mutagenesis for gene tagging and gene cloning in barley was demonstrated through production of a large number of transgenic insertion lines. Efforts to test the potential suitability of this approach in some other cereal species such as wheat and sorghum are underway. Occurrence of functionally redundant duplicated copies of many genes in rice, maize, barley, and other cereal species such as polyploid bread wheat, etc. (Gaut and Doebley, 1997; Goff et al., 2002) hamper functional analysis of all the available genes due to the following reasons. First, insertional mutation in such genes may not be expressed, as the loss of function in one copy of the gene may be compensated for by the expression of another copy. Second, many genes function at more than one stage of plant development and are critical for plant's viability; mutations in such genes may be lethal. Lastly, mutations with subtle phenotypic alterations may escape detection during phenotypic screening, thereby making it impossible to correlate the changed phenotype with the specific disrupted sequence. In order to circumvent the above problems and to identify genes, complementary techniques of activation tagging and gene trap analyses that utilize random integration of reporter gene constructs have been developed, e.g., ty1, copia, gypsy, lines, and sines (Weigel et al., 2000).

4.3 TYPES OF TRANSPOSONS

Transposable elements are divided into two major categories based on their identified speculated transposition mechanisms (Capy et al., 1998).

- **Class I** or the retrotransposons, which transposes through an RNA intermediate generated by reverse transcription elements. They usually have a large copy number, which reflects their mechanism of transposition (Hirochika et al., 2001), e.g., ty1, copia, gypsy, lines, sines.
- **Class II** or DNA-based transposons, which transpose through DNA cut and paste–type mechanisms, e.g., Ac/Ds elements, En/Spm elements, Is elements.

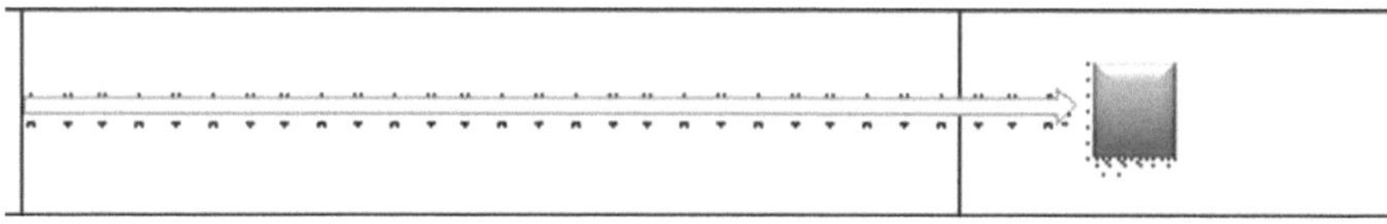

FIGURE 4.1 Timeline of the genes studied and reported from the various cereal crops.

4.4 MECHANISM OF TRANSPOSON MUTAGENESIS

The transposon system is made up of a transposon (a specific DNA sequence) and a transposase enzyme (Figure 4.1). Each strand of the target sequence is cut or nicked by the transposase enzyme. It locates the proper place along the transposon and binds there, excising the transposon. The target sequence is duplicated, and two of the copies displace the transposon from the middle. The transposon then secures itself into the two produced free ends. After that, it inserts the transposon at the proper spot in another DNA sequence. Two strands become continuous when the ligase seals the nicks.

4.5 TRANSPOSON MUTAGENESIS IN CEREAL CROPS

4.5.1 Rice

Rice (*Oryza sativa*) is a major grain crop that is consumed by more than half of the world's population (Pramanick et al., 2020). As rice genome sequences become more widely available, the reverse genetic method of finding out how genes function should become crucial. Due to its small size and economic importance as a major crop, the rice genome has undergone extensive analysis (Sasaki and Moore, 1997; Sasaki, 1999). Unfortunately, effective techniques utilizing such data to systematically analyze the functions of rice genes are still lacking.

However, Maize Ac and Ds elements were used to construct a significant amount of gene tagging resources in rice, a heterologous host. Originally, rice was directly transformed to incorporate the autonomous Ac element which was cloned between a promoter and the htp gene. Rice insertional mutagenesis has also been produced using maize En/Spm transposable elements (Kumar et al., 2005). Retrotransposons or retroelements from both endogenous and heterologous sources have been demonstrated to be randomly dispersed and effective tags (Balyan et al., 2008). Mutant lines produced by exogenous insertional mutagens are, nonetheless, transgenic plants and thus a cause of worry for the environment. Although rice plants are bigger than *Arabidopsis* plants, managing numbers of transgenic rice plants inside a greenhouse is difficult. Thus, it is preferable to use endogenous insertional mutagens for rice (Hirochika et al., 2001). When stressed by culturing, cell culture, pathogen infection, and wounding, some of the rice endogenous retroelements are activated and their transposition is triggered (Balyan et al., 2008). Rice retroelement Tos17's low copy number (1–5) could be raised to 30 by activating retrotransposons, which must be accessed via tissue culture (Balyan et al., 2008). Tos17 has unique characteristics that make it potentially useful for reverse genetics research (Hirochika et al., 2001). In the future, reverse-genetics analysis may make use of a number of other transposable elements that were previously discovered in rice (Balyan et al., 2008). Tobacco's Tnt1 and Tto1 and rice's Tos17 are suitable for use in the investigation of gene function due to the frequency of transposition found in their native or heterologous hosts (Hirochika et al., 2001).

Many problems have been discovered despite the fact that rice has undergone thorough gene-tagging analyses: firstly, the frequent background mutation, which may be a result of the tissue culture procedure, and secondly, the suppression of Ds activity in subsequent generations that inhibits the production of several populations of mutant lines population. On the other hand, Ac has been associated with very steady activity (Hirochika et al., 2001). In the future, reverse-genetics analysis may make use of a number of other transposable elements that were previously discovered in rice (Table 4.1).

4.5.2 Maize

Maize has made a great model for research on transposon genetics and mutagenesis, since the turn of the 20th century (Settles et al., 2004). The maize genome is largely composed of transposable elements (TEs). Several endogenous DNA transposon systems, including the extremely active Ac/Ds and Mu transposons (Robertson, 1978), are readily applicable for extensive mutagenesis in maize. The Mutator (Mu) element family is the most active class II transposon family in maize (Lisch, 2002). MuDR, a high copy number autonomous TE that Robertson first identified in 1978, regulates the transposition of 16 kinds of nonautonomous Mu elements in addition to itself (Lisch, 2015).

Molecular genetics and high-throughput sequencing make it comparatively simpler to find transposon insertions. However, early Mu-based mutation libraries in maize (Macron et al., 2020) faced difficulties in (1) inefficiently recovering and identifying newly transposed insertions and (2) making a distinction between germinal and the high background of somatic insertions. Both of these obstacles have been solved in recent investigations.

Nowadays, transposon-based resources are preferred for maize for a number of reasons. (1) Plant populations with several separate transpositions are relatively easy to produce in maize, compared to self-pollinating species like *Arabidopsis* and rice. (2) The relative ineffectiveness of transformation procedures for maize limits the development of several T-DNA lines. (3) Maize is a major model for transposon genetics because it has a number of genetically well-characterized transposon families. The so-called "cut and paste" DNA transposons that have been most favored for the creation of genomic resources in maize are the Ac/Ds (Cowperthwaite et al., 2002; Kolkman et al., 2005) and Robertson's Mutator (Bensen et al., 1995; May et al., 2003; McCarty et al., 2005) systems. The mechanisms in each of these systems enable genetic control of transposon mobility in the genome (McCarty et al., 2009).

TABLE 4.1
Types of Transposons, the Corresponding Genes, and Their Functions in Various Cereal Crops

CROPS	Insertion of Transposon	Gene	Function	Reference
RICE	Retrotransposon Tos17	OsABAl (Zeaxanthin epoxidase)	The epoxidation of zeaxanthin	Jiang et al., 2010
	Retrotransposon Tos17	PHYA (Phytochrome A)	Controlling photomorphogenesis	Jiang et al., 2010
	Retrotransposon Tos17	OsCHLH (The largest subunit of the rice Mg-chelatase)	Chlorophyll biosynthesis	Jiang et al., 2010
	Retrotransposon Tos17	Three different members of cellulose synthase genes (OsCesA4, OsCesA7, and OsCesA9)	Led to a dramatic decrease of cellulose content, resulting in changes in secondary cell wall formation	Tanaka et al., 2003
	Retrotransposon Tos17	BRANCHED FLORETLESS (EREBP/AP2 domain-containing protein)	Mediating the transition from spikelet to floret meristem	Jiang et al., 2010
	Retrotransposon Tos17	PAIR2	Results in homologous chromosome pairing during meiosis	Nonomura et al., 2004
	Retrotransposon Tos17	ANTHER INDEHISCENCE 1 (MYB transcription factor)	Anther development	Jiang et al., 2010
	Retrotransposon Tos17	OsGAMYB (Transcriptional regulator of gibberellin [GA] dependent)	Induction of alpha-amylase in aleurone and floral organ development	Jiang et al., 2010
	Retrotransposon Tos17	OsMT2b (Metallothionein)	A reactive oxygen species (ROS) scavenger	Jiang et al., 2010
	Retrotransposon Tos17	PAIR1 (Coiled-coil motif-containing protein)	Responsible for cytokinesis in male and female meiocytes and homologous chromosome pairing	Jiang et al., 2010
	Retrotransposon Tos17	OsTPC1 [Voltage-gated Ca(2+)-permeable channel protein]	A key regulator of mitogen-activated protein kinase activation and elicitor-induced hypersensitive cell death	Jiang et al., 2010

(Continued)

TABLE 4.1 (Continued)
Types of Transposons, the Corresponding Genes, and Their Functions in Various Cereal Crops

CROPS	Insertion of Transposon	Gene	Function	Reference
	Retrotransposon Tos17	GS1 (Glutamine synthetase)	Play important role in development and grain filling.	Jiang et al., 2010
	Retrotransposon Tos17	OsCAO1 and OsCAO2 (Chlorophyll a oxygenase)	Chlorophyll b biosynthesis	Jiang et al., 2010
	Retrotransposon-tagged mutants	starch synthase I	Plays role in amylopectin chain synthesis during grain development	Fujita et al., 2006
	Retrotransposon Tos17	FON1 (LRR receptor Kinase)	Controlling vegetative and reproductive development by regulating shoot apical meristem size	Jiang et al., 2010
	Retrotransposon Tos17	MYBGA (GARE-interacting transcription factor)	Tissue-specific regulating a-amylase expression by sugar and GA signaling interference	Jiang et al., 2010
	Retrotransposon Tos17	OsCLC-1 or OsCLC-2 (Voltage-gated chloride channel protein)	Role in the transport of chloride ions through the vacuolar membrane	Jiang et al., 2010
	Retrotransposon Tos17	PAIR2 (HORMA domain-containing protein)	Essential for homologous chromosome synapsis during meiosis I in rice	Jiang et al., 2010
	nDart1	chlorophyll protease gene (OsClpP5)	leads to the formation of mutant dwarf allele thl-m with leaf variegation	Tsugane et al., 2006
	nDart1	OSMYOXIB (Myosin protein)	Pollen development	Jiang et al., 2010
	nDart1	OsNAC2 (NAC domain-containing protein)	Regulating shoot branching	Jiang et al., 2010
	Retrotransposon Tos17	COW (A new member of the YUCCA protein family)	Required for maintaining water homeostasis and an appropriate root-to-shoot ratio	Jiang et al., 2010
	Retrotransposon Tos17	OsJar 1 (GH3 auxin-responsive promoter)	Involved in phytochrome and jasmonate signaling	Jiang et al., 2010

	Retrotransposon Tos17	OsRecQ1 (a QDE-3 homologue in rice)	Plays a role in the process that allows inverted repeat DNA to be transcribed into dsRNA, which can cause RNA silencing	Jiang et al., 2010
	Retrotransposon Tos17	OsIPD3 [a Ca(2+)/calmodulin-dependent protein kinase]	Required for root symbiosis with AM fungi in plants	Jiang et al., 2010
	Retrotransposon Tos17	OsSS12 (Fatty-acid desaturase)	Negative regulation of defense responses in rice	Jiang et al., 2010
	Retrotransposon Tos17	RIMI (NAC-domain protein)	As a host factor that is required for multiplication of Rice dwarf virus (RDV)	Jiang et al., 2010
MAIZE	Mutator transposon insertion	starch branching IIa	reduced branching activity of glucans in leaves compared to seeds	Blauth et al., 2001
	Mutator transposon insertion	inositol phosphate kinase (ZmIpk)	resulted in approximately 30% reduced phytic acid content in seed	Shi et al., 2003
	Mutator transposon insertion	sugary2	resulted in short glucan chains of amylopectin	Zhang et al., 2004
	Mutator transposon insertion	ZmPRPL35-1	encoding large subunit of plastid ribosomes resulted in alterations of embryo development without affecting endosperm development	Magnard et al., 2004
	Mutator transposon insertion	molybdopterin synthase gene (viviparous 15 locus)	The mutant exhibited viviparous phenotype due to lack of molybdenum cofactor, which is an essential component required for ABA biosynthesis	Suzuki et al., 2006
	Mutator transposon insertion	6-Phosphogluconate dehydrogenase (6PGDH) gene at the Pgd3 locus	suggesting an important role for 6PGDH enzyme in seed development	Settles et al., 2007
Barley	Retrotransposon insertion	HvAC T1	responsible for citrate secretion from the roots for external detoxification of aluminum	Kashino-Fujii et al., 2018
Sorghum	candystripe1 transposon	y1	provides resistance against anthracnose leaf blight	Ibraheem et al., 2010

4.5.3 Barley

Barley is an important industrial crop with expanding use in food and bioproducts. It is also a significant model plant for genetics study because it is a true diploid with strong similarity to other Triticeae crops, particularly wheat. It is most typically utilized in the brewing and distilling industries as feed and malt. However, malting quality is a complex feature that impedes genetic progress in barley. Thus, the Activator/Dissociation (Ac/Ds) transposon-based technique is utilized to ensure optimal utilization of barley in the brewing and distilling industries. Due to the limited effectiveness of genetic transformation and the enormous genome size of barley, it has great promise. Transposon insertion in barley has been successful in determining the gene (*HvAACT1*) for regulating differential aluminum tolerance using both retrotransposon insertion and DNA methylation.

Furthermore, new insertion locations in barley mutants show that BARE-1 and Tpo1-like transposons were involved in causing morphological alterations in the plant. At least some of the morphological aberrations in barley mutants, such as color spots on leaves, are most likely caused by an insertion or resultant point mutations. Further research pertaining to the success of barley is being carried out.

4.5.4 Sorghum

Sorghum (*Sorghum bicolor* L.) is the world's fifth most significant grain crop and a source of sustenance for millions of African and Southeast Asian people. Sorghum's stalk includes a high level of soluble sugars that can be easily turned into ethanol, in addition to its high grain yield. Bagasse (sugar extraction residue) is a good lignocellulosic source for biofuel and can also be utilized to create energy through direct combustion. Furthermore, sorghum is a good feed crop, with yield comparable to forage maize (*Zea mays*) but requiring far less water for irrigation (Rocateli et al., 2012). The sorghum genome contains minimal gene redundancy since it is a diploid plant that hasn't undergone genome duplication in over 70 million years (Paterson et al., 2004, 2009). Sorghum is a desirable model for functional genomics for maize, sugarcane, *Miscanthus*, and other C4 bioenergy crops with complicated genomes due to the small size of its genome (;730 Mb), which has been entirely sequenced (Swigonova et al., 2004; Paterson et al., 2009; Schnable et al., 2009; Hughes et al., 2014). Sorghum has potential for plant transposon tagging due to its short genome size and low concentration of repetitive DNA.

The first active transposable element identified from sorghum was *Cs1 (Candystripe1)*. According to the findings, *Cs1*-homologous sequences are found in low copy number in sorghum and other grasses such as sudangrass, maize, rice, teosinte, and sugarcane. *Cs1*'s low copy number and high transposition frequency suggest that it is an effective gene isolation method in sorghum. *Cs1* is a plant transposable element belonging to the CACTA family.

Other unique features, such as *msd* and *erl* mutants, were discovered in the mutant library and could be used to improve sorghum (Burow et al., 2014; Xin et al., 2015). Additionally, results in sorghum could be easily applied to other Poaceae species using integrated genomics techniques. The timeline of genes identified in various cereal crops is shown below (Figure 4.1).

4.6 TRANSPOSON MUTAGENESIS—AT A GLANCE

It is essential to give the thousands of newly discovered genes their proper functions as more and more sequences are made available in the databases. In an effort to solve this issue, numerous strategies are being attempted. Disrupting or creating mutations in the genes and analyzing the results is the most direct way to understand how the sequenced genes in an organism function. In plants, techniques such as insertional mutagenesis, sense and antisense suppression, and gene replacement have all been developed for this purpose (Ramachandran and Sundaresan, 2001). Random insertional mutagenesis is now the most frequently used method for detailed gene function studies in plants. Transposons or T-DNA can be used as insertional mutagens in plants (Ramachandran and Sundaresan, 2001).

Although transposon insertional mutagenesis is the oldest of these techniques, resources based on mutations brought on by specific DNA insertions are expected to be important in functional genomics for the following reasons. (1) Large insertions are especially potent in affecting gene function (as are null mutations). (2) The mutations resulting in the loss of function are usually recessive and biologically stable. (3) Large insertions are easier to find and locate in the genome than other forms of mutations (such as point mutations). Recessive, loss-of-function mutations are an important starting point for functional examination of a gene (McCarty et al., 2009).

Transposon insertional mutagenesis is a good option for T-DNA mutagenesis when tissue culture transformation is inefficient, as it is for many agricultural species. Just a few numbers of primary transformants are required for the transposon mutagenesis strategy, and insertions are produced through propagation (Qu et al., 2008). Transposable elements from maize (*Z. mays*) are key tools for developing a large number of gene knockouts in the functional genomics of monocot plants and cereals because high-throughput T-DNA transformation is not a realistic solution for these plant species. Several studies have shown that the maize Ac/Ds transposable elements work in barley (*Hordeum vulgare*) and mapped Ds launching pads have been built for the study of functional genomics (Qu et al., 2008).

Complementary methods of activation tagging and gene trap analyses, which make use of random integration of reporter gene constructs, have been developed in order to get around the aforementioned issues and discover genes. In heterologous systems, transposon-based gene tagging suffers from a number of problems, mostly from the presence of numerous copies, which causes the same issues as multiple T-DNA insertions. Mutagenesis can't always serve as a tool to determine the function of the genes. One of the factors is that many genes have redundant functions, and mutations in such genes may not result in phenotypes that are immediately noticeable. Another factor is that many genes have numerous developmental functions, and mutations in these genes may result in early lethality (Jiang et al., 2010).

The first application of plant transposons being used for gene tagging was the Ac element's use in the cloning of the bronze gene in maize (Fedoroff et al., 1983). Since then, more and more plant genes have either been cloned or discovered utilizing *Ac/Ds, En/Spm,* or *Mu*. Transposon tagging in plants has, up until recently, only been possible in species with active and well-studied transposon systems, including maize and *Antirrhinum* (Walbot et al., 1992).

4.7 FUTURE PROSPECTS

After being identified as mobile genetic elements more than fifty years ago, transposons emerged as effective instruments for molecular genetic research. The usage of tagging systems has increased in other easily transformed species like *Arabidopsis*, tomato, and rice as a result of the transposition of maize *Ac/Ds* and *En/Spm* elements. Sorghum, a cereal grain related to maize, has one of the shortest genomes of all cereal grains. As a result, finding an active transposable element for gene tagging in sorghum may open up a new path for isolating the related genes in maize in the years to come. It will take less time and be more effective to find causative genes for phenotypes that are helpful for agriculture, not just for cereal crops but also for other grasses where there are few genetic research pathways due to genome complexity. These systems will quickly become more effective for future functional genomic analysis because of the modification and improvement of transposases.

REFERENCES

Alonso, J.M., Stepanova, A.N., Leisse, T.J., Kim, C.J., Chen, H., Shinn, P., Stevenson, D.K., Zimmerman, J., Barajas, P., Cheuk, R., Gadrinab, C., Heller, C., Jeske, A., et al., 2003. Genome-wide insertional mutagenesis of Arabidopsis thaliana. *Science*. 301, 653–657.

Azpiroz-Leehan, R., Feldmann, K.A., 1997. T-DNA insertion mutagenesis in Arabidopsis: going back and forth. *Trends Genet.* 13(4), 152–156.

Baker, B., Schell, J., Lorz, H., and Fedoroff, N, 1986. Transposition of the maize controlling element "Activator" in tobacco. *Proc. Natl. Acad. Sci.* USA. 83, 4844–4848.

Balyan, H. S., Sreenivasulu, N., Riera-Lizarazu, O., Azhaguvel, P., & Kianian, S. F., 2008. Mutagenesis and High-Throughput Functional Genomics in Cereal Crops: Current Status. *Adv. in Agron.* 98, 357–414.

Bensen, R. J., G. S. Johal, V. C. Crane, J. T. Tossberg, P. S. Schnable, R. B. Meeley and S. P. Briggs, 1995. Cloning and characterization of the maize An1 gene. *Plant Cell,* 7, 75–84.

Blauth, S. L., Yao, Y., Klucinec, J. D., Shannon, J. C., Thompson, D. B., & Guiltinan, M. J., 2001. Identification of Mutator insertional mutants of starch-branching enzyme 2a in corn. *Plant Physiology*. 125(3), 1396–1405.

Burow, G., Xin, Z., Hayes, C., and Burke, J., 2014. Characterization of a multi-seeded mutant of Sorghum for increasing grain yield. *Crop Sci.* 54, 2030–2037.

Capy, P., Bazin, C., Higuet, D., and Langin, T., 1998. "Dynamics and Evolution of Transposable Elements." Landes Bioscience, Austin, Texas.

Cowperthwaite, M., W. Park, Z. Xu, X. Yan, S. C. Maurais and H. K. Dooner, 2002. Use of the transposon Ac as a gene-searching engine in the maize genome. *Plant Cell.* 14, 713–26.

Fedoroff, N. V., 1983. Controlling elements in maize. *In Mobile Genetic Elements*, J. Shapiro, ed. (New York: Academic Press). L–63.

Fedoroff, N. V., Wessler, S., and Shure, M., 1983. Isolation of the transposable maize controlling elements Ac and Ds. *Cell.* 35, 235–243.

Feng, C. P., and Mundy, J., 2006. Gene discovery and functional analysis in model plant Arabidopsis. *J. Integr. Plant Biol.* 48, 5–14.

Fujita, N., Yoshida, M., Asakura, N., Ohdan, T., Miyao, A., Hirochika, H., & Nakamura, Y., 2006. Function and characterization of starch synthase I using mutants in rice. *Plant physiology*, 140(3), 1070–1084.

Gaut, B. S., and Doebley, J. F., 1997. DNA sequence evidence for the segmental allotetraploid origin of maize. *Proc. Natl. Acad. Sci. USA*. 94, 6809–6814.

Goff, S., Ricke, D., Lan, T., Presting, G., Wang, R., Dunn, M., Galezebrook, J., Session, A., Oeller, P., Varma, H., Hadley, D., Hutchinson, D., et al., 2002. A draft of the rice genomes (*Oryza sativa* L. *spp. Japonica). Science.* 296, 92–100.

Hamer, L., Dezwaan, T. M., Montenegro-Chamorro, M. V., Frank, S. A., and Hamer, J. E., 2001. Recent advances in large-scale transposon mutagenesis. *Curr. Opin. Chem. Biol.* 5, 67–73.

Hieter, P., and Boguski, M., 1997. Functional genomics: Its all how you read it. *Science.* 278, 601–602.

Hirochika, H., 2001. Contribution of the Tos17 retrotransposon to rice functional genomics. *Curr. Opin. Plant Biol.* 4, 118–122.

Hughes, T.E., Langdale, J.A., and Kelly, S., 2014. The impact of widespread regulatory neofunctionalization on homeolog gene evolution following whole-genome duplication in maize. *Genome Res.* 24, 1348–1355.

Ibraheem, F., Gaffoor, I., & Chopra, S., 2010. Flavonoid phytoalexin-dependent resistance to anthracnose leaf blight requires a functional yellow seed1 in Sorghum bicolor. *Genetics.* 184, 915–926.

Ito, T., Seki, M., Hayashida, N., Shibata, D., and Shinozaki, K., 1999. Regional insertional mutagenesis of genes on Arabidopsis thaliana chromosome V using the Ac/Ds transposon in combination with a cDNA scanning method. *Plant J.* 17, 433–444.

Ito, Y., Eiguchi, M., and Kurata, N., 2004. Establishment of an enhancer trap system with Ds and GUS for functional genomics in rice. *Mol. Genet. Genomics.* 271, 639–650.

Jiang, S. Y., & Ramachandran, S., 2010. Natural and artificial mutants as valuable resources for functional genomics and molecular breeding. *International Journal of Biological Sciences.* 6, 228.

Kashino-Fujii, M., Yokosho, K., Yamaji, N., Yamane, M., Saisho, D., Sato, K., & Ma, J. F., 2018. Retrotransposon insertion and DNA methylation regulate aluminum tolerance in European barley accessions. *Plant Physiology.* 178, 716–727.

Koga, A., Suzuki, M., Inagaki, H., 1996. Transposable element in fish. *Nature.* 383, 30. https://doi.org/10.1038/383030a0

Kolkman, J. M., L. J. Conrad, P. R. Farmer, K. Hardeman, K. R. Ahern, P. E. Lewis, R. J. Sawers, S. Lebejko, P. Chomet and Brutnell, T. P., 2005. Distribution of Activator (Ac) throughout the maize genome for use in regional mutagenesis. *Genetics.* 169, 981–95.

Krysan, P. J., Young, J. C., and Sussman, M. R., 1999. T-DNA as an insertional mutagen in Arabidopsis. *Plant Cell.* 11, 2283–2290.

Kumar, A. and Bennetzen, J.L., 1999. *Plant Retrotransposons. Annual Review of Genetics*, 33, 479–532. https://doi.org/10.1146/annurev.genet.33.1.479

Kumar, C. H., Wing, R. A., and Sundaresan, V., 2005. Efficient insertional mutagenesis in rice using the maize En/Spm elements. *Plant J.* 44, 879–892.

Lampreht Tratar, Ursa; Horvat, Simon; Cemazar, Maja, 2018. Transgenic Mouse Models in Cancer Research. *Frontiers in Oncology*. 8, 7 doi:10.3389/fonc.2018.00268

Lisch, D., 2002. Mutator transposons. *Trends in plant science*, 7, 498–504.

Lisch, D., 2015. Mutator and MULE transposons. *Mobile DNA III*, 801–826.

Magnard, J. L., Heckel, T., Massonneau, A., Wisniewski, J. P., Cordelier, S., Lassagne, H., ... & Rogowsky, P. M., 2004. Morphogenesis of maize embryos requires ZmPRPL35-1 encoding a plastid ribosomal protein. *Plant physiology*. 134, 649–663.

Marcon, C., Altrogge, L., Win, Y. N., Stöcker, T., Gardiner, J. M., Portwood, J. L., and Hochholdinger, F., 2020. BonnMu: a sequence-indexed resource of transposon-induced maize mutations for functional genomics studies. *Plant physiology.* 184, 620–631.

Masson, P., and Fedoroff, N., 1989. Mobility of the maize Suppressor-mutator element in transgenic tobacco cells. *Proc. Natl. Acad. Sci. USA*. 86, 2219–2223.

May, B. P., H. Liu, E. Vollbrecht, L. Senior, P. D. Rabinowicz, D. Roh, X. Pan, L. Stein, M. Freeling, D. Alexander and R. A. Martienssen, 2003. *Maize-targeted mutagenesis: A knockout resource for maize*. Proc Natl Acad Sci USA. 100, 11541–6.

McCarty, D. R., & Meeley, R. B., 2009. Transposon resources for forward and reverse genetics in maize. Handbook of maize: *Genetics and genomics*. (JL Bennetzen, S. Hake, Eds.). Springer Science, Berlin, 561–584.

McCarty, D. R., A. M. Settles, M. Suzuki, B. C. Tan, S. Latshaw, T. Porch, K. Robin, J. Baier, W. Avigne, J. Lai, J. Messing, K. E. Koch and L. C. Hannah, 2005. Steady-state transposon mutagenesis in inbred maize. *Plant J.* 44, 52–61.

McClintock, B., 1950. The origin and behavior of mutable loci in maize. *Proc. Natl. Acad.Sci. USA.* 36, 344–355.

Nonomura, K. I., Nakano, M., Murata, K., Miyoshi, K., Eiguchi, M., Miyao, A., ... & Kurata, N., 2004. An insertional mutation in the rice PAIR2 gene, the ortholog of Arabidopsis ASY1, results in a defect in homologous chromosome pairing during meiosis. *Mol. Genet. Genom.* 271, 121–129.

Paterson, A.H., Bowers, J.E., and Chapman, B.A., 2004. Ancient polyploidization predating divergence of the cereals, and its consequences for comparative genomics. *Proc. Natl. Acad. Sci. USA.* 101, 9903–9908.

Paterson, A.H., et al., 2009. The Sorghum bicolor genome and the diversification of grasses. *Nature.* 457, 551–556.

Pereira, A., 1998. Heterologous transposon tagging system in Transgenic Plant Research (K. Lindsey, Ed.). Harwood Academic Publishers, UK, 91–108.

Pereira, A., and Saedler, H., 1989. Transposition of the maize En/Spm element in transgenic tobacco. *EMBO J.* 8, 1315–1321.

Pramanick, B., Brahmachari, K., Kar, S. and Mahapatra, B. S., 2020. Can foliar application of seaweed sap improve the quality of rice grown under rice–potato–greengram crop sequence with better efficiency of the system. *Environ. Boil. Fishes.* 32, 1–10.

Qu S, Desai A, Wing R, Sundaresan V., 2008. A versatile transposon-based activation tag vector system for functional genomics in cereals and other monocot plants. *Plant Physiol.* 146, 189–99. doi: 10.1104/pp.107.111427

Qu S, Jeon J-S, Ouwerkerk PBF, Bellizzi M, Leach J, Ronald P, Wang G-L, 2009. Construction and application of efficient *Ac–Ds* transposon tagging vectors in rice. *J Integr Plant Biol.* 51, 982–992.

Ramachandran, S., and Sundaresan, V., 2001. Transposons as tools for functional genomics. Plant Physiology and Biochemistry. 39(3–4), 243–252. doi:10.1016/s0981-9428(01)01243-8

Robertson, D. S., 1978. Characterization of a mutator system in maize. *Mutat. Res.* 51, 21–28.

Rocateli, A.C., Raper, R.L., Balkcom, K.S., Arriaga, F.J., and Bransby, D.I., 2012. Biomass sorghum production and components under different irrigation/tillage systems for the southeastern U.S. *Ind. Crops Prod.* 36, 589–598.

Sasaki, T., 1999. Current status of and future prospects for genome analysis in rice. In *Molecular Biology of Rice* (Shimamoto, K., ed.). Tokyo, 3–16.

Sasaki, T. and Moore, G., 1997. *Oryza: from Molecule to Plant*. Dordrecht, Kluwer Academic Publishers.

Schnable, P.S., et al., 2009. The B73 maize genome: complexity, diversity, and dynamics. *Science.* 326, 1112–1115.

Settles, A. M., Holding, D. R., Tan, B. C., Latshaw, S. P., Liu, J., Suzuki, M., ... & McCarty, D. R., 2007. Sequence-indexed mutations in maize using the UniformMu transposon-tagging population. *BMC genomics.* 8(1), 1–12.

Shi, J., Wang, H., Wu, Y., Hazebroek, J., Meeley, R. B., & Ertl, D. S., 2003. The maize low-phytic acid mutant lpa2 is caused by mutation in an inositol phosphate kinase gene. *Plant physiol.* 131(2), 507–515.

Shimamoto, K., Miyazaki, C., Hashimoto, H., Izawa, T., Itoh, K., Terada, R., Inagaki, Y., and Iida, S., 1993. Trans-activation and stable integration of the maize transposable element Ds cotransfected with the Ac transposase gene in transgenic rice plants. *Mol. Gen. Genet.* 239, 354–360.

Sundaresan, V., 1996. Horizontal spread of transposon mutagenesis: New uses of old elements. *Trends Plant Sci.* 1, 184–191.

Sundaresan, V., Springer, P., Volpe, T., Haward, S., Jones, J. D., Dean, C., Ma, H., and Martienssen, R., 1995. Patterns of gene action in plant development revealed by enhancer trap and gene trap transposable elements. *Genes Dev.* 9, 1797–1810.

Suzuki, M., Mark Settles, A., Tseung, C. W., Li, Q. B., Latshaw, S., Wu, S., ... & McCarty, D. R., 2006. The maize viviparous15 locus encodes the molybdopterin synthase small subunit. *Plant J.*, 45(2), 264–274.

Swigonová, Z., Lai, J., Ma, J., Ramakrishna, W., Llaca, V., Bennetzen, J.L., and Messing, J., 2004. Close split of sorghum and maize genome progenitors. *Genome Res.* 14, 1916–1923.

Tanaka, K., Murata, K., Yamazaki, M., Onosato, K., Miyao, A., & Hirochika, H., 2003. Three distinct rice cellulose synthase catalytic subunit genes required for cellulose synthesis in the secondary wall. *Plant Physiol.* 133(1), 73–83.

Tissier, A. F., Marillonnet, S., Klimyuk, V., Patel, K., Torres, M. A., Murphy, G., and Jones, J. D. G., 1999. Multiple independent defective Suppressor-mutator transposons insertions in Arabidopsis: A tool for functional genomics. *Plant Cell.* 11, 1841–1852.

Tsugane, K., Maekawa, M., Takagi, K., Takahara, H., Qian, Q., Eun, C. H., & Iida, S., 2006. An active DNA transposon nDart causing leaf variegation and mutable dwarfism and its related elements in rice. *Plant J.*, 45(1), 46–57.

Upadhyaya, N. M., Zhu, Q. H., Zhou, X. R., Eamens, A. L., Hoque, M. S., Ramm, K., Shivakkumar, R., Smith, K. F., Pan, S. T., Li, S. Z., Peng, K. F., Kim, S. J., et al., 2006. Dissociation (Ds) constructs, mapped Ds launch pads and a transiently-expressed transposase system suitable for localized insertional mutagenesis in rice. *Theor. Appl. Genet.* 112, 1326–1341.

Walbot, V., 1992. Strategies for mutagenesis and gene cloning using transposon tagging and T-DNA insertional mutagenesis. *Annu. Rev. Plant. Physiol. Plant Mol. Biol.* 43, 49–82.

Walbot, V., 2000. Saturation mutagenesis using maize transposons. *Curr. Opin. Plant Biol.* 3, 103–107.

Weigel, D., Ahn, J. H., Blazquez, M. A., Borevitz, J. O., Christensen, S. K., Fankhauser, C., Ferrandiz, C., Kardailsky, I., Malancharuvil, E. J., Neff, M. M., Nguyen, J. T., Sato, S., et al., 2000. Activation tagging in Arabidopsis. *Plant Physiol.* 122, 1003–1013.

Xin, Z., Gitz, D., Burow, G., Hayes, C., and Burke, J.J., 2015. Registration of two allelic erect leaf mutants of sorghum. *J. Plant Regist.* 9, 254–257.

Zhang, X., Colleoni, C., Ratushna, V., Sirghie-Colleoni, M., James, M., & Myers, A., 2004. Molecular characterization demonstrates that the *Zea mays* gene sugary2 codes for the starch synthase isoform SSIIa. *Plant Mol. Biol.* 54(6), 865.

5 Application of Nuclear Techniques in Crop Improvement: A Review

Nihar Ranjan Chakraborty,[*1] *Achal Kant,*[2] *and Sandip Debnath*[1]

[1]Department of Genetics and Plant Breeding and Crop Physiology, Institute of Agriculture, Visva-Bharati University, Sriniketan, West Bengal, India

[2]Narayan Institute of Agricultural Sciences, Gopal Narayan Singh University, Jamuhar, Sasaram, Rohtas, Bihar, India

*Corresponding author: nrchakraborty@gmail.com

5.1 INTRODUCTION

In the 19th century, Gregor Johann Mendel discovered the laws of heredity, which led to the development of genetics as a fundamental science in the field of plant breeding. The introduction of hybridization techniques, on the other hand, brought about significant developments in the field of plant breeding. The goal was to produce pure-line offspring that were genetically superior to the parental types in a number of ways by combining desirable genes that were found in two or more distinct varieties (Albert et al. 2022). The practice of plant breeding that is known as cross-breeding, also known as recombinant breeding, is one that is based on the crossing of different genotypes and is then followed by trait selection. Later on, in the late 1920s and early 1930s, Lewis John Stadler conducted research on the induction of genetic alterations through the use of X-rays. This research laid the foundation for another type of plant breeding known as mutation breeding. The variation that is produced in this manner is then amplified further by recombination of alleles on homologous chromosomes and their independent assortment during meiosis. The majority of the genetic variations that can be found in any organism, including plants, come from mutation (Goldstein and Pollock 1997). The ensuing variation is both the raw material for natural selection and the driving force behind the process of evolution. Because spontaneous mutations are so uncommon and unpredictable in terms of when they take place, it can be challenging to use them in plant breeding programs (Brunner 1995). As a result of this process, mutant forms that show large and small effects on the phenotype can emerge for a wide variety of characteristics (Saludes-Tapia et al. 2021). Along with recombinant breeding and transgenic breeding, mutation breeding is now an essential component of the modern approach to plant breeding. This method of breeding plants through sexual reproduction has become the most common one, as highlighted by

 DOI: 10.1201/9781003392897-5

Novak and Brunner(1992), and it is frequently complemented by germplasm that has been derived from induced mutation(Kharkwaland Shu 2009).

In spite of the scarcity of natural resources, there has been an increase in the demand for food and feed in recent years due to the expanding human population. Even worse crop production conditions can result from erratic rainfall, sudden and severe drought conditions, excessive flooding, and other phenomena that are frequently attributed to climate change. In order to counteract the deteriorating state of food security, there is an urgent need to significantly boost the yield potential of crop plants. In the past, induced mutants that possessed desirable qualities were utilized either directly or indirectly in the breeding process for new varieties. Induced mutations have also been gaining increasing importance in plant molecular biology over the past decade as a tool to identify and isolate genes as well as to study their structure and function (Mba 2013). This has all taken place during the last decade. Apart from traditional mutation induction techniques such as gamma-ray irradiation and chemicals, a unique approach known as space-induced mutagenesis is increasingly being used for crop improvement (Liu 2008). Significant advancements have been made in technique optimization, gene identification, and the discovery of molecular pathways associated with mutant lines, elite agricultural resource development and novel varieties, and the establishment of national research networks and platforms.

It is absolutely necessary for plant breeders to have an understanding of the genes that control important agronomic and quality traits in order for them to develop appropriate strategies and efficiently implement breeding programs. Therefore, as a result of increasing yield potential and stability, induced mutations can contribute further to increasing global food production both directly and indirectly.

5.2 UTILITY OF NUCLEAR TECHNIQUES

Inducing mutations is an established and true strategy for developing heritable variation in crops, and as such, it has enormous promise for expanding plant genetic diversity (Ellis et al. 2000). This would boost the plants' ability to adapt to the pressures caused by climate change, particularly the frequent introduction of new pests and disease strains. The genetic makeup of plants can be altered using radioisotopes, resulting in the generation of strains that are superior in a variety of ways. By inducing mutations in crops, it is possible to generate varieties that are more disease resistant, are tolerant to harsh climatic conditions, have higher yields, and have shorter growth times. This process can be done by applying small doses of gamma or neutron radiation.

This technique is not the same as genetically redesigning plants (Tsatsakis et al. 2017), which involves combining the plant's DNA with the DNA of other genetic material to create a new set of genes. This method is unique from genetically manipulating plants. Because of the changed DNA, the genetically engineered plants will gain new or different features. The traditional method, in addition to other nuclear techniques, can be a valuable answer in the process of resolving many of the challenges that arise in agricultural settings (Ahloowalia et al. 2004), and its implementation in underdeveloped countries are both examples of its real use.

The practical use of the method is to increase the effectiveness of nuclear agriculture and to improve the quality of the advanced countries in more than 50 years and in other countries in developing a 30-year history. According to Ahloowalia and Maluszynski (2001), the reservoir for the activity of plant breeding and the genetic mutation increases, which in turn leads to an increase in genetic diversity. An ongoing mutation in the DNA sequence, which results in a change in the gene and leads to improper construction of amino acids and proteins, was the cause of the problem. This phenomenon led to the formation of the fundamental organism, and many of the activities that take place within the cell went through a period of transformation (Robbelen 1990).

Over 1000 different mutant varieties of major staple crops are currently being cultivated on tens of millions of hectares worldwide. These crops have the potential to increase rural income, improve human nutrition, and contribute to the world's environmentally sustainable food security. Breeding with mutations has become the most successful field for the application of nuclear techniques in food and agriculture. Farmers have been given access to an increasing number of mutant crop varieties, most notably in the form of rice. In the field of agriculture, radiation could be used to eliminate insect pests, breed crops that are more resistant to disease, boost the nutritional value of certain crops, improve their capacity for baking or melting, or shorten the amount of time needed to prepare certain foods. The natural processes of evolution of plant DNA that have been driven by human-directed agriculture for thousands of years can be sped up through the use of a technique known as mutation induction, which involves the use of mutagens such as chemicals or gamma radiation. Plant breeders create a large number of different varieties of mutant plant strains, from which they select the ones that have desirable characteristics, such as higher yield, resistance to adverse conditions or diseases, and improved nutritional value.

Induced genetic diversity and breeding with promising mutations for farmer-preferred traits in agronomically sound genetic backgrounds have paved the way for crop improvement all over the world for over seven decades now. This has allowed for a greater variety of improved crops to be produced. This is evidenced by the fact that the Mutant Variety Database contains entries for more than 3300 mutant varieties, covering more than 220 plant species and coming from more than 70 countries (MVD 2021). Induced mutations strengthen plant germplasm pools (Suprasanna et al. 2015) and make it possible for faster genetic gain, particularly in situations where the genetic base is limited. Therefore, compared to traditional breeding, mutation breeding has the potential to facilitate greater genetic gain. This advantage, when combined with developing front-end technologies for effective and precise selection, has the potential to both quicken the rate at which crop improvements are made and increase the rate at which genetic gain is achieved. In developing Member States of the Food and Agriculture Organization of the United Nations (FAO) and the International Atomic Energy Agency (IAEA), breeding with induced genetic diversity has continued to be a highly effective avenue for the improvement of both simple and complex crop traits (Jacob et al. 2018).

It is common practice in the agricultural industry to apply fertilizer in order to increase crop production and maximize yield. However, they come at a very high price and have a significant negative impact on the surrounding environment. As a result, it is imperative that an effective use of them be made. Scientists and farmers are able to make the most efficient use of fertilizers thanks to radioactive tracers (Cimpeanu et al. 2014), which allow them to determine the proportion of the applied chemicals that are taken up by the plant and the remainder that is lost to the surrounding environment. This is useful in not one but two ways: first, it lowers costs for the farmer, and second, it reduces the amount of damage done to the environment by the overuse of fertilizers that can be caused by those fertilizers.

5.3 UNDERSTANDING OF RADIOACTIVITY AND RADIATION AND HOW IT WORKS

The fundamental unit of an element is called an atom (diameter 10^{-8} cm), where the number of protons does not have to equal the number of electrons and the number of neutrons does not have to equal the number of electrons (Figure 5.1). $^{a}_{b}X$: Isotopes of an element that differ in mass number but share the same atomic number are chemically identical, e.g., $^{1}H_{1}$, $^{2}H_{1}$, $^{3}H_{1}$. Here, "b" establishes the chemical characteristics, while "a" roughly reflects the atomic weight.

The stability of an element is determined by radioactivity, which is the ratio of neutrons to protons in the nucleus. Higher atomic number (>84) elements are more likely to be unstable, and an unstable nucleus will attempt to become stable by radiating the excess energy. Despite the fact that the term "radioactivity" was first used by Madame Curie, Henry Becquerel, a French physicist, discovered a uranium compound that was emitting an intangible radiant energy in 1896.

Gamma rays are known to affect plant growth and development by causing changes in cells and tissues that are cytological, genetic, biochemical, physiological, and morphogenetic (Jan, et al. 2012; Kovacs and Keresztes 2002). The biological impact of gamma rays causes the production of free radicals, which can alter crucial plant cell components (Hong et al. 2022; Marcu et al. 2013). Gamma ray effects are solely dose dependent and may also differ depending on the matrix of the subject matter (Marcu et al. 2013). Ionizing radiations are a form of electromagnetic radiation (Lavanya et al. 2003), similar to light that is visible to the human eye. But because their wavelengths are shorter (0.001–10 nm, as opposed to 2000–3000 nm for UV light), they are able to pass through the substances that surround them. They transfer their energy to the surrounding matter, which results in the ionization of the

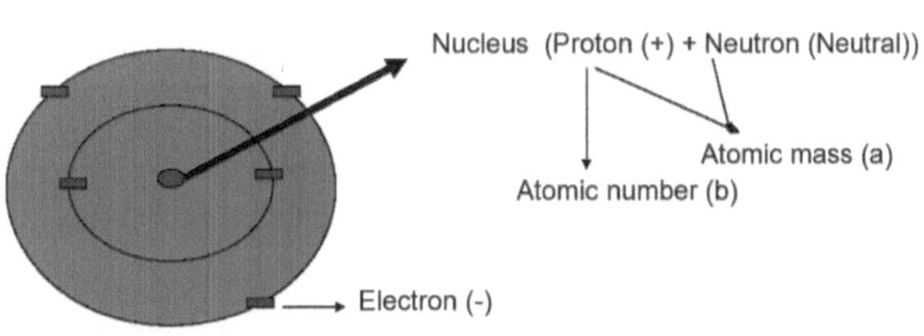

FIGURE 5.1 Basic unit of an element.

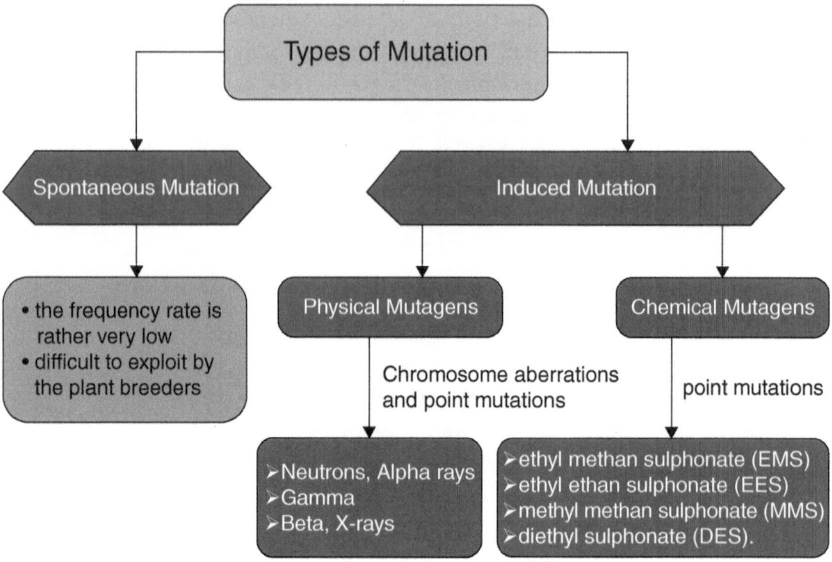

FIGURE 5.2 Different types of mutation.

atoms of the matter that they travel through. Another way to think of this is that they "knock off" electrons from the orbits of atoms. Therefore, the formation of a positive ion takes place when an electron is expelled from an atom; the formation of a negative ion takes place when the expelled electron is accepted by another atom. Atoms go through a series of chemical transformations as a result of ionization. Changes in genetic makeup can occur as a result of ionization processes that have an effect on biologically significant molecules (Reisz et al. 2014). Radiation, in addition to causing ionization, also results in the formation of free radicals that have a very short life (only a fraction of a second). These free radicals have also been linked to changes in genetic material, which can result in mutations.

Gamma rays are the type of radiation that is used most frequently as a physical mutagen (Figure 5.2). Two of the reasons for this are how simple it is to handle, as well as how readily available it is. Gamma rays are distinct from X-rays in both the amount of energy that they pack into each photon and their source. Gamma rays are derived from radioactive isotopes, whereas X-rays are manufactured in X-ray machines. Gamma rays can come from a variety of different elements, but the most common ones are cobalt 60 and cesium 137. The structure that houses the gamma radiation is referred to as the gamma cell. It is shaped like a short cylinder and contains the radioactive isotope in a cylinder-like arrangement. The sample chamber, into which the seeds are loaded, is placed inside the source cylinder, where it is then subjected to radiation coming from all directions. After the predetermined amount of time has elapsed for the sample to be exposed, the sample chamber is withdrawn. Lead shielding is used to encase the source in order to prevent personnel from being exposed to radiation.

Dose and dose rate: The required dose and the dose rate determine (Rühm et al. 2018) how long the samples are exposed to gamma rays. The duration of this exposure is based on the required dose. Calibration is performed at regular intervals on the dose rate, which is expressed as Gy/min. The unit of measure for gamma ray dose is the gray (Gy), and 100 R is equal to 1 Gy. The amount of time spent exposed to the gamma cell can range anywhere from a few minutes to an hour. This is due to the fact that the dose rate of gamma cells is quite high, while the doses used in mutation breeding are not particularly high. There is a significant amount of variety seen in the radiosensitivity of the various crop plants. As an index of radiosensitivity, the dose (D_{50}) that results in a reduction in growth of fifty percent can be used.

Mutagens produced by chemicals: The most common types of chemical mutagens are alkylating agents such as ethyl methane sulfonate (EMS) and diethyl sulfate (DES). These mutagens interact with DNA in a way that results in the alkylation of phosphate groups as well as the bases of purine and pyrimidine. The formation of 7-alkylguanine is the event that takes place the most frequently. Because the majority of alkylating agents have the potential to cause cancer, it is imperative that they be handled with extreme caution. These are capable of reacting with water and have a high level of reactivity. Since the products of hydrolysis are not mutagenic, the treatment solutions must be prepared immediately before use and must not be stored. There are a variety of half-lives for alkylating agents (time needed for degradation of half the initial amount). The half-life is significantly affected by temperature variations as well.

For chemical mutagens, the dose is determined by combining the concentration of the chemical with the length of time that the treatment is administered. The type of seed, as well as the overall health of the seed, will determine the optimal concentration of the chemical that causes mutations (dry or soaked). The half-life of the chemical is another factor that influences the dose. Therefore, the concentration for EMS can range from 0.3 to 1.5%, and the treatment time can be anywhere from 2 to 6 hours. Because of its short half-life, the concentration of DES can be anywhere from 0.1 to 0.6% for 2 hours, but the solution must be replaced with a new one every half-hour. After the stipulated amount of time has elapsed for the treatment, the seeds are subjected to the necessary procedure of being washed for a total of 4 hours in moving water. Blotting paper can be used to remove excess moisture from the seeds before they are sown in the field. In the event that the seeds cannot be planted right away, they should be kept in a dry place until their moisture content reaches 13%. It is essential that the used mutagen solution be disposed of in a secure manner. The next step explains how this can be accomplished. After mixing it together with additional sodium bicarbonate powder in a bucket, transfer the mixture to a large drum and fill it with approximately 50 L of water. After 48–72 hours, pour the mixture down the drain along with any excess water.

The fast development of DNA sequencing technologies and related data analysis techniques is transforming plant breeding (Moorthie et al. 2011). Similarly, developments in these fields will greatly aid plant mutant breeding. For example, sequencing technology combined with accompanying informatics tools can aid in the resolution of a bottleneck in mutation breeding, or the selection of mutants at the population level. This usually necessitates progressing and phenotyping mutant populations for up to five or six generations. Advanced DNA sequencing and bioinformatics techniques, in general, can improve precision and accelerate plant mutation breeding efforts. DNA sequencing is currently available in two varieties: (1) short-read sequencing and (2) long-read sequencing. The resequencing of genomes to discover mutations, primarily SNPs and InDels, is one application of short-read DNA sequencing (Wurtzel et al. 2010). The first steps in the analysis entail converting raw sequence reads from the sequencer into relevant output.

5.4 HOW TO DETECT AND ISOLATE MUTATION?

Ionizing radiation causes various kinds of mutations across the genomes of individual plant cells that have been exposed to radiation or genetically modified. The analyses of self-pollinated generations (M_2 or later generations) of the mutagenized plants, however, are the basis for current knowledge on radiation-induced genome-wide mutations in plants (M_1 generation) (Kitamura et al. 2022). Therefore, mutations that are rarely passed on to offspring and those that occur in nongermline cells could not be examined.

5.4.1 Detection of Natural Mutations

Various tools have been developed to detect polymorphism in plants. Through restriction enzyme digestion of DNA and selective amplification of DNA fragments, AFLP (Amplified Fragment Length Polymorphisms) has been widely used to detect DNA polymorphisms in a variety of crops (Agarwal et al. 2008). SSRs (also known as microsatellites) are "2-6" nucleotide repeats that are highly polymorphic and found throughout the genome (Miah et al. 2013). GBS (Genotyping-by-Sequencing) and RAD (Restriction-site Associated DNA) use high-throughput sequencing to investigate SNP diversity in different crop genotypes (Elshire et al. 2011). EcoTILLING is a high-throughput method for detecting variation in natural populations that employs a technology similar to TILLING (Comai et al. 2011).

5.4.2 Detection of Induced Mutations

With the forward genetics system, point mutations can be discovered using either the traditional map-based approach or the recently established MutMap technique (Abe et al. 2012). MutMap has been demonstrated to discover mutations in uncommon genomic regions in rice semidwarfism. In this case, a desirable mutant line was crossed with the parent line first, and then self-pollinated to produce F2 progenies for SNP identification. When insertional mutagenesis occurs, the gene responsible for the changed phenotype can be identified using the Thermal Asymmetric Interlaced PCR (TAIL-PCR) method, which uses nested PCR with degenerate primers (Liuand Whittier 1995).

The M_1 generation is made up of the plants that emerge from the treated seeds. Cross-pollination should be prevented at all costs in order to ensure self-pollination.

It is simple to accomplish this by cultivating it alone. The use of selfing bags is costly, time-consuming, and detrimental to seed set. It is beneficial to plant seeds more closely together to ensure that fewer branches or tillers are produced, as this will lessen the sectorial effect (Jurka and Kapitonov 1999). Even though many different factors can be researched to evaluate the harm brought on by mutagen treatment, these factors might not be significant if the goal is to find desirable mutations or boost variability. Furthermore, most crop plants now have access to this information. It might not be possible to publish this information. The parameters of M_1 injury (Singh and Mohapatra 2004) that are typically studied are the following: number of tillers, pollen sterility, number of seeds, seed sterility, abnormal leaves, leaves with chlorophyll-deficient sectors, survival and plant height at harvest, etc. There is no genetic basis for the effects observed in M_1 plants. In diploids, heritable changes first manifest in the second generation (M_2) (Brunner 1995). Mutants may also show up in subsequent generations in polyploids. The M_1's morphometric data are irrelevant. Different techniques for harvesting and raising the M_2 may be used, depending on the goal of the experiment and the ontogeny of the plant. Because primary tillers in monocotyledonous plants develop from the primordia in the seed at the time of mutagenic treatment, they are more likely to experience mutations. In dicotyledonous plants, it is anticipated that there will be more mutations in the seeds on the main branch.

5.4.3 Harvesting Techniques

5.4.3.1 Bulk Method

The M_1 plants' seeds can be bulked dose-wise and planted to produce the M_2 generation (Albokari 2014) when the goal is to isolate specific mutations with a desirable trait. Bulking one or a few seeds from the panicles (or fruits) of the primary tillers will increase the effectiveness of this method (or main and secondary branches). Greater chances of mutant isolation result from representation of seeds from all primary and secondary branch regions.

5.4.3.2 M_1 Plant to Row

The M_2 generation is grown as M_1 plant to row (Shadakshari et al. 2001), with each row of 30–40 plants representing one M1 plant progeny. All or a sample of the seeds from each M_1 plant are harvested individually. The benefit is that the mutant can be recovered from the sister plants in the following generation if it is lost before harvest.

5.4.3.3 M_1 Panicle, Branch, Pod to Row

Using this technique, each M_1 plant will produce a number of offspring since they all grow from various panicles or pods (Imam and Chakraborty 2019; et al. Ahirwar et al. 2014). Given that planting and harvesting will require a lot of labor, this method is obviously more expensive. Depending on how many people one can manage In the M_2 generation, 30,000–40,000 plants may be raised. To compare the control plants with the mutants easily, controls should be grown at regular intervals.

5.4.4 Mutation Detection in the M_2

Because lethal chlorophyll mutants only survive for 7–12 days, screening for these mutations should begin at the seedling stage. The expression of the mutant trait can be studied at various growth stages and can be harvested easily if all morphological mutants are labeled. It is possible to remove the neighboring plants that look normal in order to guarantee the survival of weak mutants. The mutation frequency (Russell et al. 1958) can be expressed as the number of mutants per 100 M_2 plants using the total number of mutants in each dose. By observing the mutant characters at various growth stages, the spectrum of induced mutants can be evaluated. Gather every mutant for research on the M_3 generation's breeding habits.

5.4.5 Effectiveness and Efficiency of Mutagenesis

The level of a mutagen's ability to cause mutations is known as its mutagenic effectiveness. This is calculated as the percentage of the dose of mutagen that caused the mutation to occur divided by the frequency of the mutations (Ambavane et al. 2015). The dose for chemical mutagens is calculated as concentration times treatment time. Mutagenic efficiency, which is calculated as the ratio of mutation frequency to M_1 injury (% mutation/lethality in M_1), illustrates the relationship between the damage caused by a mutagen and the frequency of mutations (Russell et al. 1979). Mutagenic effectiveness and efficiency were found to be greatest in nonbasmati aromatic rice at 300 Gy and lowest at 350 Gy. Efficiency and effectiveness of mutagens are used to identify the frequency of favorable mutations (Imam et al. 2019). The mutagenic efficiency enhanced with increasing irradiation dose, but no association between mutagen doses and mutagenic efficacy was reported. With 200–300 Gy dose of gamma rays, less damage was detected in biological features such as mortality, injury, and sterility. At 200 and 300 Gy, desirable mutants were established.

The adoption of a visible marker allowed for the decrease of genomic complexity in M_1 plants as well as the discovery of radiation-induced genome-wide alterations (anthocyanin pigmentation). Because the M_1 population is directly exposed, evaluating the M_1 genome sequence (Kitamura et al. 2022) rather than the genomes of M_2 or later generations may expose the impacts of mutagens more clearly. The M_1 genome analysis also eliminates the possibility of mutation loss during gametogenesis. The mutational assessment of the M_1 generation may potentially provide crucial information regarding the genetic progression of certain ornamental species propagable using vegetative clonal methods. However, because of chimerism, which occurs when multiple types of altered cells coexist in various amounts, identifying mutations across the M_1 is difficult.

5.5 TILLING (TARGETING INDUCED LOCAL LESIONS IN GENOMES)

It is a molecular biology technique for detecting gene mutations. TILLING and mutant breeding have played critical roles in the introduction of mutations, as well as the discovery and application of these mutations in genetic improvement. TILLING requires careful consideration of the genotype and portion of the plant to be used as an explant, the kind and concentration of the mutagen, the size of the population to be checked, and the mutation detection method. TILLING relies on the formation of DNA heteroduplexes, which occur when several alleles are amplified by PCR, heated, and slowly cooled.

EcoTILLING is a way of searching for natural mutations in humans using TILLING processes, generally for population genetics research. DEcoTILLING is a hybrid of TILLING and EcoTILLING that employs a low-cost technique for fragment identification. There are several methods for detecting mutations following heteroduplex cleavage. The most common method is to run a denaturing polyacrylamide gel through a LI-COR DNA analyzer (also known as the LI-COR). PCR products are amplified using primers labeled with infrared dye (IRD) in this approach. Both forward and reverse primers have a dye labeled at the 5′ end that can be recognized in one of the "LI-two COR channels." Objects are denaturized at a rate of 5–6%.

The majority of TILLING systems depend on high-throughput genotyping technologies like LI-COR gene analyzers, which use fluorescently labeled primers and are quite expensive for single laboratory setups. TILLING may be challenging to execute because of the necessary expenditure and technical skills. Detection approaches based on agarose have recently been presented as low-cost alternatives to current technology-intensive platforms (Sato et al. 2006), and their utility for identifying EMS-induced mutations in large libraries has recently been shown (Caldwell et al. 2004).

There are now numerous TILLING platforms available for a wide range of model and crop plants. Mutants with favorable features have been established and are now being introduced into breeding programs as a result. TILLING is a high-throughput reverse genetics technique that can precisely pinpoint mutations in qualities that are difficult to target using front genetics; yet, mutations in visibly recognizable attributes like plant height can be investigated using the front genetics approach.

TILLING has the following advantages: (i) it generates a variety of allelic mutations pertaining to genetic study; (ii) mutations that are hard to identify via forward genetics may be revealed because TILLING can target a specific gene of interest; (iii) it is appropriate to any organism with a different genome size and ploidy level; (iv) it results in stable mutation; and (v) no exogenous DNA is introduced into the plant (Tadele et al. 2010).

TILLING is a procedure that consists of the following major steps: (i) mutagenesis; (ii) nonchimeric population advancement; (iii) germplasm asset preparation; (iv) DNA filtration and sample accumulating; (v) population testing to discover mutations in the target gene; and (vi) mutant line recognition and target gene sequencing (Till et al. 2006). TILLING has grown in popularity as a method of enabling mutation or variety in crop plants and exploiting this variability to benefit agricultural plants. The TILLING population, once generated, may be utilized to screen for a variety of desirable characteristics.

5.6 CROP IMPROVEMENT THROUGH SPACE MUTAGENESIS

Since the 1930s, gamma and X-rays have been the most common agents employed to induce genetic diversity for plant mutant breeding. Heavy ion beams, electron beams, proton beams, and cosmic radiation are all becoming more popular. In November 2022, Seed Irradiation in Space for Induced Genetic Diversity and Plant Mutation Breeding was launched at the International Space Station (ISS) (Liu et al. 2009). Gravitation and radiation are two physical phenomena that differ dramatically in outer space from the Earth's surface. Photons emitted by the sun, as well as energetic particles from the solar wind and galactic cosmic rays (GCRs), constitute space radiation. The electromagnetic radiation emitted by the sun falls into several spectral categories, ranging from gamma rays to radio waves. With 95% protons, 4% helium, and around 1% heavier nuclei, electrons, and positrons, GCRs dominate the nonelectromagnetic radiation environment in interstellar space. For many astrobiology experiments, radiation information is the critical differentiation between the multiple possible Earth orbits and other locations. There is a rising interest in understanding the impact of the space environment on producing mutations in plant genomes and affecting plant physiology in order to improve plants' ability to withstand poor growth conditions on Earth. Substantial progress is expected in the area, with a sustained interest in investigating plant biology in space for both feeding astronauts and creating mutations with space exposure to breed resistant agricultural varieties on Earth. Shijian 8, a plant-breeding satellite, was launched and retrieved in September 2006 to conduct basic and applied research in space breeding (Mohanta et al. 2021). Ground-based simulation algorithms and high-throughput mutant gene screening approaches were also devised and optimized to efficiently develop new resources and varieties for higher performance via space-induced mutagenesis (Chen et al. 2020).

Eight mutant populations for wheat, maize, rice, soybean, and cotton were created using established space-induced or simulated mutation approaches. In wheat, for example, the primary related features include biotic and abiotic stress resistance, plant height, tiller number, spike length, and thousand-grain weight. Hundreds of elite mutant lines with improved critical agronomic features, such as high anther culture frequency, high yield, excellent quality, and strong disease or stress tolerance, were produced and employed for crop breeding from these mutant resources. Many wheat mutant varieties, such as Hangmai 2566 and Hangmai 501, have been produced using a high anther culture frequency and high yield potential wheat mutant line SPLM2. Suinong 86 and Suidou 5 have been developed using a disease-resistant soybean mutant line (Liu et al. 2023). Novel mutant germplasm can be selected directly for new kinds or cultivated indirectly as a donor parent of superior genes. Luo et al. discovered that space mutagenesis occurs more frequently than gamma-ray radiation in a genome-wide analysis of rice space-induced mutations (Luo et al. 2014).

The study of mutagenic effects on plant species in space is a critical topic that crosses many fields, including space biology, genetics, mutations, and hybridization. Although many studies have demonstrated the mutagenic effects of space (Horneck et al. 2010), and a variety of types have been chosen by space mutation, there is still a scarcity of research on the molecular properties and genetic pathways of space-induced mutations, particularly in an important food crop like rice.

The following issues need to be addressed further: (1) investigation of the combination mutagenic effects of microgravity and space radiation; (2) single-factor assessment and ground simulation of space radiation mutagenic features; (3) development of a single-cell mutation map of space radiation-induced mutations and genetic network research; and (4) high-efficiency identification and rapid fixation of space-induced variations.

5.7 HANDLING OF MUTANT POPULATIONS

5.7.1 Population Size

Since mutations are induced in the treated seeds the frequency of mutations depends on the size of the sample treated; the greater the number of seeds treated, the greater the mutation frequency. Taking several factors into account it was estimated that about 6000 seeds need to be treated to get about the minimum number (10) of desired mutants. However, this estimation may not have direct application in practice due to various factors. Three doses of mutagen may be used. For gamma rays the doses that cause 30–50% growth reduction in the laboratory have been proven to be useful for mutation induction. For chemical mutagens the levels of injury should be low at 10–30%. An M_1 population of 1000–1500 plants may be grown so that the M_1 population will not become too large to study effectively.

5.7.2 M_1 Generation

The M_1 generation consists of the plants that sprout from the treated seeds. To achieve self-pollination and avoid cross-pollination, extreme caution should be exercised. This is simply accomplished by growing it in isolation. The use of selfing bags is costly and time intensive, and it has a negative impact on seed set. Closer seed planting to promote the formation of fewer branches or tillers is beneficial in reducing the sectorial effect.

5.7.2.1 M_1 Injury Parameters

While several metrics can be evaluated to assess the injury produced by mutagen treatment, they may be insignificant if the goal is to isolate beneficial mutations or increase variability.

Furthermore, such data is currently available for the majority of crop plants. It is conceivable that this information will not be published. The following M_1 damage parameters are commonly studied: seed emergence, seedling height, root length, aberrant leaves, leaves with chlorophyll deficient sectors, survival and plant height at harvest, number of tillers, pollen sterility, number of seeds, seed sterility, and so on. The impacts shown on M_1 plants are not inherited. In diploids, heritable alterations manifest in the next generation (M_2). Mutants may appear in successive generations of polyploids. Morphometric data in the M_1 is meaningless.

5.7.3 M_1 Harvest and M_2 Population

Depending on the experiment's goal and the plant's ontogeny, multiple ways of harvesting and raising the M_2 may be used. The risks of obtaining mutations in monocotyledonous plants are higher in the primary tillers because these tillers develop from the primordia present in the seed at the time of mutagenic treatment. The seeds on the main branch of dicotyledonous plants are likely to have a higher incidence of mutations.

5.8 MUTATION BREEDING STRATEGY FOR OBTAINING MUTANTS

Induced mutagenesis, particularly using gamma (γ) irradiation, plays an important role in generating genetic variability in crops including cereals, pulses, and oilseeds for desirable traits of economic importance and these variants were used further in the cross-breeding program of several seed-propagated crops (Oladosu et al. 2016; Rampure et al. 2017; Mahla et al. 2018; Chaudhary et al. 2019) and also meets up challenges including world food and nutritional security (Celik & Atak 2017) (Figure 5.3). Induced mutation has successfully been employed in rice breeding worldwide (Oladosu et al. 2016; Roy et al. 2018). The number of mutant varieties through γ-irradiation officially released and recorded in the FAO/IAEA Mutant Varieties Database (MVD) is 1665 including 533 mutant rice varieties, among them 22 in India (MVD 2021). Since the 1960s, breeders have most frequently used

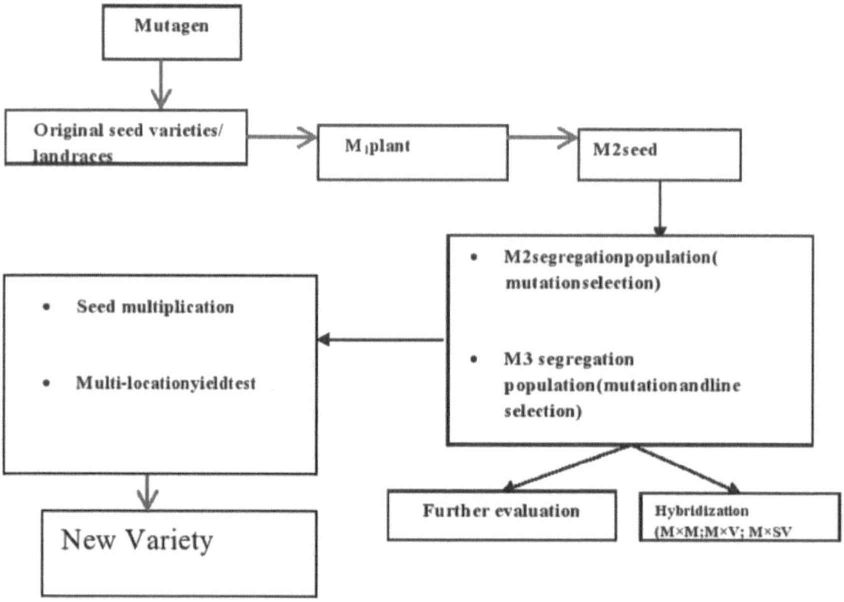

FIGURE 5.3 Procedure for utilization of potent mutant variety.

gamma rays to increase various qualities in plant mutation breeding programs (Beyaz & Yildiz 2017). Gamma rays are the most penetrating electromagnetic radiation and operate more simply than other physical mutagens.(Kovacs & Keresztesa 2002; Li et al. 2019). Gamma rays are a type of nonparticulate ionizing radiation that is created by the decay of radioisotopes such as cobalt 60 (^{60}Co) and cesium 137 (^{137}Cs). It has an effect on plant growth and development by causing cytological, genetic, biochemical, physiological, and morphological changes in cells and tissues (Gunckel & Sparrow 1967).

In India, Jagannath is the first mutant variety developed through induced mutagenesis in 1975. Some important rice mutants such as HUR-36 (1990), ADT-41 (1994), Malviya Dhan-36 (1997), MO 15 (1999), PNR-519 (2000), Dhanu (2000), Pusa-NR-546 (2003), CRM2007-1 (2006), Anashwara (2006), and TCDM-1 (2019) have been developed through induced mutagenesis up to 2020 in India by using gamma irradiation (MVD 2021). Primary research centers that make an important contribution to the field of induced mutations, development, and release of mutant varieties are the Bhabha Atomic Research Centre (BARC) in Mumbai, the Indian Agricultural Research Institute (IARI) in New Delhi, the International Crops Research Institute for Semi-Arid Tropics (ICRISAT) in Hyderabad, the Tamil Nadu Agricultural University (TNAU) in Coimbatore, and the National Botanical Research Institute (NBRI) situated in Lucknow.

Any successful mutation breeding program depends on the knowledge of the radiosensitivity of any mutagens in crops. Radiosensitivity of crops varies according to species, cultivars, plant portion of the mutation, and material water content (Spencer-Lopes et al. 2018). Free radicals, which can change or harm essential plant cell components, are created as a result of interactions with molecules in plant cells, particularly water (Gowthami et al. 2017). According to one set of studies, the maximum probability of creating useful mutations is discovered at a level where 50% of the irradiated individuals die (Angeles-Espino et al. 2013). Some scientists, on the other hand, have pointed out that another dose, ahead of LD_{50}, with the highest likelihood of causing beneficial mutations is when 30% or 50% of a growth decrease (GR_{30}/GR_{50}) is detected (Khalil et al. 2014; Kant et al. 2020).

The appropriate radiation dose for crop development projects is established by applying the LD_{50} and the GR_{30}/GR_{50} to induce useful mutations with low influence on the genome, but large doses may have many impacts on the genome, resulting in aberrations or undesirable modifications (Thole et al. 2012). Furthermore, low irradiation doses may stimulate cell division and enzymatic activity, as well as increase the number of photosynthetic pigments (Aparna et al. 2013), whereas high irradiation doses may inhibit enzyme and hormonal activities, resulting in physical damage to organelles or chemical changes in the cells (Alvarez-Holguin et al. 2019) and also causes higher pollen and spikelet sterility in M_1 plants (El-Keredy 1990; El-Degwy 2013) due to various chromosomal aberrations and meiotic abnormalities including chromosome univalent, multivalent, stickiness and laggards, and bridges (Ata et al. 2017; Wani & Bhat 2017). These forms of reactions were not caused by genetic changes and will not be inherited. According to this viewpoint, low and high irradiation doses may generate plant phenotypic modifications that may be misunderstood as genetic changes (Alvarez-Holguin et al. 2019). As a result, before beginning a mutagenesis-based genetic improvement program, the ideal radiation dose must be identified; this increases the likelihood of finding useful mutations that help to avoid excessive losses in terms of time, human cost, field area, and experimental materials.

The mutagenic effectiveness (mutation per dose of mutagen) and mutagenic efficiency (mutation in connection to undesirable changes/damages such as lethality, injury, and sterility) also play a pivotal role in induced mutagenesis to producing a higher spectrum with high frequency of desirable mutations such as short stature, early maturity, bold or cylinder type grains, dark green leaves, high tillering with improved grain yield, etc., which is free from undesirable ones (Ramesh et al. 2019). Moreover, an effective and efficient dose of physical mutagens also generates different kinds of undesirable mutations like sterile- and fizzy-type panicles (Komatsu et al. 2001; Zhu et al. 2003), extrasterile glumes (Yoshida et al. 2009: Ren et al. 2013; Yoon et al. 2018), and narrow leaves (Wei et al. 2013), which is only used for collection of genetic information in rice crop. The desirable and undesirable mutations result from changes in the DNA sequences, viz., transitions, transversions, insertions, and deletions (In-Dels) of nucleotide base pairs within the coding region of a functional gene (Nielen et al. 2018). This mutagenic defaults in a functional gene, ultimately blocking or deactivating a particular biosynthesis pathway and signaling, which affects the expression of a specific phenotype or trait (Nielen et al. 2018; Ferrero-Serrano et al. 2019).

The semidwarf plant height is one of the most preferred agronomic traits in cereal crops, including rice. Recently, at least 76 plant height mutants and 85 plant height reduction genes have been identified in rice (Liu et al. 2009; Ferrero-Serrano et al. 2019; Bhuvaneswari et al. 2020). However, the most common plant height reduction gene, "*sd1* (*semidwarf1*)," has been widely utilized in developing high-yielding varieties across the world. The semidwarf gene provides short and thick culms with improved lodging resistance, is well responsive to nitrogen fertilizer, has a higher harvest index, and results in higher grain yields without affecting grain quality. (Futsuhara et al. 1997; Gaur et al. 2020). The rice *sd1* is also known as the "green revolution gene," because it has contributed to the significant increase in crop production during the green revolution in the 1960s, especially in Asian countries. The *sd1* allele is a recessive mutant form of *Sd1* functional gene. It was originally identified from the Taiwanese rice cultivar "Dee-Geo-Woo-Gen" (DGWG) by using positional cloning strategy (Monna et al. 2002). The *sd1* gene from DGWG generated through spontaneous mutation showed a 383-bp deletion starting from the middle of *exon1* to the upstream of *exon2* (Monna et al. 2002; Spielmeyer et al. 2002; Bhuvaneswari et al. 2020). Another radiation-induced *sd1* mutant gene was identified from Calrose76, a 1-bp substitution in *exon2* (Monna et al. 2002; Spielmeyer et al. 2002).

In rice, the *Sd1* functional gene is located at 38.38Mb on the long arm of chromosome 1 and possesses a 3123-bp-long open reading frame (ORF) consisting of 3 exons (557, 321, and 291bp) and 2 introns (103 and 1472bp) (Bhuvaneswari et al. 2020). It encodes *Gibberellin 20-oxidase* which is the key enzyme in gibberellin biosynthesis pathway (Spielmeyer et al. 2002). The *Gibberellin 20-oxidase*, a 389 amino acid chain enzyme involved in the catalytic conversion of GA_{12}/GA_{53} to bioactive gibberellic acid precursors GA_9/GA_{20} (Hedden& Phillips 2000; Ferrero-Serrano et al.

2019). Expression of *Gibberellin 20-oxidase* is found to occur in stems, unopened flowers, and leaf blades (Sasaki et al. 2002; Sakamoto et al. 2004). Semidwarfism, viz., poor internodal elongation in rice stems, is often associated with defective gibberellin biosynthesis pathway that reduces intercalary meristem cell division and elongation (Wang et al. 2009; Andrew-Peter-Leon et al. 2021).

The single use of semidwarf gene *sd1* solely confers the semidwarfism in rice cultivars that commonly grow worldwide due to its nondetrimental effects on grain yield (Hedden 2003; Hedden & Sponsel 2015). At present, no other semidwarf gene, specially in nonbasmati aromatic rice, has been identified, which has such a broad utility in aromatic rice improvement programs. This narrow genetic base of the present *sd1* gene in semidwarf varieties has also led to a loss of genetic diversity in rice (Tomita & Ishii 2018; Andrew-Peter-Leon et al. 2021). Hence, there is an urgent need to characterize a novel semidwarf gene alternative to *sd1* and use it to increase genetic diversity in semidwarf varieties and nonbasmati aromatic rice breeding programs.

5.9 BENEFICIAL USE OF MUTATION BREEDING

Mutation breeding is especially useful when normal hybridization methods are challenging or ineffective. One of the most important elements of mutant breeding has been the rapid correction of weaknesses in varieties and advanced breeding lines, as well as the creation of polygenic mutations and the development of ideotypes for different agroclimatic situations. Induced mutation has been shown to boost yield as well as other agronomic characteristics such as straw stiffness, maturity time, adaptability, shattering resistance, disease resistance, protein content, baking quality, malting quality, and a variety of other traits (Borojevic 1990; Brunner 1995). Mutation breeding often focuses on features that either were not favored by natural selection during the evolutionary process or were not enhanced during prior plant breeding efforts. Sigurbjornsson and Micke (1974) demonstrated the growing importance of induced mutation in plant development.

One of the most serious disadvantages of mutation breeding is the unfavorable effect caused by the mutant gene's pleiotropic action or the simultaneous mutation of closely related genes. Although the majority of induced mutations are harmful, valuable mutants can be recovered using an effective selection approach. Yet, the occurrence of even a few favorable mutations in high-yielding variety has the significant benefit of becoming homozygous and exhibiting its superiority within a few generations after induction in M_2 or M_3 generations in comparison to F_6 or F_7 generations in the case of hybridization.

Hence, produced mutations can serve as an alternative or supplement to natural diversity and hybridization. Artificial mutations may be comparable to those that occurred naturally, although many of them are unlikely to have occurred normally or have been lost from the natural population. Desirable mutants fit for the present agricultural system could be retained using appropriate selection strategies (Brock 1971).

Popular cultivars Badshabhog, Bahadurbhog, and Blackjoha were subjected to gamma irradiation ranging from 200 Gy to 400 Gy with a 50 Gy interval, with an unexposed treatment serving as a control. The experiment was designed in a randomized block design with three replications for in vitro and in vivo conditions. Duncan's test and regression analysis were used to analyze the data. At increasing dosages of gamma rays, most of the traits of each cultivar were drastically reduced. The ideal dose of gamma rays was determined to be 358.37 Gy and 346.10 Gy in Badshabhog; 331.19 Gy and 319.17 Gy in Bahadurbhog; and 314.55 Gy and 314.05 Gy in Blackjoha based on the weighted mean of LD_{50} and GR_{30}/GR_{50} with "40% and 10%" weighted, respectively, under in vitro and in vivo circumstances. This ideal dose of gamma rays can be utilized to create desirable mutants of these cultivars while causing the least amount of harm. Blackjoha was the most radiosensitive, whereas Badshabhog was the least.

Kole et al. (2008) investigated 18 morphologically distinct mutants in M_4 generation along with their two mother genotypes (IET 14142 and IET 14143) derived from Tulaipanja, an aromatic nonbasmati rice cultivar from West Bengal, and discovered that panicle number had the greatest positive direct effect, followed by grain number, test weight, plant height, days to flower, and straw weight. Overall, the results showed that larger panicle number per plant, test weight, straw weight, and medium plant height were selected, with a fair balance for moderate grain. Chakraborty and Kole (2014) reported that the increase in values of GCV and PCV in M_3 over M_2 generation for flag leaf length and spikelet number was due to "release of additional variability" and also noticed the scope for isolating promising segregants in micromutational population even in early generations.

The appropriate radiation dose for crop development projects is established by applying the LD_{50} and the GR_{30}/GR_{50} to induce useful mutations with low influence on the genome, but large doses may have many impacts on the genome, resulting in aberrations or undesirable modifications (Thole et al. 2012). Furthermore, low irradiation doses may induce cell division and enzymatic activity or increase the number of photosynthetic pigments (Aparna et al. 2013), whereas high irradiation doses may restrict enzyme and hormonal activities, resulting in physical damage to organelles or chemical changes in the cells (Alvarez-Holguin et al. 2019). These types of responses are not connected to genetic changes and will not be gained. According to this viewpoint, low and high irradiation doses may generate plant phenotypic alterations that may be misinterpreted as genetic changes (Alvarez-Holguin et al. 2019). As a result, before beginning a mutagenesis-based genetic improvement program, the optimal radiation dose must be identified; this increases the likelihood of finding useful mutations that help to avoid excessive losses in terms of time, human cost, field area, and experimental materials. Kanta et al. (2020) estimated the appropriate radiation dose by balancing both in vitro and in vivo data: 350–360 Gy for Badshabhog, 320–330 Gy for Bahadurbhog, and 315 Gy for Blackjoha.

5.10 ACHIEVEMENTS THROUGH APPLICATION OF MUTATION BREEDING

Some of the advantages of mutant breeding are increased yield, earlier maturity, stress resilience, salt tolerance, waterlogging tolerance, and larger seed size (Yali and

Mitiku 2022). Mutation breeding has resulted in improved varieties of wheat, rice, and barley. Many crop types have been developed with high yield, increased quality, earliness, dwarfness, disease resistance, and reduced toxin content. Mutation has also been used to produce male sterility, which reduces the cost of hybrid seed production, broadens the genetic variety available, and helps crop response. Rigorous testing of desirable mutants in M_4 generation the following important agronomical characters were isolated by authors:

- Semidwarf mutants having desirable plant type with up to 50 cm. reduction in plant height as compared to their parent.
- Early maturity mutants having 30 days early.
- Mutants having photo- and thermoinsensitivity properties.
- Mutants having good ratooning habit.

This mutagenesis study produced a significant quantity of genetic variety for economically and agronomically important traits. After rigorous testing under various agroclimatic conditions, these mutant lines may be offered as varieties for commercial cultivation or used as prospective parents for future cross-breeding programs.

The vast majority of mutant varieties that have been released (Table 5.1) (including grains, pulses, oil, root and tuber crops, and ornamentals) have been released in developing nations, resulting in favorable economic consequences. Hence, plant mutant breeding is an effective strategy for protecting and improving global food security.

TABLE 5.1
Some Popular Mutant Varieties Cultivated in India

Common name	Latin name	Mutant cultivar
Rice	*Oryza sativa*	PNR-381, PNR-102
Wheat	*Triticum aestivum*	NP-836, Sharboti Sonora
Sugarcane	*Saccharum officinarum*	Co-6608
Castor	*Ricinus communis*	Aruna
Blackgram	*Vigna mungo*	TAU 1,TU 94-2
Cowpea	*Vigna unguiculata*	V-16 (Amba)
Mung bean	*Vigna radiate*	TAP-7, MUM-2, BM-4, LGG-409, LGG-450, Co 4, Dhauli, Pant moong-1, TM 96-2
Chickpea	*Cicer arietinum*	Pusa-547, Pusa-408 (Ajoy), Pusa-413 (Atul), Pusa-417 (Girnar)
Jute	*Corchorus capsularis*	Shymali

Source: Chakraborty and Paul (2013).

5.11 INDUCED MUTATIONS FOR QUALITY AND NUTRITION IMPROVEMENT

Induced mutation has been employed to increase crop quality for many years. Many parts of these two characteristics were debated. The study focused on the addition of mineral elements (biofortification) and amino acids required by humans and animals, the alteration of protein and fatty acid profiles for nutritional and health purposes, the adjustment of physicochemical properties of starches for various end-uses, the enhancement of phytonutrients in fruits, and the reduction of antinutrients in staple foods. Numerous mutant genes have been successfully integrated into commercial crop varieties, boosting their nutritional value significantly. Mutant variants created and released in major crops have been widely planted by farmers, leading to enhanced food production and contributing to food security. Some examples are listed below:

The protein in maize endosperm is lacking in two critical amino acids, lysine and tryptophan. QPM cultivars were created using the opaque 2 mutant gene as well as endosperm and amino acid modifier genes. QPM contains nearly twice as much lysine and tryptophan as regular maize and 30% less leucine and has been demonstrated to have dramatic benefits on human and animal nutrition, growth, and performance. Hundreds of thousands of hectares of QPM cultivars are now grown.

The majority of phosphorus in seeds is accumulated as phytic acid and its salts (phytate). Because phosphorus and mineral elements such as iron and zinc cannot be digested by humans and monogastric animals in the form of phytate, lowering phytic acid would increase phosphorus and micronutrient mineral element bioavailability. A number of LPA mutations have been created in barley, rice, wheat, and soybean, according to findings presented at the meeting, and unique LPA barley types have been introduced for commercial usage. LPA variations are projected to be produced in a wide range of crops, with the potential to reduce phosphorus pollution while boosting phosphorus and micronutrient mineral bioavailability in cereal grains and legume seeds. The end application of plant oils determines their ideal composition. While unsaturated fatty acids (oleic, linolenic) are appealing for salad and cooking oils, higher concentrations of saturated fatty acids (stearic, palmitic) are prioritized for oils used in the food industry, because high-temperature processes (frying) necessitate oils that are thermooxidation resistant. Several studies have been undertaken on the change of fatty acid composition using mutation induction in soybean, sunflower, and other oil crops. By utilizing the changed genes, new types have been created for a variety of purposes.

5.12 IMPACT OF MUTANT CULTIVARS

Mutants are either introduced as varieties or used as a starting point in breeding efforts to generate genetic change. The newly released mutant varieties have larger yields, are disease resistant, have increased quality, and are tolerant to changes in the environment. A large number of these mutant cultivars have been distributed in poor nations, enhancing their economic position (Mir et al. 2020). These cultivars cover hundreds of millions of acres of agricultural land, while the influence on these countries' national economies is assessed in billions of dollars. Although the major goal of mutation breeding has been to create novel cultivars, the genetic stocks that have been developed can be employed in a variety of plant breeding applications, such as a donor parent in conventional breeding programs or as a parent in hybrid breeding programs. The approach of lowering phenotypic expression using induced mutagenesis to locate a gene is now a significant part of molecular genetics and genomics research (Roychowdhury and Tah 2013). The IAEA has classified its mutant variants under four breeding procedures, notably:

1. direct use of a mutant line obtained by physical or chemical mutagenesis or somaclonal variation;
2. indirect use of a mutant line/lines as a parental variety/varieties in cross-breeding (crossing between mutant lines or with a commercial variety/ varieties);
3. usage of mutant gene allele (trait); and
4. utilization of wild species' genes.

More progress has been made in the other classes by employing mutant lines as parents in breeding programs, as well as mutagenesis in breeding facilities.

The detection of millions of mutations in a short amount of time has grown more convenient and cost-effective as next-generation sequencing (NGS) techniques have advanced. High-throughput next-generation sequencing–based mutation mapping approaches, such as MutMap, MutMap-Gap, MutChromSeq, Mut-Ren-Seq, and whole-genome sequencing–based mutation mapping, have recently been created. These methods have high applicability as well as cost-effective tools to identify desired genes in less time (Sanchez-Martin et al. 2016). Under situations of exponentially rising human population and climate change, the merging of cutting-edge mutant breeding technologies with powerful selection and speed breeding tools broadens its application in meeting food security issues.

REFERENCES

Abe, A., Kosugi, S., Yoshida, K., Natsume, S., Takagi, H., Kanzaki, H., Matsumura, H., Mitsuoka, C., Tamiru, M., Innan, H., Cano, L., Kamoun, S., Terauchi, R. 2012. Genome sequencing reveals agronomically important loci in rice using MutMap. *Nature Biotechnology, 30*(2), 174–178.

Agarwal, M., Shrivastava, N. and Padh, H., 2008. Advances in molecular marker techniques and their applications in plant sciences. *Plant Cell Reports, 27*, pp.617–631.

Ahloowalia, B.S. and Maluszynski, M., 2001. *Induced mutations – a new paradigm in plant breeding. Euphytica*, 118, 167–173.

Ahloowalia B S, Maluszynski M, Nichterlein K., 2004. *Global impact of mutation-derived varieties. Euphytica*, 135(2), 187–204.

Albert, D., Dumonceaux, T., Carisse, O., Beaulieu, C. and Filion, M., 2022. Combining Desirable Traits for a Good Biocontrol Strategy against *Sclerotinia sclerotiorum. Microorganisms*, 10(6), 1189.

Albokari, M., 2014. Induction of mutants in durum wheat (*Triticum durum* Desf cv. Samra) using gamma irradiation. *Pakistan Journal of Botany*, 46(1), pp.317–324.

Alvarez-Holguin, A., Raul Morales-Nieto, C., Hugo Avendano-Arrazate, C., Corrales-Lerma, R., Villarreal-Guerrero, F., Santellano-Estrada, E., & Gomez-Simuta, Y. 2019. Mean lethal dose (LD_{50}) and growth reduction (GR (50)) due to gamma radiation in Wilman lovegrass (*Eragrostis superba*). *Revista Mexicana De Ciencias Pecuarias*, 10(1): 227–238. DOI: 10.22319/rmcp.v10i1.4327

Ambavane, A.R., Sawardekar, S.V., Sawantdesai, S.A. and Gokhale, N.B., 2015. Studies on mutagenic effectiveness and efficiency of gamma rays and its effect on quantitative traits in finger millet (*Eleusine coracana* L. Gaertn). *Journal of Radiation Research and Applied Sciences*, 8(1), 120–125.

Andrew-Peter-Leon, M. T., Ramchander, S., Kumar, K. K., Muthamilarasan, M., & Pillai, M. A. 2021. Assessment of efficacy of mutagenesis of gamma-irradiation in plant height and days to maturity through expression analysis in rice. *PLoS ONE*, 16(1): 01–20. DOI: 10.1371/journal.pone.0245603

Angeles-Espino, A., Valencia-Botín, A. J., Virgen-Calleros, G., Ramirez-Serrano, C., Paredes-Gutierrez, L., & Hurtado-De la Pena, S. 2013. Lethal doses (LD_{50}) determination using Co^{60} on *Agave tequilana* var. *Azul vitroplantlets. Revista Fitotecnia Mexicana*, 36(4): 381–386.

Aparna, M., Chaturvedo, A., and Sreedhar, M. 2013. Impact of gamma rays on the seed germination and seedling parameters of groundnut (*Arachis Hypogaea* L.). *Asian Journal of Experimental Biological Sciences*, 4(1): 61–68.

Ata, A.M., Mahmoud, M.A.H., Gamea, A.M. and Abu-El-Soud, K.A.R.A.M., 2017. Cytological effects of gamma ray on onion plants and their M1 generation. Egyptian Journal of Plant Breeding, 21, 785–799.

Bhuvaneswari, S., Gopala Krishnan, S., Ellur, R. K., Vinod, K. K., Bollinedi, H., Bhowmick, P. K., Bansal, V.P., Nagarajan, M. & Singh, A. K. 2020. Discovery of a Novel Induced Polymorphism in *SD1* Gene Governing Semi-Dwarfism in Rice and Development of a Functional Marker for Marker-Assisted Selection. *Plants*, 9(9): 01–14. DOI: 10.3390/plants9091198

Beyaz, R., Yildiz, M. 2017. The use of gamma irradiation in plant mutation breeding. *Plant Eng*. https://doi.org/10.5772/intechopen.69974

Borojević, S., 1990. *Principles and methods of plant breeding*. Elsevier.

Brunner, H., 1995. Radiation induced mutations for plant selection. *Applied radiation and isotopes*, *46*(6–7), 589–594.

Caldwell DG, McCallum N, Shaw P, Muehlbauer GJ, Marshall DF, Waugh R. 2004.A structured mutant population for forward and reverse genetics in Barley (L.). *Plant J.* 2004, 40, 143–150. 10.1111/j.1365-313X.2004.02190.x.

Celik, O., & Atak, C. 2017. Applications of ionizing radiation in mutation breeding. *New Insights on Gamma Rays, Ahmed M. Maghraby, IntechOpen*, 111–132. DOI: 10.5772/66925

Chakraborty N R and Kole P C. 2014. Gamma ray induced early generation polygenic variability in medium grain aromatic non-basmati rice. *International Journal of Plant Breeding and Crop Science* 1: 28–35.

Chakraborty, N.R. and Paul, A. 2013. Role of induced mutations in enhancing nutrition quality and production of food. *International Journal of Bio-Resource and Stress management* 4(1): 014–019.

Chaudhary, J., Deshmukh, R., & Sonah, H. 2019. Mutagenesis approaches and their role in crop improvement. *Plants,* 8(11): 1–4. DOI: 10.3390/plants8110467

Chen, Y., Wu, B., Zhang, C., Fan, Z., Chen, Y., Xin, B. and Xie, Q., 2020. Current progression: application of high-throughput sequencing technique in space microbiology. *BioMed Research International, 2020*(3): 1–13.

Cimpeanu, C., Barna, C. and Iliescu, M., 2014.32P–Radioactive Tracer for the Evaluation of Fertilizers Influence on Nutrients Translocation Process from Soil to the Plants. *Rom. Journ. Phys , 59*(9–10), p.1048.

Comai, L.; Young, K.; Till, B. J.; Reynolds, S. H.; Greene, E. A.; Codomo, C. A.; Enns, L. C.; Johnson, J. E.; Burtner, C.; Odden, A. R.; Henikoff, S. 2004. Efficient discovery of DNA polymorphisms in natural populations by Ecotilling. *Plant J., 37*(5), 778–786.

El-Degwy, I. S. 2013. Mutation induced genetic variability in rice (*Oryza sativa* L.). *Egyptian Journal of Agronomy*, 35(2): 199–209. DOI: 10.21608/AGRO.2013.87

El-Keredy, M. S. (1990). Induced useful mutations of some field crops in Egypt. In *fourth Conference of Agronomy, University of Cairo, Egypt,* 1, 237–253.

Ellis, J., Dodds, P. and Pryor, T., 2000. Structure, function and evolution of plant disease resistance genes. *Current Opinion in Plant Biology*, *3*(4), 278–284.

Elshire, R. J.; Glaubitz, J. C.; Sun, Q.; Poland, J. A.; Kawamoto, K.; Buckler, E. S.; Mitchell, S. E. 2011. A robust, simple genotyping-bysequencing (GBS) approach for high diversity species. *PLoS One, 6* (5), e19379.

Ferrero-Serrano, A., Cantos, C., and Assmann, S. M. 2019. The role of dwarfing traits in historical and modern agriculture with a focus on rice. *Cold Spring Harbor Perspectives in Biology*, 11(11): 01–30. DOI: 10.1101/cshperspect.a034645

Futsuhara, Y. & Kikuchi, F. 1997. Science of the Rice Plant. *Matsuo, T., Futsuhara, Y., Kikuchi, F. and Yamaguchi, H.* (eds.), vol. 3: pp. 300–318. Food and Agriculture Policy Research Center, Tokyo.

Gaur, V. S., Channappa, G., Chakraborti, M., Sharma, T. R., & Mondal, T. K. 2020. "Green revolution" dwarf gene *sd1* of rice has gigantic impact. *Briefings in Functional Genomics*, 19(5&6): 390–409. DOI: 10.1093/bfgp/elaa019

Goldstein, D.B. and Pollock, D.D., 1997. Launching microsatellites: a review of mutation processes and methods of phylogenetic inference. *Journal of Heredity*, *88*(5), 335–342.

Gowthami, R., Vanniarajan, C., Souframanien, J., & Pillai, M. A. 2017. Comparison of radiosensitivity of two rice (*Oryza sativa* L.) varieties to gamma rays and electron beam in M_1 generation. *Electronic Journal of Plant Breeding*, 8(3), 732–741. DOI: 10.5958/0975-928X.2017.00111.9

Gunckel, J. K., & Sparrow, A. H. 1967. Ionizing radiations: biochemical, physiological and morphological aspects of their effects on plants. *Encyclopedia of Plant Physiology,* 16, 556–611

Hedden, P. 2003. Constructing dwarf rice. *Nature biotechnology*, *21*(8), 873–874. DOI: 10.1038/nbt0803-873

Hedden, P., & Phillips, A. L. 2000. Gibberellin metabolism: new insights revealed by the genes. *Trends in plant science*, 5(12): 523–530. DOI: 10.1016/S1360-1385(00)01790-8

Hedden, P., & Sponsel, V. 2015. A century of gibberellin research. *Journal of Plant Growth Regulation*, 34(4): 740–760. DOI: 10.1007/s00344-015-9546-1

Hong, M.J., Kim, D.Y., Jo, Y.D., Choi, H.I., Ahn, J.W., Kwon, S.J., Kim, S.H., Seo, Y.W. and Kim, J.B., 2022. Biological effect of gamma rays according to exposure time on germination and plant growth in wheat. *Applied Sciences*, *12*(6), p.3208.

Horneck, G., Klaus, D.M. and Mancinelli, R.L., 2010. Space microbiology. *Microbiology and Molecular Biology Reviews*, *74*(1), 121–156.

Imam Z., Chakraborty, N. R. and Gadi J. 2019. Gamma ray induced effectiveness and efficiency of chlorophyll mutants in non-basmati aromatic rice. *Journal of Biotechnology and Crop Science* 8(12): 4–9.

Jacob, P., Avni, A. and Bendahmane, A., 2018. Translational research: exploring and creating genetic diversity. *Trends in Plant Science*, *23*(1), 42–52.

Jan, S., Parween, T. and Siddiqi, T.O., 2012. Effect of gamma radiation on morphological, biochemical, and physiological aspects of plants and plant products. *Environmental Reviews*, *20*(1), 17–39.

Jurka, J. and Kapitonov, V.V., 1999. Sectorial mutagenesis by transposable elements. *Genetica*, *107*(1–3), 239–248.

Kant A and Chakraborty N R. 2021. Induction of mutation through gamma irradiation in non-basmati aromatic "Badshabhog" rice (*Oryza sativa* L.). *Applied Biological Research* 23(1): 50–59. DOI:10.5958/0974-4517.2021.00007.0

Kant A, Chakraborty NR, Das BK. 2020. Immediate radiation effects and determination of optimal dose of gamma rays on non-basmati aromatic rice (*Oryza sativa* L.) of eastern India. *Journal of Experimental Biology and Agricultural Sciences*, 8(5): 586–604. ISSN No. 2320 – 8694. DOI: http://dx.doi.org/10.18006/2020.8(5).586.604

Khalil, S. K., Rehman, S., Afridi, K., & Jan, M. T. 1986. Damage induced by gamma radiation in morphological and chemical characteristics of barley. *Sarhad Journal of Agriculture*, 2(1): 45–52.

Kharkwal, M.C. and Shu, Q.Y., 2009. The role of induced mutations in world food security. *Induced plant mutations in the genomics era.* Food and Agriculture Organization of the United Nations, Rome, pp. 33–38.

Kitamura, S., Satoh, K. and Oono, Y., 2022. Detection and characterization of genome-wide mutations in M1 vegetative cells of gamma-irradiated Arabidopsis. *PLoS Genetics*, *18*(1), e1009979.

Kole, P.C., Chakraborty, N.R. and Bhatt, J.S. 2008. Analysis of variability, correlation and path coefficients in induced mutants of aromatics non-basmati rice. *Tropical Agricultural Research* 11: 61–64.

Komatsu, M., Maekawa, M., Shimamoto, K., & Kyozuka, J. 2001. The *LAX1* and *FRIZZY PANICLE 2* genes determine the inflorescence architecture of rice by controlling rachis-branch and spikelet development. *Developmental Biology*, 231(2): 364–373. DOI: 10.1006/dbio.2000.9988

Kovacs, E., & Keresztes, A. 2002. Effect of gamma and UV-B/C radiation on plant cells. *Micron*, 33(2): 199–210.

Lavanya, A.B., 2003, December. Effects of electromagnetic radiation on biological systems: A short review of case studies. In *8th International Conference on Electromagnetic Interference and Compatibility* (pp. 87–90).IEEE.

Li, F., Shimizu, A., Nishio, T., Tsutsumi, N., & Kato, H. 2019. Comparison and characterization of mutations induced by gamma-ray and carbon-ion irradiation in rice (*Oryza sativa* L.) using whole-genome resequencing. *G3: Genes, Genomes, Genetics*, *9*(11): 3743–3751. DOI: 10.1534/g3.119.400555

Liu, L.X., Guo, H.J., Zhao, L., Gu, J. and Zhao, S., 2008. Advances in crop improvement by space mutagenesis in China. In *ICSC* (Vol. 4, No. 274, p. 3).

Liu, L.X., Guo, H.J., Zhao, L.S., Wang, J., Gu, J.Y. and Zhao, S.R., 2009. Achievements and perspectives of crop space breeding in China. *Induced plant mutations in the genomics era*, Food and Agriculture Organization of the United Nations (FAO), pp.213–215.

Liu, T., Ji, J., Cheng, Y., Zhang, S., Wang, Z., Duan, K. and Wang, Y., 2023. CRISPR/Cas9-mediated editing of GmTAP1 confers enhanced resistance to Phytophthora sojae in soybean. *Journal of Integrative Plant Biology*.

Liu, Y. G.; Whittier, R. F. 1995. Thermal Asymmetric Interlaced PCR – Automatable Amplification and Sequencing of Insert End Fragments from P1 and Yac Clones for Chromosome Walking. *Genomics, 25* (3), 674–681.

Luo W. L., Guo T., Yang Q. Y., Wang H., Liu Y. Z., Zhu X. Y., et al. 2014. Stacking of five favorable alleles for amylase content, fragrance and disease resistance into elite lines in rice (*Oryza sativa*) by using four HRM-based markers and a linked gel-based marker. *Mol. Breed.* 34, 805–815. DOI: 10.1007/s11032-014-0076-5.

Mahla, H. R., Sharma, R., & Bhatt, R. K. 2018. Effect of gamma irradiations on seed germination, seedling growth and mutation induction in Cluster bean [*Cyamopsis tetragonoloba* (L.) Taub.]. *Indian Journal of Genetics and Plant Breeding*, 78(2): 261–269.

Marcu, D., Damian, G., Cosma, C. and Cristea, V., 2013. Gamma radiation effects on seed germination, growth and pigment content, and ESR study of induced free radicals in maize (*Zea mays*). *Journal of Biological Physics, 39*, pp.625–634.

Mba, C., 2013. Induced mutations unleash the potentials of plant genetic resources for food and agriculture. *Agronomy, 3*(1), pp. 200–231.

Miah, G.; Rafii, M. Y.; Ismail, M. R.; Puteh, A. B.; Rahim, H. A.; Islam, K. N.; Latif, M. A. 2013. A Review of Microsatellite Markers and Their Applications in Rice Breeding Programs to Improve Blast Disease Resistance. *Int. J. Mol. Sci., 14* (11), 22499–22528.

Mir, A.S., Maria, M., Muhammad, S. and Ali, S.M., 2020. Potential of mutation breeding to sustain food security. In *Genetic Variation*: Rafael Trindade Maia and Magnólia de Araújo Campos, eds., Intech Open, pp. 1337–1357, 1–15.

Mohanta, T.K., Mishra, A.K., Mohanta, Y.K. and Al-Harrasi, A., 2021. Space breeding: the next-generation crops. *Frontiers in Plant Science, 12*, p.771985.

Monna, L., Kitazawa, N., Yoshino, R., Suzuki, J., Masuda, H., Maehara, Y., Tanji, M., Sato, M., Nasu, S. & Minobe, Y. 2002. Positional cloning of rice semidwarfing gene, *sd1*: rice "green revolution gene" encodes a mutant enzyme involved in gibberellin synthesis. *DNA research*, 9(1): 11–17. DOI: 10.1093/dnares/9.1.11

Moorthie, S., Mattocks, C.J. and Wright, C.F., 2011. Review of massively parallel DNA sequencing technologies. *The HUGO Journal, 5*, pp.1–12.

MVD (2021) Mutant variety database. Available on http://mvd.iaea.org/. Accessed on 15 July 2021.

Nielen, S., Forster, B. P., & Badigannavar, A. 2018. *Types of Mutations. Manual on mutation breeding (Spencer-Lopes, M. M. Forster, B.P. & Jankuloski, L., eds. 3rd).* Food and Agriculture Organization of the United Nations (FAO), *Rome, Italy*: 083–104.

Novak, F.J. and Brunner, H., 1992. Plant breeding: Induced mutation technology for crop improvement. *IAEA Bull, 4*, pp.25–33.

Oladosu, Y., Rafii, M. Y., Abdullah, N., Hussin, G., Ramli, A., Rahim, H. A., Miah, G. & Usman, M. 2016. Principle and application of plant mutagenesis in crop improvement. *A Review. Biotechnology & Biotechnological Equipment*, 30(1): 01–16. DOI: 10.1080/13102818.2015.1087333

Ramesh, M., Vanniarajan, C., Ravikesavan, R., Aiyanathan, K. E. A., & Mahendran, P. P. 2019. Mutagenic effectiveness and efficiency in barnyard millet (*Echinochloa frumentacea*) using physical, chemical and combination of mutagens. *Electronic Journal of Plant Breeding*, 10(2): 949–956. DOI: 10.5958/0975-928X.2019.00122.4

Rampure, N. H., Choudhary, A. D., Jambhulkar, S. J., & Badere, R. S. 2017. Isolation of desirable mutants in safflower for crop improvement. *Indian J. Genet*, 77(1): 134–144. DOI: 10.5958/0975-6906.2017. 00018.9

Reisz, J.A., Bansal, N., Qian, J., Zhao, W. and Furdui, C.M. 2014. Effects of ionizing radiation on biological molecules – mechanisms of damage and emerging methods of detection. *Antioxidants & redox signaling, 21*(2), pp. 260–292.

Ren, D., Li, Y., Zhao, F., Sang, X., Shi, J., Wang, N., Guo, S., Ling, Y., Zhang, C., Yang, Z., & He, G. 2013. *MULTI-FLORET SPIKELET1*, which encodes an AP2/ERF protein, determines spikelet meristem fate and sterile lemma identity in rice. *Plant Physiology*, 162(2): 872–884. DOI: 10.1104/pp.113.216044

Robbelen, G., 1990. Mutation breeding for quality improvement – a case study for oilseed crops. *Molecular Breeding Review*, 6: 1–44.

Roy, B., Kumar, V., Tulsiram, S. D., & Das, B. K. 2018. Development of high yielding aromatic mutants of rice (*Oryza sativa* L.) from a local aromatic cultivar, Tulaipanji by using gamma radiation. *Indian Journal of Genetics and Plant Breeding*, 78(4): 409–416. DOI: 10.31742/IJGPB.78.4.2

Roychowdhury, R. and Tah, J., 2013. Mutagenesis – A potential approach for crop improvement. *Crop improvement: new approaches and modern techniques*. In: Hakeem, K.R., Ahmad, P., Ozturk, M., eds. Crop improvement: new approaches and modern techniques. New York (NY): Springer, pp.149–187.

Rühm, W., Azizova, T., Bouffler, S., Cullings, H.M., Grosche, B., Little, M.P., Shore, R.S., Walsh, L. and Woloschak, G.E., 2018. Typical doses and dose rates in studies pertinent to radiation risk inference at low doses and low dose rates. *Journal of Radiation Research*, 59(suppl_2), pp.ii1–ii10.

Russell, W.L., Russell, L.B. and Kelly, E.M., 1958. Radiation dose rate and mutation frequency: The frequency of radiation-induced mutations is not, as the classical view holds, independent of dose rate. *Science*, 128(3338), pp.1546–1550.

Russell, W.L., Kelly, E.M., Hunsicker, P.R., Bangham, J.W., Maddux, S.C. and Phipps, E., 1979. Specific-locus test shows ethylnitrosourea to be the most potent mutagen in the mouse. Proceedings of the National Academy of Sciences, 76(11), pp.5818–5819.

Sakamoto, T., Miura, K., Itoh, H., Tatsumi, T., Ueguchi-Tanaka, M., Ishiyama, K., Kobayashi, M., Agrawal, G.K., Takeda, S., Abe, K., Miyao, A., Hirochika, H., Kitano, H., Ashikari, M., & Matsuoka, M. 2004. An overview of gibberellin metabolism enzyme genes and their related mutants in rice. *Plant Physiology*, 134(4): 1642–1653. DOI: 10.1104/pp.103.033696

Saludes-Tapia, M., Gonzalez, M.B., Campabadal, F., Suñé, J. and Miranda, E., 2021. A simple, robust, and accurate compact model for a wide variety of complementary resistive switching devices. *Solid-State Electronics*, *185*, p.108083.

Sánchez-Martín, J.; Steuernagel, B.; Ghosh, S.; Herren, G.; Hurni, S.; Adamski, N.; Vrána, J.; Kubaláková, M.; Krattinger, S.G.;Wicker, T. 2016. Rapid gene isolation in barley and wheat by mutant chromosome sequencing. *Genome Biology*, 17, 221.

Sasaki, A., Ashikari, M., Ueguchi-Tanaka, M., Itoh, H., Nishimura, A., Swapan, D., Ishiyama, K., Saito, T., Kobayashi, M., Khush, G. S., Kitano, H., & Matsuoka, M. 2002. A mutant gibberellin-synthesis gene in rice. *Nature*, 416(6882): 701–702. DOI: 10.1038/416701a

Sato Y, Shirasawa K, Takahashi Y, Nishimura M, Nishio T. 2006. Mutant selection from progeny of gamma-ray-irradiated rice by DNA heteroduplex cleavage using Brassica petiole extract. *Breeding Science*, 56: 179–183. DOI: 10.1270/jsbbs.56.179.

Shadakshari, Y.G., Chandrappa, H.M., Kulkarni, R.S., and Shashidhar, H.E. 2001. Induction of beneficial mutants in rice (*Oryza sativa* L.). *Indian Journal of Genetics and Plant Breeding*, 61: 274–276.

SigurbjörnSSon, B. and Micke, A., 1974. Polyploidy and induced mutations in plant breeding. *Vienna: IAEA*, pp.303–343.

Singh, B. and Mohapatra, B.K., 2004. Prediction of M2 mutation frequency based on M1 estimates in blackgram. *Legume Research-An International Journal*, *27*(2), pp.137–139.

Spencer-Lopes, M.M., Jankuloski, L., Mukhtar Ali Ghanim, A., Matijevic, M., & Kodym, A. 2018. Physical mutagenesis. *Manual on mutation breeding (Spencer-Lopes, M. M., Forster, B. P., & Jankuloski, L. eds.)*. Food and Agriculture Organization of the United Nations (FAO), Rome, Italy, pp. 05–49.

Spielmeyer, W., Ellis, M. H., & Chandler, P. M. 2002. Semidwarf (*sd-1*), "green revolution" rice, contains a defective gibberellin 20-oxidase gene. *Proceedings of the National Academy of Sciences*, 99(13): 9043–9048. DOI: 10.1073/pnas.132266399

Suprasanna, P., Mirajkar, S.J. and Bhagwat, S.G., 2015. Induced mutations and crop improvement. *Plant Biology and Biotechnology: Volume I: Plant Diversity, Organization, Function and Improvement*, In Plant Biology and Biotechnology. Bahadur, B., Venkat Rajam, M., Sahijram, L., Krishnamurthy, K. Eds. Springer, New Delhi, pp.593–617. https://doi.org/10.1007/978-81-322-2286-6_23

Tadele, Z., Mba, C., Till, B. J. 2010.TILLING for mutations in model plants and crops. In *Molecular Techniques in Crop Improvement:2nd Edition* Jain, S. M., Brar, D. S., Eds. Springer, pp 307–332.

Thole, V., Peraldi, A., Worland, B., Nicholson, P., Doonan, J. H., & Vain, P. (2012). T-DNA mutagenesis in *Brachypodium distachyon. Journal of Experimental Botany*, 63(2): 567–576.

Till, B. J., Zerr, T., Comai, L., Henikoff, S. 2006. A protocol for TILLING and Ecotilling in plants and animals. *Nature Protocols, 1*(5), 2465–2477.

Tomita, M. and Ishii, K., 2018. Genetic performance of the semidwarfing allele sd1 derived from a Japonica rice cultivar and minimum requirements to detect its single-nucleotide polymorphism by MiSeq whole-genome sequencing. *BioMed Research International*, 2018, Article ID 4241725, 7 pages. https://doi.org/10.1155/2018/4241725.

Tsatsakis, A.M., Nawaz, M.A., Kouretas, D., Balias, G., Savolainen, K., Tutelyan, V.A., Golokhvast, K.S., Lee, J.D., Yang, S.H. and Chung, G., 2017. Environmental impacts of genetically modified plants: a review. *Environmental Research, 156*, pp.818–833.

Wang, L., Wang, Z., Xu, Y., Joo, S. H., Kim, S. K., Xue, Z., Xu, Z., Wang, Z., & Chong, K. (2009). *OsGSR1* is involved in crosstalk between gibberellins and brassinosteroids in rice. *The Plant Journal*, 57(3): 498–510. DOI: 10.1111/j.1365-313X.2008.03707.x

Wei, X. J., Tang, S. Q., Shao, G. N., Chen, M. L., Hu, Y. C., & Hu, P. S. 2013. Fine mapping and characterization of a novel dwarf and narrow-leaf mutant dnl1 in rice. *Genetics and Molecular Research*, 12(3): 3845–3855. DOI: 10.4238/2013.september.23.2

Wurtzel, O., Dori-Bachash, M., Pietrokovski, S., Jurkevitch, E. and Sorek, R., 2010.Mutation detection with next-generation resequencing through a mediator genome. *PLoS One*, *5*(12), p.e15628.

Yali, W. and Mitiku, T., 2022. Mutation breeding and its importance in modern plant breeding. *Journal of Plant Sciences*, *10*(2), pp.64–70.

Yoon, H., Yang, J., Liang, W., Zhang, D., & An, G. (2018). *OsVIL2* regulates spikelet development by controlling regulatory genes in *Oryza sativa. Frontiers in Plant Science,* 9: 01–12. DOI: 10.3389/fpls.2018.00102

Yoshida, A., Suzaki, T., Tanaka, W., & Hirano, H. Y. 2009. The homeotic gene long sterile lemma (G1) specifies sterile lemma identity in the rice spikelet. *Proceedings of the National Academy of Sciences*, 106(47): 20103–20108. DOI: 10.1073/pnas.0907896106

Zhu, Q. H., Hoque, M. S., Dennis, E. S., & Upadhyaya, N. M. 2003. Ds tagging of *BRANCHED FLORETLESS 1 (BFL1)* that mediates the transition from spikelet to floret meristem in rice (*Oryza sativa* L). *BMC Plant Biology*, 3(1): 01–13. DOI: 10.1186/1471-2229-3-6

6 Applications of DNA Marker Techniques in Plant Mutation Research

Shilpa Gupta and Diksha Singla*
Department of Biochemistry, Punjab Agricultural University, Ludhiana, Punjab, India
*Corresponding Email: shilpagupta@pau.edu

6.1 INTRODUCTION

Any alteration to an organism's DNA that is not the result of normal recombination and segregation is referred to as a mutation. For plant breeding, the direct usage of mutation is an immensely useful supplementary approach. The key benefit of this strategy is that it takes less time than hybridization to develop a crop with better character(s). In 1920s, when it was discovered that heritable mutations could be created in plants using physical (irradiation) or chemical treatments, globally, plant breeders have used mutation breeding (Stadler, 1928). This approach to crop variety improvement remained in interest all over the 1950s and 1960s, and it did result in the release of a sizable number of varieties, such as those from Scandinavian barley breeding (Lundqvist, 2014). The gene pool used for modern barley breeding still contains altered genes from these ancient mutant species. But contrarily, mutation-based breeding has one major limitation of managing the burden of undesirable mutations that accompany the selected ones and hampers the development of high-yielding varieties (Mba, 2013). The development of new genetic modification technologies (GM traits) in 1980s further curtailed the breeders' interest in using mutation-based breeding. The two main techniques for creating mutations in plants are irradiation (physical) and chemical mutagen therapy (Mba et al., 2011; Leitao, 2011). Chemical mutagens (such as EMS, MNU and NaN_3) usually result in single base substitutions (G/C to A/T transition), whereas physical mutagens viz. X-rays and γ-radiations lead to a mixture of point mutations and larger chromosome deletions (single base deletions/substitutions). Chemical mutagens have the advantage of creating mutant populations with high mutation density, which makes it simpler to explore specific mutations in a population (Szarejko et al., 2017).

For the effective use and conservation of plants, understanding of the genetic makeup and level of genetic variation within and between plant populations is a basic

 DOI: 10.1201/9781003392897-6

prerequisite. Evolutionary history, mating system, population density, and gene flow mechanisms are the major determinants that determine the extent and structure of genetic variation within plant species (Amom and Nongdam, 2017). Morphological traits and ethnological and cytological traits have traditionally been exploited to describe the levels and patterns of diversity, but these parameters only account for a small fraction of the plant genome and are dependent on environmental variables (Nongdam and Nirmala, 2007). This limits their usefulness in elucidating the potentially complex genetic structures that can exist between and within taxa (Avise, 2004).

Advancements in molecular biology and biochemistry over the past few decades have prevailed over these limitations by making many potential DNA marker techniques available to plant breeders (Horst and Wenzel, 2005). This milestone in plant genome research led to the accomplishment of PCR-based molecular marker techniques a decade later. Since then, many applications of numerous marker techniques in several areas of genomics and molecular plant breeding have been reported (Nadeem et al., 2018). Currently, DNA marker techniques are one of the most valuable tools available for crop improvement studies. Using specific or arbitrary designed oligonucleotide primers, PCR methods conveniently amplify specific DNA sequences from segments of genomic DNA. These techniques often use polymorphic nucleic acids between individuals or populations (Collard et al., 2005). Point mutations in oligonucleotide priming sites cause genotypes to show a differential pool of fragments. In some instances, polymorphism could be caused by insertion or deletion mutational events that changed the distance between termini sequences.

In order to analyze plant genomes, DNA marker techniques mediated by PCR applications have been widely used. The success of molecular breeding has also been tremendously aided by advancements in plant genetics, which have given greater insights into DNA markers and the genetic variety of plant species. DNA marker technology has advanced quickly in recent years, resulting in new methods that considerably simplify plant development and improvement research. In addition to being extremely helpful in the preservation and management of various plant germplasm, DNA markers have been shown to be very beneficial in determining the degree and distribution of diversity across a wide range of crop species (Hailu and Asfere, 2020). Also, by using marker-assisted selection (MAS), DNA marker technologies have accelerated plant molecular breeding and made it easier to collect genetic maps (Collard et al., 2005; Jiang, 2013). DNA markers also provide important phylogenetic information by helping to introduce desirable traits in cultivated species from wild relatives. In addition, the application of DNA marker techniques makes it possible to directly observe and describe differences in gene sequence. Ultimately, the availability of gene sequence information makes it easier to clone and modify desirable genes for study of plant mutations and crop improvement. The development of DNA marker techniques has brought a level of precision previously unattainable in plant breeding. Thus, the present chapter aims to describe different DNA marker techniques along with their applicability for plant mutation studies.

6.2 DNA/MOLECULAR MARKER TECHNIQUES

Differences in DNA base sequences between individual organisms or species that are closely related to the target gene expressing a trait are known as DNA/molecular markers. The expressed characteristic or biological function, the target gene, and closely related molecular markers are typically all inherited simultaneously (Collard et al., 2005). The locus or loci, which refer to the precise gene location of the marker in the chromosomes, may or may not be known. In particular, a molecular or DNA marker does not affect a trait related to the function or expression of the linked gene(s). Whether the gene(s) involved are known/not known, or whether that gene(s) may/may not be identified, the close relationship of DNA markers to gene or parameter/trait of a specific biological function suggests that the markers may correspond to a particular gene locus and help in recognition of related traits (Reddy et al., 2021). When polymorphism (i.e., differences in the base sequences of markers) occurs between individuals or species, molecular marker approaches can be helpful for detecting the individual genotypic differences in same/different species. These polymorphisms are caused by insertions, deletions, point mutations, duplications, and translocations, although they don't always influence gene activity (Nadeem et al., 2018). The following characteristics should be met by ideal DNA markers for effective usage in plant mutation studies:

- Codominant in expression (heterozygotes-homozygotes differentiation)
- Evenly distributed throughout the whole genome (no clustering in particular regions)
- Highly reproducible
- Ability to detect higher level of polymorphism
- Single copy and no pleiotropic effect
- Clear distinct allelic traits (easy identification of different alleles)
- Cost-effective marker development and genotyping
- Easy assay/detection and automation
- High availability and suitable for duplication/multiplication (easy data accumulation and data sharing)
- Genome-specific (particularly in polyploids)
- No negative phenotypic effects

On the basis of detection method, DNA marker techniques are grouped into different groups: hybridization-based markers, polymerase chain reaction (PCR)–based markers, and DNA sequence-based markers. The genotyping of many plant genomes is currently possible using a wide range of DNA marker approaches. DNA or molecular markers are progressively more used for experimental and applied plant mutagenesis. Considerations when deciding which particular marker to use will depend on particular plant species under study, the purpose of the study, and the accessibility of necessary resources.

6.2.1 Hybridization-based Marker (RFLP)

Based on the hybridization, the foremost and the only DNA marker known was restriction fragment length polymorphism (RFLP). It was developed by Botstein and coworkers in 1980. Due to insertions/deletions (also known as InDels), translocations, point mutations, inversions, and duplications, individuals of the same species exhibit polymorphism. In RFLP, firstly, pure DNA is extracted, followed by mixing up of isolated DNA with restriction enzymes (cut DNA at specific site) isolated from bacteria that result in generation of many variable-sized fragments/bands. Gel electrophoresis (horizontal agarose/polyacrylamide gel electrophoresis [PAGE]) helps in separation of these bands of variable lengths. Deletions, inversions, mutations, transpositions, and translocations in base pair are the major causes of the variation that leads to RFLP pattern. These mutations lead to acquisition (gain) or loss of recognition sites, resulting in fragments of different lengths and polymorphisms. When even a single base pair is changed at the recognition site, restriction enzymes do not cleave fragments. In contrast, the marker is referred to as being heterozygous if a point mutation occurs on only one chromosome but not on the other because both bands are present (Madhumati, 2014). RFLP markers are also very common and randomly distributed in the genome. However, in practice, analysis of RFLP is a tedious, technically demanding protocol and time-consuming. Any DNA digestion requires large amounts of high molecular weight purified DNA. Additionally, analysis of RFLP markers is not suitable for automation. Also, the cost impact on probe design can be a drawback for plant species without suitable probes. The invention of less complex and more affordable PCR-based marker profiling techniques has rendered RFLP obsolete. RFLP identifies human genetic variation and phylogenetic relations and has been widely used in genetic mapping studies.

6.2.2 PCR-based DNA Markers

PCR technique was developed by Kary Mullis in 1983 which uses Taq DNA polymerase, which was obtained from *Thermus aquaticus*, to exponentially amplify small amounts of target DNA. To create the ideal primer for the amplification of a certain nucleotide segment, the flanking sites of the target DNA must first have their sequence information known. In addition to Taq polymerase and primers, the deoxynucleotide triphosphates (dNTPs), Mg^{2+}, and a suitable buffer are vital to the basic procedure of PCR.

The PCR-based DNA markers have more advantages than the ones based on hybridization because of the smaller quantity of required DNA, higher reproducibility, absence of radioisotopes, reliability, and higher polymorphism in a quick time.

6.2.2.1 Amplified Fragment Length Polymorphism (AFLP)

This technique is a combination of RFLP- and PCR-based markers (Lynch and Walsh, 1998). In this, the primer recognition sequences are ligated to the DNA fragments obtained from restriction digestion. An essential characteristic of AFLP technique is the concomitant screening of representative DNA regions that are arbitrarily distributed across the genome. Mutations/changes in the sequences surrounding the restriction sites and duplications, insertions, or deletions inside the amplification segments can result in AFLP polymorphisms. For AFLP analysis, both partially degraded and high-quality DNA can be used, but the DNA must be free of PCR inhibitors and restriction enzymes. Certain enzymes, which combine frequent cutters (like Taq1 or Mse1) and rare cutters (Pst1 or Eco RI), can digest genomic DNA.

In order to provide a recognized sequence for PCR, the double-stranded oligonucleotide adapters are constructed and ligated to both ends of the fragments. These adapters do not have any initial restriction sites following ligation. The first round of PCR is executed using primer combinations with a single base pair (bp) extension, and the last round of amplification is carried out using primer combinations with up to 3 bp extensions. Denaturing polyacrylamide gel or agarose gel with autoradiography or $AgNO_3$ staining is used to observe the generated fragments. Due to its great repeatability and capacity to identify excessive polymorphism in a single reaction, the AFLP approach is among most crucial molecular tools for genetic investigation (Vos et al., 1995). The constraints of AFLP are the need for highly purified and DNA with high molecular weight and the probable comigration of nonhomologous fragments belonging to distinct loci.

6.2.2.1.1 Methylation Sensitive Amplified Polymorphism (MSAP)

Fundamentally, MSAP is a moderation to AFLP technique, which involves digestion of genomic DNA using methylation-sensitive restriction endonucleases, amplifying the resulting fragments, and analyzing the results. In this process, the extracted gDNA is separated into two parts and every part is digested with *Eco*RI, an indifferent cutter that is negligibly influenced by DNA cytosine methylation and identifies the GAATTC target site. In the next step, the content is next digested by the methylation-sensitive *MspI* or *HpaII* isoschizomers, respectively ("methyl-sensitive cutter"). These enzymes share an identical recognition of the restriction site (CCGG), but response to cytosine methylation is different. The digested DNA samples are then ligated to two dsDNA adapters compatible with *EcoRI* and *MspI/HpaII*-generated ends. After that, non-selective or pre-selective primers complementary to the adapters are used to pre-amplify the ligated fragments, which are then pre-amplified using a pair of selective primers (these are one- to three-base extended variants of the non-selective or preselective primers at the 3′ ends). A limited population of fragments from such amplification are segregated to compare the various band patterns (Schulz et al., 2013; Fulneček and Kovařík, 2014).

Hao and Deng (2003) used the MSAP method to preserve shoot tips of apple (*Malus pumila* Mill.) for a year at 4°C and a 12-hour photoperiod in a medium that included 2% mannitol. No genetic difference could be found between the preserved samples and the shoot before preservation (control) using AFLP markers. However, the MSAP analysis found that 6 of the 389 examined markers had changed. The scientists attributed this to alterations from hemimethylation to complete methylation status rather than *de novo* methylation or demethylation. The plants' varying DNA methylation status was thought to be a reaction to various stresses brought on by the *in vitro* preservation settings. The authors justified the use of this conservation procedure as being superior to field collections despite the substantial variation that was discovered.

6.2.2.2 Random Amplified Polymorphic DNA (RAPD)

RAPD is based on differential PCR amplification of genomic DNA. It uses short random oligonucleotide sequences (mostly 10 bases in length) to infer DNA polymorphisms created by deletions or rearrangements at or between oligonucleotide primer binding sites in the genome (Williams et al., 1990). Because this approach does not require prior knowledge of the genome for analysis, it can be used species-wide with universal primers. The fundamental drawback of this method is that the profiling depends on the reaction conditions, i.e., it can vary across two different laboratories, and that the profiling cannot distinguish between heterozygous and homozygous individuals because each primer amplifies multiple distinct loci in the genome (Bardakci, 2001). High-density genetic mapping in numerous crop species, including faba bean (Torress et al., 1993), alfalfa (Kiss et al., 1993), and apple (Hemmat et al., 1994), has been created in a relatively short period of time due to the effectiveness and rapid speed of RAPD analysis. The markers linked to disease-resistant genes in lettuce (*Lactuca* sp.) (Paran et al.1991), common bean (*Phaseolus vulgaris*) (Adam-Blondon et al., 1994), and tomato (*Lycopersicon* sp.) (Martin et al., 1991) were successfully discovered by RAPD analysis of nonisogenic lines (NILs).

Both DNA amplification fingerprinting (DAF) and randomly primed PCR are separately developed methods that are variations of RAPD. PCR using a single primer is randomly primed (around 1015 nucleotides in length). It contains low stringent amplification during the first two PCR cycles. The remaining cycles are then performed with a higher annealing temperature to increase the rigidity (Welsh and McClelland, 1990). Because it required autoradiography, this RAPD variant was not very popular; however, because fragments can now be separated by agarose gel electrophoresis, it has been made simpler. In the DAF method, the amplicons are resolved using polyacrylamide gel and stained through silver staining, and a single random primer shorter than 10 nucleotides is used for amplification (Caetano-Anolles and Bassam, 1993).

6.2.2.3 Minisatellites

The genomes of higher organisms contain three forms of multiple copies of simple repetitive DNA sequences (minisatellites, microsatellites, and satellite DNAs) arranged in arrays of widely varying sizes (Litt and Luty, 1989). They form the smallest class of simple repetitive DNA sequences and are also known as short tandem repeats (STRs), simple sequence repeats (SSRs), or simple sequence length polymorphisms (SSLPs). Like RFLP, it also involves the use of restriction endonucleases for genomic DNA digestion, but minisatellites are conceptually an entirely different class of markers. It has chromosomal regions with tandem repetitions of 10–50 base motifs that are surrounded by conserved DNA restriction sites. A multiband minisatellite outline, typically within a size range of 4–20 kb, is created using standard multilocus probes capable of hybridizing to the sequences of minisatellite in various species (Jeffreys et al., 1985). Through molecular cloning of DNA restriction fragments, subsequent screening with multilocus minisatellite probes, and separation of specific fragments, the locus-specific probes may be created. The primary cause of length polymorphisms is assumed to be variation in the number of repeat units brought on by uneven crossover or genetic rearrangement. The degree of polymorphism is considerable because of the excessive mutation rate in minisatellites, which usually results in unique multilocus profiles for different individuals in the population. Variable number tandem repeat (VNTR) loci is another name for minisatellite loci (Nakamura et al., 1987). Minisatellites have excessive polymorphism and good reproducibility, which are their main advantages. The disadvantages of minisatellites are much like RFLPs as the methodological procedures are very similar. When using multilocus probes, highly informative profiles are usually observed because many informative bands are formed per reaction. In this situation, it is impossible to interpret the band profile in terms of alleles and loci, and homologous fragments cannot exist. The random distribution of minisatellites across the genome has also raised questions (Schlotterer, 2004). Minisatellites gave rise to the term "DNA fingerprinting," which is now more extensively used as a DNA-based technique to uniquely identify individuals. Studies of genetic identity, the identification of cultivars and variations, and the study of clonal development and structure all benefit greatly from the use of minisatellites (Jeffreys et al., 1985).

6.2.2.4 Microsatellites

Microsatellites display tandem repeats; however, their repeat motifs are shorter, i.e., 1–6 bp. Specific primers (usually 20–25 bp) may be designed to amplify the microsatellite by PCR if nucleotide sequences in the flanking regions are known. By creating a genomic library with short inserts, screening the library with synthetically labeled oligonucleotide repeats, and sequencing the positive clones, microsatellites and their flanking sequences can be identified. As an alternative, microsatellites can be identified by searching sequence databases for microsatellite sequence motifs, from which adjacent primers can then be designed. Moreover, it is possible to use primers that have already been developed for closely similar species. It is believed that changes in the number of microsatellite repeat units are primarily caused by slippage of DNA polymerase during DNA replication, or mismatch and nonuniform crossover of the slipped strands, which can be detected by gel electrophoresis (Matsuoka et al., 2002). Allelic codominance, high genomic frequency, and random distribution throughout the genome are advantages of microsatellites (Morgante et al., 2002). High levels of polymorphism in microsatellites make them extremely useful markers that can be utilized for genetic mapping and investigations, from the individual level (for clone and variant identification) to the level of closely related species, greatly lowering the analytical work (Jarne and Lagoda, 1996). When suitable primer sequences for the species of interest are unavailable, the principal drawback of microsatellites is the high production cost, which makes their use to unstudied groups difficult. Null alleles (alleles that do not amplify the desired PCR product) may emerge as a result of mutations in the primer annealing sites, which can result in errors in genotype evaluation. Using microsatellite primers made from germplasm unrelated to the test species enhances the ability for null alleles to be present (weak cross-species amplification). Null alleles can lead to an overestimation of heterozygosity and a biased estimation of genotypic and allelic frequencies. Microsatellites' underlying mutation paradigm (infinite allele model or stepwise mutation model) is still up for debate. Sometimes, numerous forward and reverse mutations in microsatellite sites might cause homoplasia, underestimating genetic diversity. The emergence of stuttering bands due to mismatch of DNA slipping strands during PCR amplification is a very common finding in microsatellite analysis (Freudenreich et al., 1997). Because fragment sizes are more complicated and heterozygotes and homozygotes can be mistaken for one another, this can make it more difficult to evaluate band profiles. However, interpretation can be improved by including an appropriate reference genotype in which the band size is known for the experiment. Conversely, high mutation rates make them unsuitable for studies at higher taxonomic levels.

6.2.2.5 Intersimple Sequence Repeats (ISSRs)

Zietkiewicz et al. (1994) were the first to describe this method, which involves amplifying DNA segments that are situated between two identical regions of microsatellite repeats that are orientated in opposition to one another. The amplification of simple repetitions of various sizes often uses microsatellites with di-, tri-, tetra-, or pentanucleotide core sequences as primers. The length (15–35 mers) and higher annealing temperatures of ISSR primers lead to higher stringency. The GC content of the primer, however, affects the annealing temperature. A product between 200 and 2000 bp in length results from PCR amplification. The ISSR band pattern can be seen by autoradiography or $AgNO_3$ staining following separation by agarose or polyacrylamide gel electrophoresis. To generate an ISSR polymorphism, no prior knowledge of the template DNA sequence is necessary. ISSR markers have a dominant mode of inheritance, are simple, are randomly distributed throughout the genome, and need only a small quantity of DNA. They are used to identify plant species and cultivars, study taxonomy and genetic diversity, map genetic variation, and evaluate the clonal fidelity of *in vitro*–grown plants (Nikanta et al., 2017). A major limitation of ISSR markers is the low reproducibility and homology of related amplicons (Semagn et al., 2006).

6.2.2.6 Simple Sequence Repeats (SSRs)

Several species' nuclear, mitochondrial, and chloroplast genomes contain a large number of SSRs that are randomly distributed and polymorphic (Kalia et al., 2011). In tandem repeats of mono-, di-, tri-, tetra-, penta-, and hexa-nucleotide motifs, these short sequences are constructed (Koelling et al., 2012). The mono-, di-, tri-, and tetra-nucleotide motifs are the most prevalent. SSRs are caused by single-stranded DNA slippage, double-stranded DNA recombination, the mobility of mobile DNA genetic molecules, and nucleotide variations from an evolutionary perspective (Koelling et al., 2012). ESTs and protein-coding genes have both been shown to contain SSRs (Yan et al., 2017). A subset of microsatellites are SSRs. Simple sequences that generate short tandem repeats and polymorphism changes in microsatellite length are the two more forms of microsatellites. SSRs among these microsatellite classes show fewer repeats per locus but more polymorphisms

(Padmakar et al., 2015). The ability to detect higher-level polymorphisms of SSRs in microsatellite regions is improved by using PCR in the analysis of the SSR technique (Padmakar et al., 2015).

SSRs, markers with codominant inheritance and locus specificity, have many benefits for effective plant mapping and population genetics investigations (Kumar et al., 2021; Padmakar et al., 2015). SSR is very informational; however, discovery of DNA sequencing-based marker or creation process can be pricey. Since SSR markers which are locus specific proved to be more informative, their development could act as an important genetic resource. Therefore, the separation and characterization of individual loci is necessary for the development of these SSR-specific loci. Microsatellite probe DNA library building, library screening, and positive clone sequencing are the procedures to be followed in order to achieve this. Finally, PCR primer synthesis and optimization are carried out. The very low recovery of functional SSRs, typically brought on by the production of nonspecific PCR products and monomorphic loci, is another challenge. Furthermore, for effective genotyping, a high-resolution gel instrument system is needed. However, the use of gel or capillary electrophoresis for profiling and recording limits large-scale investigation of SSR markers.

6.2.2.7 Sequence-related Amplified Polymorphism (SRAP)

SRAP, a simple and affordable dominant marker technique, was developed by Li and Quiros (2001). They created markers that precisely amplified coding sections of the *Brassica oleracea* genome by targeting AT-rich promoter, spacers, and introns and GC-rich exons. The reverse primer has an AT-rich sequence near the 3′ end, while the forward primer has a GC-rich region there. The forward primer's core contains the CCGG sequence, whereas the reverse primer's core has the AATT sequence (Lin et al., 1999). PCR amplification products are separated by agarose gel electrophoresis, and bands are then identified by autoradiography. Like with ISSR and RAPD markers, DNA fragments are scored based only on the presence or absence of bands. In population genetic analysis of closely related hybrids, SRAP has been used to assess genetic variation at the species level, to construct linkage maps, and to identify quantitative trait loci (Uzun and Cagirgan, 2009).

6.2.2.8 Conserved DNA-derived Polymorphism (CDDP)

To identify rice varieties, Collard and Mackill developed this technique in 2009. Moreover, *Solanum dulcamara*'s genetic diversity has been evaluated using it (Poczai et al., 2011). Primers designed to bind to these genes and produce polymorphic band patterns are targeted at short conserved regions found in many copies of plant genomes. Moreover, common plant genes related to biotic and abiotic stress or for plant development are targeted by primers. Highly conserved DNA regions share the same priming sites but have different genomic distributions, so variations within these regions can be detected as length polymorphisms. Amplification of a single long primer at high annealing temperature improves CDDP reproducibility. The resulting DNA fragments are separated by electrophoresis in the size of 200~1500 bp, and the band pattern is observed after autoradiography.

6.2.2.9 Single-strand Conformational Polymorphism (SSCP)

This is a mutation recognition and screening technique that allows the detection of mutated variants in many different species. It is an easy and technically simple method based on the differences of electrophoretic mobility in the single-stranded DNAs for the detection of unknown mutations and precise analysis of mutational and allelic sequence variations.

It is an effective method that is widely used to find small insertions and deletions, inversions, and single nucleotide substitutions. Single-stranded DNA in nondenaturing gels is altered and has differential mobility as a result of single nucleotide changes. This property of single-stranded DNA (ssDNA) is used when SSCP is used to detect mutations. The difference in migration patterns of wild-type and a mutated single-stranded PCR fragment during gel electrophoresis is the basis for the detection of sequence variants or mutations. This method is based on the differential electrophoretic gel mobility of denatured ssDNA that has unique folding.

6.2.2.10 Expressed Sequence Tag (EST)

EST is complementary DNA (cDNA) used to convert messenger RNA (mRNA) into DNA using reverse transcriptase. EST development is necessary to improve mRNA stability. DNA or cDNA molecules are more stable than RNA molecules, according to a comparison of nucleic acid chemistry. cDNA consists only of exon-coding portions of RNA of DNA sequences transcribed from mRNA with excised or spliced intronic sequences that do not encode proteins. Because it is so useful for identifying new genes and providing important information about how the genes work, EST is widely used. EST has helped develop DNA microarrays and related probes to study gene function. In addition, many EST-based molecular markers, such as RFLPs, ISSRs, CAPS, and SNPs, are considered valuable (Xiao et al., 2020).

6.2.2.11 Sequence-tagged Site (STS)

STS are characteristically short DNA stretches with unique sequences. The exact sequence of an STS is usually located only at one point throughout the genome and is determined by the primer sequence. In 1989, Olson and coworkers developed and published the first STS markers as DNA markers used in the Human Genome Project's physical mapping. Later, it proved helpful and was applied to numerous studies of the plant genome. If the locus can be cloned and sequenced, STS can be created for any genomic region. STS markers are simple to use, dominant, and highly reproducible (Olson et al., 1989) and become highly useful when combined with robust, reliable, and stable DNA markers associated with genes encoding specific features of agronomic interest.

6.2.2.12 Start Codon Targeted (SCoT)

This new marker system was created using the short conserved region of the plant genome that borders the ATG start codon. This includes using 18-mer primers at an annealing temperature of 50°C (Collard and Mackill, 2009). PCR uses a single primer, meaning that the forward and reverse primers are the same primers as in case of RAPD and ISSR markers. These marker techniques are usually reproducible, and annealing temperature and annealing period are not the primary factors in reproducibility (Gorji et al., 2011). These are dominant markers used in quantitative trait loci (QTL) mapping, mass segregation analysis, and plant genetics (Collard and Mackill, 2009). The resulting PCR products can be subjected to conventional agarose gel electrophoresis to separate the fragments which can be detected by autoradiography.

6.2.2.13 Cleaved Amplified Polymorphic Sequence (CAPS) Markers

CAPS marker was initially known as a PCR-RFLP marker developed by combining RFLP and PCR (Maeda et al., 1990). This method involves PCR amplification of the target DNA followed by restriction enzyme digestion (Jarvis et al., 1994; Michaels and Amasino, 1998). As a result, CAPS products are seen using agarose gel or acrylamide gel electrophoresis. The sequence information in a database of genomic or cloned RAPD bands or cDNA sequences is used to generate the primers used in this approach. These markers are adaptable, and by combining CAPS with single-strand conformational polymorphism, SCAR, AFLP, or RAPD, the probability of discovering DNA polymorphism can be effectively increased. These are codominant markers that have been widely used in studies on molecular identification, map-based cloning, and genotyping (Spaniolas et al., 2006).

6.2.2.14 Retrotransposons-based Markers

Retrotransposons are highly conserved sequences which have long terminal repeats (LTRs) at their ends (Wu, 2017). These are the most prevalent mobile genetic elements in eukaryotes and are often found in regions of the genome adjacent to known genes (Kalendar et al., 2010). A number of retrotransposon analyses between and within plant species have revealed that these markers have substantial degrees of polymorphisms and heterogeneity (Kalendar et al., 2017). Furthermore, it is known that plant genomes contain a large number of retrotransposons. For molecular marker studies, the prevalence and wide-ranging random distribution of retrotransposons in various plant genomes have been examined (Kalendar et al., 2010). The design of primers for retrotransposon marker analysis is guided by LTRs. Retrotransposon insertions are distinctive in that they are irreversible, which makes them particularly useful for phylogenetic studies.

6.2.2.15 Interprimer Binding Site (iPBS)

The need for LTRs to design element-specific primers restricts the use of retrotransposons as DNA markers. To attain any good primer sequences, numerous cloning and sequencing processes are needed. For direct PCR amplification, most LTR sections lack conserved motifs (Kalendar et al., 2011). To distinguish between various LTRs and immediately see variations between plant cultivars, Kalendar et al. (2010) developed iPBS markers. Primer binding sites (PBSs), a conserved sequence found adjacent to the 5′LTR in all retrotransposons, are used in this method. By creating complementary base pairing between the conserved region of the PBS and the terminal sequence of the transfer RNA (tRNA), the reverse transcription is initiated (Kelly et al., 2003). Primers can be designed to match conserved regions in PBS, and PCR amplification can be used to produce DNA fragments of various LTRs (Kalendar et al., 2011). Cloning of LTRs previously relied on conserved protein-coding domains, which limited screening for autonomous elements (Monden et al., 2014). However, with iPBS technology, various LTR sequences can be screened while performing efficient DNA fingerprinting. As effective as IRAP and RBIP markers, iPBS amplification produces polymorphism levels. Due to different retrotransposon activity and retrotransposon recombination, the iPBS markers are extremely effective and reliable in detecting polymorphisms and clonal differentiation (Kalendar et al., 2010). iPBS is also an extremely reliable and robust DNA fingerprinting technique that does not require knowledge of sequence information (Baranek et al., 2012). It is used in the study of genetic diversity and phylogeny of many plant species. The use of iPBS markers in studies of genetic diversity in peas, lentils, and apricots has also been successfully performed (Baranek et al., 2012; Baloch et al., 2015).

6.2.3 Sequence-based DNA Markers

DNA markers based on the identification of specific DNA sequences from an unknown pool of DNA are known as sequence-based markers. Sequencing is a technique for identifying nucleotide bases and their sequence along a DNA strand. This method was developed as a result of the less reliable and polymorphic hybridization-based markers. The emergence of single nucleotide polymorphisms (SNPs), leading to high polymorphisms, has transformed plant breeding with the advent of sequencing technologies like next-generation sequencing (NGS) and genotyping by sequencing (GBS). Heather and Chain (2016) provide a quick overview of the many sequencing methods that have been created so far. But the following is also a description of some significant sequencing techniques.

6.2.3.1 Sanger Method of Sequencing

The first technique for DNA sequencing was plus-and-minus which was used by Sanger and Coulson (Sanger and Colson, 1975). The fundamental idea behind this technique is that single-stranded DNA molecules can be distinguished from one another using PAGE when they differ in length by just one nucleotide. Sangers' dideoxy sequencing method is another name for this technique, which makes use of modified bases called dideoxy nucleotides (Heather and Chain, 2016). In the initial research, DNA polymerase I from *Escherichia coli* and bacteriophage T4 DNA polymerase were employed in this technique (Englund, 1971; Englund, 1972). The products of the subsequent polymerization were put onto acrylamide gels and resolved using ionophoresis. This technology had several drawbacks; thus Sanger and his coworkers produced a new sequencing method after 2 years, in which oligonucleotides were sequenced by polymerization by enzymes (Sanger et al., 1977). The greatest number of variations can be measured with this method. It is very reproducible and uses a small amount of DNA. Although this approach does not cover the entire genome, it is expensive, is time-consuming, and detects less polymorphism below the species level (Franca et al., 2002).

6.2.3.2 Pyrosequencing

Pyrosequencing is a sequencing method based on the principle of synthesis (Hyman, 1988), where phosphate is found during the synthesis of DNA (Ronaghi, 2001). In this technique, a single-stranded DNA template that has been biotin-labeled and coupled with certain enzymes is hybridized with the sequencing primer (Gharizadeh et al., 2006). Four cycles are used to introduce deoxynucleoside triphosphates one at a time to the reaction mixture. The polymerization of nucleic acid, which releases the inorganic compound Ppi (pyrophosphate), initiates the reaction. Inorganic phosphate is continuously released as the nucleotides are introduced to the process, and this released Ppi is in an amount equivalent to the nucleotide that has been incorporated. Initially, this method was used to monitor the DNA polymerase activity. Pyrosequencing is of two types: liquid phase sequencing and solid phase sequencing (Ronaghi et al., 1996).

6.2.3.3 Next-generation Sequencing (NGS)

The demand for high-throughput, low-cost sequencing increased as sequencing techniques advanced. Due to this demand, NGS was created, and it currently generates millions of sequences. This process has the capacity to generate between several hundred million and several hundred billion DNA bases in a single run (Shendure and Ji, 2008). Numerous companies, including Roche 454 FLX Titanium (Thudi et al., 2012), Illumina MiSeq and HiSeq 2500 (Bentley et al., 2008), and Ion Torrent PGM (Rothberg et al., 2011), have successfully developed this technique and offer their services commercially. These NGSs have affordable costs and more accurate whole-genome coverage (Deschamps et al., 2012). All NGS methods begin with the same process of randomly shearing and ligating DNA fragments at both ends using universal adapters to create the template DNA. One or more nucleotides are incorporated during this constant channel sequencing, which releases a signal that is picked up by a sequencer (Metzker, 2010). NGS has several advantages: (i) it is of higher accuracy compared to earlier sequencing techniques; (ii) it is low cost and high throughput; (iii) this method is currently being used in whole genome sequencing to examine the greatest number of SNPs and to take into account the variety existing within the species, as well as in genome-wide association studies (GWAS) and the construction of linkage/halophyte maps (Elshire et al., 2011); and (iv) the area of metagenomics has been strengthened by the use of NGS to sequence older DNA samples (Mardis, 2008).

6.2.3.4 Genotyping by Sequencing (GBS)

GBS is a straightforward and versatile method that is nowadays used successfully. The Buckler lab used the Illumina NGS platform to develop this method. Modernization of the NGS method has reduced the cost of sequencing, ensuring the effective use of GBS for large genome species with significant levels of diversity (Elshire et al., 2011). There are two categories of GBS approaches based on the use of ion PGM systems. (i) Restriction enzyme digestion: This technique is primarily used in marker-assisted selection (MAS) programs to find new markers, and no specific SNPs are found here. In this procedure, DNA is digested with one or two particular restriction enzymes before the ligation of adapters. (ii) Multiplex enrichment PCR: This approach uses certain PCR primers to amplify specific points of interest. A whole set of SNPs are found for a genomic segment, as opposed to the restriction enzyme method's digestion. GBS is now used for many other species with complicated genomes; however, it was initially created to study the high-resolution association in maize. The main benefits of GBS are (a) lower costs when compared to other approaches, which makes it a cutting-edge method for identifying SNPs in various animals and crops; (b) this method yields acceptable outcomes in population research, crop breeding, and germplasm characterization (Poland and Rife, 2012); (c) a large number of SNPs generated by GBS are employed for genotyping and genetic analysis (Beissinger et al., 2013); (d) it reduces sample handling; and I it contains fewer PCR and purification sets (Davey et al., 2011).

The following sequence-based markers have been developed using the abovementioned sequencing method single nucleotide polymorphisms (SNPs)

SNPs are genetic polymorphisms that develop when single base pairs are positioned differently in DNA sequences between individuals of a species or different species (Manivannan et al., 2021). At a frequency of one SNP per 100–300 base pairs of DNA, they are normally restricted to noncoding, coding, and intergenic regions of the genome, with varying frequencies within each of these genomic regions (Fusari et al., 2008). When a recognition site is present on one allele and that specific sequence is lacking on another allele, SNP variants are generated. SNPs are either present or absent as a result of mutational events like insertions, deletions, transversions, or transitions brought on by nucleotide substitutions. Moreover, after enzyme digestion and gel electrophoresis, various DNA fragments with polymorphisms of varying lengths result from nonrestriction caused by the absence of restriction enzyme recognition sites or restriction caused by the presence of these sites in the DNA region.

In genetic linkage analysis, the genes recombine during meiosis (Tanksley, 1993). Genetic mapping studies include creating genetic linkage maps after genotyping people in segregating populations using DNA markers that cover the organism's genome. DNA-based markers have been widely employed for the mapping and finding of genes/QTL as well as for the development of saturation genetic linkage maps since their discovery in the 1980s. SNP markers have been used for QTL mapping investigations on a larger scale as a result of the development of HTP technology for SNP genotyping and the widespread availability of sequence information. This is primarily due to SNPs' high genome abundance, which enables them to produce maps with the highest resolution when compared to other marker systems (Bhattramakki et al., 2002; Jones et al., 2007).

SNP markers (based on genes) providing resistance to leaf rust (*Lr34* locus), stripe rust (*Yr18*), and powdery mildew (*Pm38*) diseases were developed (Lagudah et al., 2009). In canola, for mapping *fad2* and *fad3* genes, SNP mutations were found by sequencing their genomic clones, and SNP markers were designed thereafter (Hu et al., 2006). To enable direct selection of favorable *fad2* and *fad3* alleles in marker-assisted trait introgression and breeding, allele-specific PCR tests were created. Using the high-resolution melting (HRM) approach, SNP markers in barley that were connected to the covered smut resistance gene *Ruh.7H* were discovered (Lehmensiek et al., 2008). In sugar beet, an anchored linkage map based on AFLP, SNP, and RAPD markers was created to map QTL for *Beet necrotic yellow vein virus* resistance genes, *Rz4*(Grimmer et al., 2007) and *Rz5* (Grimmer et al., 2008). EST-derived SNPs were used to create a consensus genetic map for cowpea, which will be useful for genomic and QTL mapping research in this crop (Muchero et al., 2009). The *VTC2* gene in *Arabidopsis* was cloned using fine mapping and map-based cloning procedures in one of the postgenomic era investigations in 2002 (Jander et al., 2002). Using SNP and InDel markers, the authors precisely localized the gene interval from approximately 980 kb region to 20 kb interval. In that time, an additional nine potential genes were found, and the underlying mutation was later found. Several other studies have been published in this field, even if just a few instances that illustrate the use of SNP markers in QTL mapping and genomic investigations have been included here. Current developments in HTP genotyping technology and sequencing data will make it easier to quickly identify causal variants and clone QTL of interest for MAB application.

Discovery of SNPs is quite easy in plants with simple genomes as compared to complex genomes. The major reason behind this is the highly repetitive genome of the complex plants (Meyers et al., 2001). Recent developments in NGS technologies, such as 454 Life Sciences (Roche Applied Science, Indianapolis, IN), HiSeq (Illumina, San Diego, CA), SOLiD, and Ion Torrent (Life Technologies Corporation, Carlsbad, CA) (Mardis, 2008) have eliminated the hurdles like low throughput and high cost of SNP discovery. By avoiding highly repetitive portions of a genome, transcriptome resequencing utilizing NGS technologies enables quick and affordable SNP detection inside genes (Morozova and Marra, 2008). Many plant genomes, including those of maize, canola, eucalyptus, sugarcane, tree species, wheat, avocado, and black currant (Barbazuk et al., 2007; Trick et al., 2009; Novaes et al., 2008; Bundock et al., 2009; Parchman et al., 2010; Lai et al., 2012; Kuhn, 2012; Russell et al., 2011 respectively), have successfully used this technology. The NimbleGen sequence capture method (Roche Applied Science, Madison, IN) (Hodges et al., 2007) was first created for human illness diagnostic research, and it increased the throughput and coverage level of gene-based SNP detection in plants (Springer et al., 2009). This method involves microarray-based exon sequence capture and enrichment, followed by targeted resequencing using NGS. For investigations including genome/exome mining, comparable in-solution target capture technologies, such as Agilent SureSelect, are also commercially available. Nevertheless, because the design of capture probes requires these reference resources, this method would only be effective for crops with available reference genome sequences or big transcriptome (EST) databases.

Despite the allure of finding SNPs using transcriptome or exome resequencing, this method is narrowly focused, concentrating only on coding areas. It is clear that the availability of SNPs within coding sequences provides molecular geneticists with a very effective tool for identifying a causal mutation (Varshney, 2009). Though they may be many megabases away from genes in intergenic spaces, noncoding regulatory sequences like enhancers or locus control regions frequently include quantitative trait loci (QTL) (Dean, 2006). It is difficult to find SNPs inside those regulatory regions using transcriptome or exon sequencing. It is crucial to use genome complexity reduction strategies in conjunction with NGS technologies in order to find SNPs in a genome-wide manner while avoiding repetitive and duplicated DNA. Throughout time, a number of methods for reducing genome complexity have been created, such as High Cot selection (Yuan et al., 2003), methylation filtering (Emberton et al., 2005), and microarray-based genomic selection (Okou et al., 2007). These strategies generally lower the quantity of repeated sequences but are unable to identify and eliminate duplication, which results in false-positive SNPs identification. Contrary to these methods, newly created genome complexity reduction technologies, such as Restriction Site Associated DNA (RAD) (Floragenics, Eugene, OR) (Baird et al., 2008) and Complexity Reduction of Polymorphic Sequences (CRoPS) (Keygene N.V., Wageningen, Netherlands) (van Orsouw et al., 2007), are computationally powerful and capable of removing duplicated SNPs. These techniques have been used to find SNPs in crops both with and without reference genome sequences (Mammadov et al., 2010; Chutimanitsakun et al., 2011).

(B) Diversity arrays technology (DarT)
The basis of DarT technique is microarray hybridization, in which several hundred polymorphic loci distributed throughout the genome are simultaneously genotyped (Jaccoud et al., 2001). Genomic representations are constructed by restriction digestion of genomic DNA, followed by ligation of restriction fragments to adapters.

In order to reduce the complexity of the genome, PCR amplification uses adapter-specific primers and selective overhangs. DNA fragments derived from the genomic representations are cloned, amplified using primers specific to the vector, purified, and finally arrayed on a microarray (Semagn et al., 2006). A discovery array is hybridized with labeled genomic representations formed from individual pooled genomes. Later, for routine genotyping, polymorphic clones with various hybridization signal intensities are set up in a genotyping array (Jaccoud et al., 2001). Easy extensibility, high reproducibility, speed, and no need for prior sequence information are the advantages of this marker. Since it is a microarray-based technology, it is technically more challenging, has lengthy and laborious stages, and requires skilled workers. The use of DarT markers in plant research is constrained by significant cost, time, physical energy, and modern laboratory equipment. Table 6.1 shows the comparison of most widely used DNA markers.

6.3 APPLICATIONS OF DNA MARKER TECHNIQUES IN PLANT MUTATION RESEARCH

DNA marker techniques are used in a variety of applications, starting from identification of gene location to crop improvement through marker-assisted breeding. In particular, the application of phylogenetic analysis adds a new dimension to the theory of evolution. In the modern period of molecular biology research, DNA marker-assisted selection is a great achievement for research scientists and plant breeders.

6.3.1 Cultivar Identity/Assessment of "Purity"

Due to the challenges of handling the numerous seed samples used in various plant breeding programs, seeds from various strains are frequently mixed. Markers may be

TABLE 6.1
Comparison of Most Widely Used DNA Markers

	RFLP	RAPD	AFLP	SSR	SNP
Presence/availability	Average	Abundant	Abundant	High	Abundant
Polymorphism type	Single bp, InDel, inversion	Single bp, InDel, inversion	Single bp, InDel, inversion	Repeat length variability	Single bp
Level of polymorphism	Medium	High	High	High	High
Repeatability/ reproducibility	High	Low	Medium	High	High
Prior DNA sequence information	Not required	Not required	Not required	Required	Required
DNA quality	High	Medium	High	Medium	Medium
Type of inheritance	Codominant	Dominant	Dominant	Codominant	Codominant
Automation	Low	Medium	Medium	High	High
Technical complexity	High	Low	Medium	Low	Medium
Developmental cost	High	Low	Low	High	High
No. of polymorphism per loci	1.0–3.0	1.5–5.0	2.0–10.0	1.0–3.0	1.0
Primary application	Genetic	Diversity	Diversity and genetic	All purposes	All purposes

Source: Semagn et al., 2006.

used to determine the true identities of particular plants. In order to take advantage of heterosis, high genetic purity and its preservation are increasingly crucial in cereal hybrids development. In hybrid rice, STS and SSR markers were exploited to validate purity, which was proved to be much easier and time saviors compared to usual "grow-out tests," that require plant growth up to the stage of maturity for evaluation of morphophysiological traits (Dar et al., 2013).

6.3.2 Evolution and Phylogeny

Early evolutionary theories were primarily on the basis of geographic and morphological variations among species/organisms. However, the new biological methods are becoming increasingly obvious due to detailed genetic structural data (Slatkin, 1987). DNA sequencing, RFLP, and several PCR-based DNA markers are widely used to reconstruct the phylogeny of different species. These methods are expected to provide breakthrough information about the subtle timescales at which closely related species diverge and which genetic variation is associated with speciation.

Also, these studies promise to shed additional light on the type of genetic variation that occurs within species (Avise, 1994). In order to comprehend how plants evolved from their wild predecessors and classify them into the proper groupings, several attempts have been made to assess levels of plant genetic variation. To classify the test germplasm as a component of the primary, secondary, or tertiary gene pool, taxonomic classification is a crucial first step. This is much more crucial when morphological markers are ambiguous and subject to error. For instance, strains PR 304 and Azucena, which are morphologically classed as *indica*, exhibit Japanese features in crossbreed tests. Yet, when examined using RAPD markers, these strategies are specifically recognized as *japonicas* (Virk et al., 1997).

6.3.3 Diversity Analysis of Exotic Germplasm

For genetic diversity estimation, identification, elimination of duplicates, and protection of the elite genotypes, it is crucial to characterize the germplasm resources present in crop species. After domestication, the genetic crop variation expanded as a result of the ongoing selection process that further lowered selection for particular criteria, such as yield. This increased susceptibility to diseases jeopardized the possibility of long-term, sustainable genetic crop improvement. In 1970s, the southern maize leaf blight outbreak in US caused significant reduction in corn yield, and it was due to substantial adoption of a single genetic component in male sterility that was unfortunately linked/associated to disease susceptibility. Thus, it is important to explore the genetic makeup of existing recent varieties in contrast to their predecessors and related species is therefore crucial. The varieties with the most diverse DNA profiles could have the larger number of novel alleles, which will increase the possibility of discovering novel and valuable genes in addition to providing information on their phylogenetic relationship. For the majority of crops, DNA profiling is now being used to guide such sampling decisions. Exotic germplasm is chosen for breeding based on specific aspects, such as: (a) it must be genetically dissimilar and (b) it must have more number of unique DNA polymorphisms (across the genome) than recent varieties (Brown and Kresovich, 1996).

Understanding the genetic variations among existing cultivars, among cultivars and their wild ancestors, and naturally available wild species is crucial. Several kinds of germplasm have previously been DNA fingerprinted using both specified and random DNA markers (Virk et al., 1997). Moreover, STMS markers can be used to assess how well breeders are able to access existing variant forms, clarify domestication processes, and enrich crop gene pools. Particularly in species where polymorphism is extremely rare, AFLP has gained appeal for researching genetic polymorphisms. The relationship between variance in wild barley's tolerance to salt and related ecogeography was assessed using AFLP (Pakniyat et al., 1997). Likewise, Blair et al. (1999) used ISSR markers to analyze rice, wheat, and pine diversity. Such investigations have contributed to plant biodiversity classification, which can be applied further in projects including wild gene introgressions.

6.3.4 Study of Heterosis

It has been reported that maize and sorghum hybrids use molecular markers to identify heterotic group, which can further be utilized to determine hybrid vigor. The generation of hybrid lines requires more time and money because they are intended to be used for high-quality hybrid production. Yet, there have been instances of matching parental lines to the appropriate heterogeneity group, despite the fact that it is presently impossible to precisely depict the degree of heterosis from DNA marker data (Bidhendi et al., 2012). It has also been suggested that heterotic hybrids and hybrids can be chosen using small subsets of molecular/DNA marker data in conjunction with phenotypic data (Boeven et al., 2016). Previously, Melchinger et al.(1991) assessed molecular marker diversity in 32 maize inbred lines. In another study, Zhang (1995) observed that the specific heterozygosity was highly correlated to midparent heterosis in rice. QTLs contributing to heterosis were mapped in elite maize inbred lines B73 X Mo17 (Stuberet al.1992 and Bains et al.1999) and highest yielding *indica* × *japonica* hybrids (Xiao et al., 1995).

6.3.5 Identification of Genomic Region Selection

Changes in the allele frequency in the genome may be of importance to breeders as it directs to screen haplotypes/alleles and may be used to configure suitable breeding procedures (Steele et al., 2006). Various uses of recognizable evidence of the selected genomic regions include QTL mapping: the selected regions can be focused for QTL examination or used to confirm recently identified marker trait associations (Fang et al., 2014). Finally, information about genomic loci selection can further be used for improving haplotypes with particular allelic mixes using MAS designs, viz., early generation selection and marker-assisted backcrossing (Herzog and Frisch, 2011).

6.3.6 Marker-assisted Introgression

Introgression basically refers to the frequent backcrossing and hybridization for transferring a highly required characteristic from one species to another. Hospital and coworkers (2009) described it as the process in which an identified QTL of a plant in one population is introgressed within a plant of another population by more than one backcrossing. DNA markers are useful in this case to regulate how near the QTL is. Moreover, it aids in background genome recovery of the recipient. Landrace genes or QTLs can be linked by introgression using DNA markers, which speeds up the process of developing a superior assortment and reducing the problem of linkage resistance (Dwivedi et al., 2015).

6.3.7 Gene Mapping and Tagging

Large-scale gene manipulation is frequently necessary for plant improvement, whether through natural selection or breeders' efforts, which always depend on the creation, assessment, and selection of the ideal allele combination (Flavell, 1995). Finding the desirable alleles/genes in a segregating population and the mapped markers permits a more precise breakdown of complicated features into their individual genetic components and offers novel ideas for effective breeding programs. Using RFLP markers, the maize genome map was the first one in plants to be defined (Gardiner et al., 1993). Maps have also been produced for barley, banana, potato, and many Brassicaceae family members (Winter and Kahl, 1995). Sequence-tagged microsatellite (STMS) markers in particular have been proven to be quite helpful for mapping of genomes. Construction of index maps, which is based on Mendelian inheritance, can easily be utilized for providing a reference point for particular genomic regions. Zhao and Kochert (1992) created the first microsatellite maps in plants for rice using (GGC)n, and Tanksley et al. (1995) using (GA), (GT)n, (GA/AG)n, (ATC) 10, and (ATT) 14. It is presently proposed that, like microsatellites, STMS are perfect for integrating genetic maps in addition to genetic variability because of the pattern of variation produced by retrotransposons (Ellis et al., 1998).

After mapping of the markers, it can be effectively used to identify numerous individual traits, including yield, seed quality, stress tolerance, and disease resistance, that are crucial to a breeding program. Breeders and molecular biologists are identifying several monogenic as well as polygenic lines for various traits to determine the potential of marker-assisted selection. A key objective is to tag useful genes, including those involved in plant disease resistance, hormone synthesis, drought tolerance, and genes involved in several developmental pathways. Tagged genes can also be used to identify beneficial genes in novel genotypes produced by hybrid programs or other techniques like transformation. The first record of gene marking was from tomatoes (Williamson et al., 1994), where the technique was used to find molecular markers linked to genes responsible for a variety of traits, but mainly for water use efficiency (Martin et al., 1989), resistance to *Fusarium oxysporum* (12 genes), leaf rust resistance genes *LR9* and *LR24*, and resistance to root knot nematodes (*Meliodogyne* sp.). In the wild rice relative *Oryza rufipogon*, Xiao et al. (1998) exploited RFLPs in locating the trait-enhancing QTL alleles. Moreover, allele-specific linked primers have proven useful for genotyping loci with allelic variations brought on by both size variations and point mutations. The waxy gene locus in maize (Shattuck-Eidens et al., 1991), the *Glu D1* complex locus associated with breadmaking quality in wheat (D'Ovidio and Anderson, 1994), the *Lr1* leaf rust resistance locus in wheat (Feuillet et al., 1995), the *Gro1* and *H1* alleles conferring resistance to the root cyst nematode *Globodera rostochiensis* in potatoes (Niewohneret al., 1995), etc., are some of the examples for this. Table 6.2 shows the application of DNA markers associated with disease resistance in various crops.

6.3.8 Mapping of Mutations

Genetic mapping of a mutation-defined gene is the requisite for isolation and cloning of a gene and afterward for identification of its protein product. SNP detection, localization of mutations, and potential identification of causative sequence variants are all combined in next-generation sequencing for genetic mapping (Leshchiner et al., 2012). An intuitive tool called CandiSNP can help with novel discoveries from forward genetic mutant screens (Etherington et al., 2014).

TABLE 6.2
DNA Marker Techniques for Disease Resistance in Various Crops

DNA marker	Crop	Trait	References
CAPS	Rice	Rice blast (*Pyricularia oryzae*)	(Jeon et al., 2003)
SCAR		Gall midge (*Orseolia oryzae*)	(Sardesai et al., 2002)
SCAR and CAPS	Maize	Sugarcane mosaic virus (SCMV)	(Dussle et al., 2002)
SSR	Barley	Barley yellow mosaic virus	(Graner et al., 1999; Williams, 2003)
CAPS		Leaf rust (*Puccinia hordei*)	(Graner et al., 2000)
SSR	Wheat	Leaf rust (*Puccinia recondita* f. sp. tritici)	(Suenaga et al., 2003)
STS and CAPS			(Seyfarth et al., 1999)
STS		Stem rust (*Puccinia graminis* f. sp. tritici)	(Mago et al., 2002)
RAPD and SSR		Yellow rust (*Puccinia striiformis* f. sp. tritici)	(Chague et al., 1999)

6.3.9 Map-based Cloning of Genes

By connecting a mutant phenotype to markers whose precise locations in the genome are known, a procedure called map-based cloning, also known as positional cloning, can determine the genetic cause of the condition. The gene linked to the soybean ripening locus E3 was cloned using a map (Watanabe et al., 2009). BPH26 encodes a protein known as CCNBSLRR, which stands for coiled-coil nucleotide-binding site leucine-rich repeat (Tamura et al., 2014). Using a map-based cloning technique, *MOC-1* is the first gene for rice tiller that has been discovered (Li et al., 2003).

6.3.10 Marker-assisted Selection (MAS)

Marker-assisted selection (MAS) is a method in which DNA marker detection and selection are integrated into a traditional breeding program (Collard and Mackill, 2008). Some important steps of MAS are shown in Figure 6.1. Once markers tightly associated with the genes or QTLs have been discovered, they can be exploited for MAS. Compared to pure selection, MAS has a number of benefits. Single large-scale MAS refers to the inclusion of MAS in an early generation, as proposed by Ribaut and Betran (1999). MAS for simply inherited features is becoming significantly more important in plant breeding programs as it speeds up the breeding process (Francia et al., 2005).

Several economically significant traits in agricultural plants are governed by significant genes or QTLs. These traits include self-incompatibility, resistance to pests and diseases, male sterility, and many morphological traits, viz. shape, color, and architecture of whole plants and/or plant parts. In nature, these features frequently have mono- or oligogenic inheritance. Even for some quality features, one or a small number of important QTLs or genes can be responsible for a significant amount of the trait's phenotypic variance (Bilyeu et al., 2006; Pham et al., 2012).

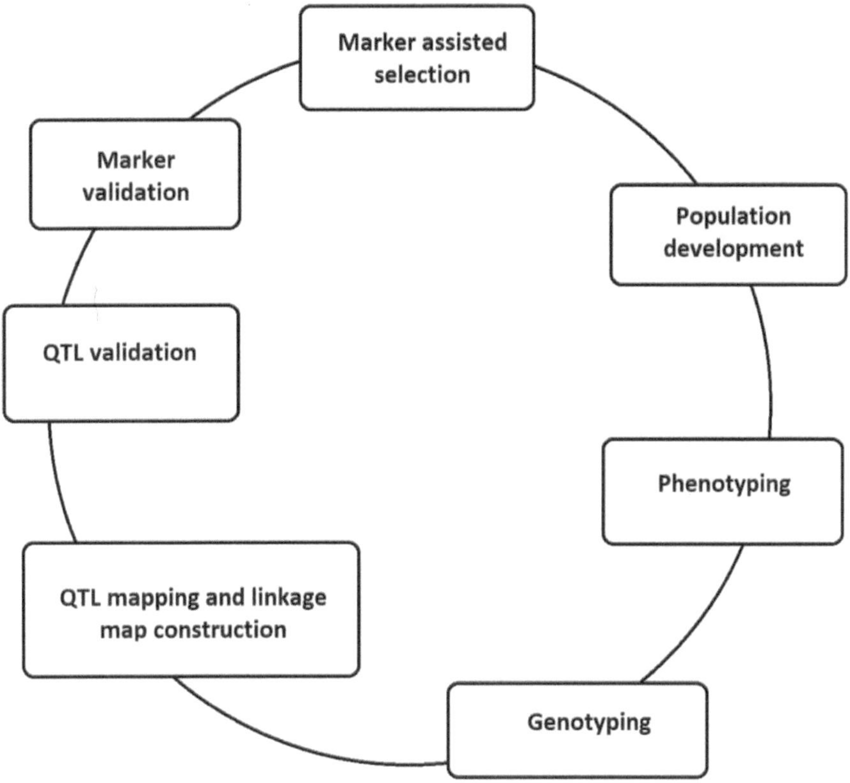

FIGURE 6.1 Some important steps involved in marker-assisted selection.

The attribute of the cultivar being developed can be significantly improved by transferring this gene to a particular line. In some cases, indirect selection for the target genes is more effective than using marker loci that are closely related to important genes for selection. When a marker is actually located within a gene, these benefits in efficiency may be the result of higher expression of the marker mRNA. However, the most practical solution may be to use an external marker whose polymorphism is simpler to detect when the target gene of interest differs between two alleles by a difficult-to-detect SNP.

The soybean cyst nematode (SCN) (*Heterodera glycines* Inchinoe) is one of the most destructive pests, resulting in huge yield damage. To curtail this damage, resistant cultivar development is the best strategy (Cregan et al., 1999), but resistant segregants identification is quite difficult and costly affair. Phenotypic assessment is time consuming and labor intensive (Young, 1999). Cregan et al. (1999) identified Satt309, the SSR marker, which is 1–2 cM away from the *rhg1* resistance gene, which was later used for several public and commercial breeding efforts. Genotypic selection with Satt309 was proved to be 80%–99% accurate in predicting susceptible lines (Young, 1999). Shi et al. (2009) developed five homozygous $F_{4:5}$ lines for all eight marker alleles linked to the soybean mosaic virus (SMV) resistance genes/loci using molecular markers. The homozygous lines demonstrated resistance to SMV strains G1 and G7 and were believed to carry all three resistance genes (*Rsv1*, *Rsv3* and *Rsv4*) that would preferably confer broad and persistent resistance to SMV.

6.3.10.1 Advancement in Marker-assisted Selection

6.3.10.1.1 Targeting Induced Local Lesions in the Genome (TILLING)

The majority of crop plants can benefit from the nontransgenic reverse genetics technique known as targeted induced local lesions in the genome (TILLING). In *Arabidopsis* plants, functioning of two genes was comprehended by TILLING (McCallum, 1990; McCallum et al., 2000). When using TILLING procedures, the mutagenic population is initially established by treating seeds with a common chemical mutagen like methyl methanesulfonate (MMS) or ethyl methanesulfonate (EMS). The most crucial methods, including liquid chromatography, enzymatic mismatch cleavage, mass spectroscopy, and array-based technologies, are used to identify the changes in nucleotide sequences of target mutant individuals (Kurowska et al., 2011). Eventually, more significant bioinformatics methods are used to analyze mutations brought on by certain mutagens, such as project aligned related sequences and evaluated SNPs (PARSESNP). While TILLING is applicable to all plant species, ploidy level and genome size should not have an impact on it. The most remarkable advantage of TILLING is the discovery of gene mutations at higher rate. Thus it can be better exploited in molecular biology as TILLING can help determine novel alleles more economically and in a lesser time span.

6.3.10.1.2 Virus-induced Gene Silencing (VIGS)

A virus-based technique called virus-induced gene silencing (VIGS) is used to take advantage of a defense mechanism that is RNA mediated. During siRNA synthesis nucleotides direct RNase complex where it focuses a certain ssRNA that resembles dsRNA molecules (Comai and Henikoff, 2006). The siRNA of infested cells must be processed to match specific regions in the genome of the virus vector, including any nonviral inserts, using double-stranded RNA intermediate molecules. Accordingly, the genome of the virus vector inserts in infected plant cells. After that, siRNa RNases complexes corresponding mRNA of the host and, ultimately, the symptoms in the plant reveal how protein-coding genes in infected plants have stopped functioning (Gupta et al., 2012). Plant reverse genomics has effectively used VIGS in recent years. It is a relatively straightforward, economical, and high-throughput technique. It is mostly employed in the detection of a gene's functional loss (Comai and Henikoff, 2006; Lu et al., 2003). VIGS has further been exploited to explore how genes in many species of plants and animals function under abiotic stresses. In order to broaden the VIGS technique's uses in a number of crops for gene investigation corresponding to abiotic stresses, a wide range of vectors have been created with high silencing efficacy. The value of VIGS has been depicted in the capability of diverse agricultural plants to withstand various abiotic stresses. A satellite DNAmβ-based VIGS vector was used to silence the *SlGRX1* gene in tomato under salinity stress that led to leaf yellowing in comparison to vector control plants, showing that GRX1 is involved in salt tolerance (Guo and Lu, 2010). Moreover, TRV-VIGS has investigated the function of CaRAV1 and CaOXR1 in the chili pepper (Lee et al., 2010). Recently, it has been used to examine oxidative stress tolerance. Oxidative stress harm in plants with these genes silenced has also been described in previous studies (Lee et al., 2010; Choi and Hwang, 2012) that investigated the function of chili pepper genes, such as CaOXR1, CaRAV1 and CaPO2. Increased lipid peroxidation under stress was the outcome of CaOXR1, CaRAV1, and CaPO2 silencing alone or in combination in the chili pepper.

6.3.11 Marker-assisted Pyramiding

When numerous genes or QTLs integrate simultaneously in one plant, this is known as pyramiding. Selection can be facilitated without phenotyping using DNA markers. Pyramiding is most frequently used to integrate genes corresponding to disease resistance into a single plant that has a stable or long-lasting resistance to a disease. Via STS and RFLP markers, Hittalmani et al. (2000) integrated genes for rice blast resistance from three parents. Castro et al. (2003) used SSR markers for combining one stripe rust gene with two QTLs for quantitative resistance.

6.3.12 Marker-assisted Backcrossing (MAB)

Markers can be used in backcross breeding for frontal area determination, recombinant, and foundational determination. Markers are used to choose the desired trait at the primary stage of foreground choice. There are several mechanisms and passive alleles known for selection of phenotypes, but their effects have been suppressed by the dominant genes; the usage of markers could be strong in this situation. The second stage, known as recombinant selection, involves backcrossing of offspring with the desired trait and strongly related flanking markers in order to lessen linkage drag. The third stage, known as the background selection, entails choosing backcross descendants using the background markers, such as offspring who have already been chosen for the desired features. In any instance, the background markers should be ideal markers but not firmly attached to the markers (Ashkani et al., 2011; Jena and Mackill, 2008). This implies that markers could be exploited to select against the donor genome, e.g., by reducing donor parent's hereditary commitment while increasing the share of intermittent parent genome (Kumar and Rao, 2014) (Table 6.3).

6.3.13 Genotyping of Cultivars

When cultivars are genotyped, the random and repetitive DNA markers are the first option. Minisatellites (Ramakrishna et al., 1995) and microsatellites, viz. $(AAGG)_4$, $(GATA)_4$, $(CAC)_5$, $(AAT)_6$, and $(CT)_{10}$ (Gupta et al., 1994), have been utilized in DNA fingerprinting to detect cultivar identification, genetic variation, and genotyping (Yang et al., 1994; Sant et al., 1999). The generated data is helpful for taxonomic investigations, defining accessions in plant germplasm repositories, and assessing genetic diversity. STMS markers created with the help of microsatellites indicate

TABLE 6.3
MAB in Different Crops

Crop	Trait	Gene/QTLs	Foreground selection	Background selection	References
Rice	Early blight	Xa21	STS	AFLP	(Chen et al., 2001)
	Tolerance, disease, resistance, quality	Subchr9 QTL, Xa21, Bph and blast QTLs, and quality loci	SSR and STS	Not performed	(Toojinda et al., 2005)
	Deep roots	QTLs on chromosomes 1, 2, 7, and 9	RFLP and SSR	SSR	(Shen et al., 2001)
	Blast	Pi1	SSR	ISSR	(Liu and Anderson, 2003)
Maize	Corn borer resistance	QTLs on chromosomes 7, 9, and 10	RFLP	RFLP	(Willcox et al., 2002)
Barley	Leaf rust	Rphq6 A	AFLP	AFLP	(van Berloo et al., 2001)
	Yield	QTLs on 2H and 3HL	RFLP	RFLP	(Schmierer et al., 2004)
Wheat	Powdery mildew	22 Pm genes	Phenotyping	AFLP	(Zhou et al., 2005)

Source: Hassan et al., 2021.

polymorphisms in closely related cultivars (Morjane et al., 1994). Microsatellites were originally used in plants to identify cultivars, and then they were used to specifically genotype other crops like rice, wheat, grapevines, and soybeans. For the protection of proprietary germplasm, this is particularly crucial. Similar to microsatellite markers, which represent a single locus, pedigree analysis benefited from microsatellite markers. These markers' multiallelism makes it possible to reliably detect comparative allelic variability over a wide variety of germplasm and makes it possible to genotype individuals ubiquitously, enabling the determination of gene flow and paternity. The microsatellite probe $(GATA)_4$ was discovered to indicate sex differences in Southern hybridization, making sex identification of dioecious plants one of the latest applications of these markers. When male and female plants don't exhibit sex-specific morphological variations prior to flowering, it can be used as a diagnostic marker. The pseudoautosomal plant sex chromosome was also found to have a random amplified polymorphic (RAPD) DNA marker in *Silene dioica* (L.) by Di Stilio et al., in 1998.

6.4 GENOME EDITING (CRISPR)

The CRISPR genome editing method has improved several crop plants (Feng et al., 2013). Because of its many advantages, including ease of use, cleaving methylation loci, etc., Cas9 technology is quickly replacing other methods (Lozano-Juste and Cutler, 2014; Hsu et al., 2013). The two most crucial components of the CRISPR method are CRISPR RNAs and the Cas protein. Two short-length RNA molecules, transencoded CRISPR RNA (tracrRNA) and CRISPR RNA (crRNA), can cleave a specific target site with the aid of Cas9 endonuclease. The hybrid formed when tracrRNA and crRNA are artificially fused together is known as single guide RNA (sgRNA) (Qi et al., 2013). The sgRNA and Cas proteins come together to produce the RNA-guided endonuclease, which mediates the cleave of a specific genome sequence (Niewoehner et al., 2013). On the basis of Cas protein, the CRISPR-Cas system is divided into three types—I, II, and III. Two distinct proteins called Cas1 and Cas2 are frequently found in all three types. Type III is most frequently found in archaea along with in some bacteria, whereas type I is in both bacteria and in archaea, and type II is in bacteria only (Makarova et al., 2011). Several economically significant crops and model plants, viz. maize, wheat, *Nicotiana tabacum*, and *Arabidopsis*, have all undergone effective genome editing (Svitashev et al., 2015; Ma et al., 2015; Shan et al., 2013; Ali et al., 2013).

6.5 PROBABLE CAUSES FOR LESSER EXPLOITATION OF DNA MARKERS IN CROP PLANTS

In 1980s, DNA marker techniques were first evolved, and PCR-based markers were first discovered in 1990s. After this, several markers were evolved and used for a variety of purposes (Xu and Crouch, 2008; Andeden et al., 2015; Baloch et al., 2015; Baloch et al., 2016). Many research articles are released every year as a result of the growth and development of marker technology. Many of these articles, meanwhile, don't have an impact on breeding at the practical level (Xu and Crouch, 2008). Similar to this, QTL mapping leads to the production of several articles that describe newly discovered QTLs. These QTLs were created as a result of research projects, and it is now necessary to utilize them after rigorous validation and to create functional diagnostic markers that could result in breeding programs that are beneficial to farmer fields (Hospital, 2005).

6.6 CONCLUSION

DNA marker technology development, from RFLP to SNPs to DArT, has been ongoing for the past 30 years. The development of NGS stages with low effort and high throughput has been motivated by advances in sequencing technology. This is done primarily for incorrect phenotyping reasons. There have been numerous initiatives to create precise, novel, and highly effective markers in crop breeding with agricultural importance (such as wheat, maize, rice potato), but for lesser exploited crops, markers have not been properly developed. There is still a lack of knowledge about molecular marker techniques in areas like molecular ecology and phylogenetics. Analysts are focusing on genome alterations of commercially valuable plants since the CRISPR technology has altered the way that plants reproduce and pass along genetic traits. The main drawback of several MAS techniques is the requirement for genome-level information in some circumstances, which necessitates extralaborious and time-consuming experimentation. Most of the techniques covered in this chapter are likely to provide a structured database that can be used on its own or in combination with sequence-level characters in particular areas of plant science where they have not yet been applied.

REFERENCES

Adam-Blondon, A.F., Sévignac, M., Bannerot, H., Dron, M., 1994. SCAR, RAPD and RFLP markers linked to a dominant gene (Are) conferring resistance to anthracnose in common bean. *Theor. Appl. Genet.* 88, 865–870.

Ali, Z., Eid, A., Ali, S. and Mahfouz, M.M., 2018. Pea early-browning virus-mediated genome editing via the CRISPR/Cas9 system in *Nicotiana benthamiana* and *Arabidopsis*. *Virus Res.* 244, 333–337.

Amom, T., Nongdam, P., 2017. The use of molecular marker methods in plants: a review. *Int. J. Curr. Res. Rev.* 9, 1–7.

Andeden, E.E., Baloch, F.S., Çakır, E., Toklu, F. and Ozkan, H., 2015. Development, characterization and mapping of microsatellite markers for lentil (*Lens culinaris* Medik.). *Plant Breed.* 134, 589–598.

Ashkani, S., Rafii, M.Y., Sariah, M., Siti, N.A.A., Rusli, I., Harun, A.R., Latif, M.A., 2011. Analysis of simple sequence repeat markers linked with blast disease resistance genes in a segregating population of rice (*Oryza sativa*). *Genet. Mol. Res.* 10, 1345–1355.

Avise, J.C., 1994. *Molecular Markers, Natural History and Evolution*, Chapman and Hall.

Avise, J.C. 2004. *Molecular Markers, Natural History and Evolution*. Sinauer Publication.

Bains, N.S., Singh, K. and Basra, A.S., 1999. Genetic basis of heterosis: Search for a new paradigm. In Basra A S (ed.) "*Heterosis and Hybrid Seed Production in Agronomic Crops.*" pp.1–24. Food Products Press.

Baird, N.A., Etter, P.D., Atwood, T.S., Currey, M.C., Shiver, A.L., Lewis, Z.A., Selker, E.U., Cresko, W.A. and Johnson, E.A., 2008. Rapid SNP discovery and genetic mapping using sequenced RAD markers. *PLoS One.* 3, e3376.

Baloch, F.S., Derya, M., Andeden, E.E., Alsaleh, A., Cömertpay, G., Kilian, B., Ozkan, H., 2015. Inter-primer binding site retrotransposon and inter-simple sequence repeat diversity among wild Lens species. *Biochem. Syst. Ecol.* 58, 162–168.

Baloch, F.S., Alsaleh, A., Andeden, E.E., Hatipoğlu, R., Nachit, M. and Özkan, H., 2016. High levels of segregation distortion in the molecular linkage map of bread wheat representing the West Asia and North Africa region. *Turk. J. Agric. Forest.* 40, 352–364.

Baránek, M., Meszaros, M., Sochorova, J., Čechova, J., Raddova, J., 2012. Utility of retrotransposon-derived marker systems for differentiation of presumed clones of the apricot cultivar Velkopavlovická. *Scientia Horticulturae.* 143, 1–6.

Barbazuk, W.B., Emrich, S.J., Chen, H.D., Li, L. and Schnable, P.S., 2007. SNP discovery via 454 transcriptome sequencing. *Plant J.* 51, 910–918.

Bardakci, F., 2001. Random amplified polymorphic DNA (RAPD) markers. *Turk. J. Biol.* 25, 185–196.

Beissinger, T.M., Hirsch, C.N., Sekhon, R.S., Foerster, J.M., Johnson, J.M., Muttoni, G., Vaillancourt, B., Buell, C.R., Kaeppler, S.M. and de Leon, N., 2013. Marker density and read depth for genotyping populations using genotyping-by-sequencing. *Genetics.* 193, 1073–1081.

Bentley, D.R., Balasubramanian, S., Swerdlow, H.P., Smith, G.P., Milton, J., Brown, C.G., Hall, K.P., Evers, D.J., Barnes, C.L., Bignell, H.R. and Boutell, J.M., 2008. Accurate whole human genome sequencing using reversible terminator chemistry. *Nature.* 456, 53–59.

Bhattramakki, D., Dolan, M., Hanafey, M., Wineland, R., Vaske, D., Register, J.C., Tingey, S.V. and Rafalski, A., 2002. Insertion-deletion polymorphisms in 3′ regions of maize genes occur frequently and can be used as highly informative genetic markers. *Plant Mol. Biol.* 48, 539–547.

Bidhendi, M.Z., Choukan, R., Darvish, F., Mostafavi, K., Majidi, E., 2012. Classifying of maize inbred lines into heterotic groups using diallel analysis. *Environments.* 7, 2252–2250.

Bilyeu, K., Palavalli, L., Sleper, D.A. and Beuselinck, P., 2006. Molecular genetic resources for development of 1% linolenic acid soybeans. *Crop Sci.* 46: 1913–1918.

Blair, M.W., Panaud, O., McCouch, S.R., 1999. Inter-simple sequence repeats (ISSR) amplification for analysis of microsatellite motif frequency and fingerprinting in rice (*Oryza sativa* L.). *Theor. Appl. Genet.* 98, 780–792.

Boeven, P.H., Longin, C.F.H., Würschum, T., 2016. A unified framework for hybrid breeding and the establishment of heterotic groups in wheat. *Theor. Appl. Genet.* 129, 1231–1245.

Brown, S.M., Kresovich, S., 1996. Molecular markers in plant genome analysis. In: Paterson, A.H. (ed), *Genome Mapping in Plants.* Clandes, pp. 85–93.

Bundock, P.C., Eliott, F.G., Ablett, G., Benson, A.D., Casu, R.E., Aitken, K.S. and Henry, R.J., 2009. Targeted single nucleotide polymorphism (SNP) discovery in a highly polyploid plant species using 454 sequencing. *Plant Biotechnol. J.* 7, 347–354.

Caetano-Anolles, G., Bassam, B.J., 1993. DNA amplification fingerprinting using arbitrary oligonucleotide primers. *Appl. Biochem. Biotechnol.* 42, 189–200.

Castro, A.J., Capettini, F.L.A.V.I.O., Corey, A.E., Filichkina, T., Hayes, P.M., Kleinhofs, A.N.D.R.I.S., Kudrna, D., Richardson, K., Sandoval-Islas, S., Rossi, C., Vivar, H., 2003. Mapping and pyramiding of qualitative and quantitative resistance to stripe rust in barley. *Theor. Appl. Genet.* 107, 922–930.

Chague, V., Fahima, T., Dahan, A., Sun, G.L., Korol, A.B., Ronin, Y.I., Grama, A., Röder, M.S. and Nevo, E.1., 1999. Isolation of microsatellite and RAPD markers flanking the Yr15 gene of wheat using NILs and bulked segregant analysis. *Genome.* 42, 1050–1056.

Chen, S., Xu, C.G., Lin, X.H. and Zhang, Q., 2001. Improving bacterial blight resistance of "6078," an elite restorer line of hybrid rice, by molecular marker-assisted selection. *Plant Breed.* 120, 133–137.

Choi, H.W. and Hwang, B.K., 2012. The pepper extracellular peroxidase $CaPO_2$ is required for salt, drought and oxidative stress tolerance as well as resistance to fungal pathogens. *Planta,* 235, 1369–1382.

Chutimanitsakun, Y., Nipper, R.W., Cuesta-Marcos, A., Cistué, L., Corey, A., Filichkina, T., Johnson, E.A. and Hayes, P.M., 2011. Construction and application for QTL analysis of a Restriction Site Associated DNA (RAD) linkage map in barley. *BMC Genom.* 12, 1–13.

Collard, B.C. and Mackill, D.J., 2008. Marker-assisted selection: an approach for precision plant breeding in the twenty-first century. *Philosophical Transactions of the Royal Society B: Biological Sciences, 363*(1491), 557–572.

Collard, B.C., Jahufer, M.Z.Z., Brouwer, J.B., Pang, E.C.K., 2005. An introduction to markers, quantitative trait loci (QTL) mapping and marker-assisted selection for crop improvement: the basic concepts. *Euphytica.* 142, 169–196.

Collard, B.C.Y., Mackill, D.J., 2009. Conserved DNA-derived polymorphism (CDDP): a simple and novel method for generating DNA markers in plants. *Plant Mol. Biol. Report.* 27, 558–562.

Comai, L. and Henikoff, S., 2006. TILLING: practical single-nucleotide mutation discovery. *Plant J.* 45, 684–694.

Cregan, P.B., Mudge, J., Fickus, E.W., Danesh, D., Denny, R. and Young, N.D., 1999. Two simple sequence repeat markers to select for soybean cyst nematode resistance conditioned by the rhg1 locus. *Theor. Appl. Genet.* 99, 811–818.

D'Ovidio, R., Anderson, O.D., 1994. PCR analysis to distinguish between allele of a member of multi-gene family correlated with wheat bread making quality. *Theor. Appl. Genet.* 88, 759–763.

Dar, S.H., Hussain, W., Sanghera, G.S., 2013. Advances in hybrid rice technology through applications of novel technologies. *Crop Improvement: An Integrated Approach.* MD Publications Pvt Ltd, New Delhi, 61–67.

Davey, J.W., Hohenlohe, P.A., Etter, P.D., Boone, J.Q., Catchen, J.M. and Blaxter, M.L., 2011. Genome-wide genetic marker discovery and genotyping using next-generation sequencing. *Nature Rev. Genet.* 12, 499–510.

Dean, A., 2006. On a chromosome far, far away: LCRs and gene expression. *Trends. Genet.* 22, 38–45.

Deschamps, S., Llaca, V. and May, G.D., 2012. Genotyping-by-sequencing in plants. *Biology.* 1, 460–483.

Di Stilio, V.S., Kesseli, R.V., Mulcahy, D.L., 1998. A pseudoautosomal random amplified polymorphic DNA marker for the sex chromosomes of *Silene dioica. Genetics.* 149, 2057–2062.

Dussle, C., Quint, M., Xu, M., Melchinger, A. and Lübberstedt, T., 2002. Conversion of AFLP fragments tightly linked to SCMV resistance genes Scmv1 and Scmv2 into simple PCR-based markers. *Theor. Appl Genet.* 105, 1190–1195.

Dwivedi, N., Kumar, R., Paliwal, R., Kumar, U., Kumar, S., Singh, M., Singh, R.K., 2015. QTL mapping for important horticultural traits in pepper (*Capsicum annuum* L.). *J. Plant Biochem. Biotechnol.* 24, 154–160.

Ellis, T.H.N., Poyser, S.J., Knox, M.R., Vershinin, A.V., Ambrose, M.J., 1998. Ty1-copia class retrotransposon insertion site polymorphism for linkage and diversity analysis in pea. *Mol. Gen. Genet.* 260, 9–91.

Elshire, R.J., Glaubitz, J.C., Sun, Q., Poland, J.A., Kawamoto, K., Buckler, E.S. and Mitchell, S.E., 2011. A robust, simple genotyping-by-sequencing (GBS) approach for high diversity species. *PLoS One.* 6, e19379.

Emberton, J., Ma, J., Yuan, Y., SanMiguel, P. and Bennetzen, J.L., 2005. Gene enrichment in maize with hypomethylated partial restriction (HMPR) libraries. *Genome Res.* 15, 1441–1446.

Englund, P.T., 1971. Analysis of nucleotide sequences at 3'termini of duplex deoxyribonucleic acid with the use of the T4 deoxyribonucleic acid polymerase. *J. Biol. Chem.* 246, 3269–3276.

Englund, P.T., 1972. The 3′-terminal nucleotide sequences of T7 DNA. *J. Mol. Biol.*, 209–224.

Etherington, G.J., Monaghan, J., Zipfel, C., MacLean, D., 2014. Mapping mutations in plant genomes with the user-friendly web application CandiSNP. *Plant Methods.* 10, 1–11.

Fang, D.D., Jenkins, J.N., Deng, D.D., McCarty, J.C., Li, P., Wu, J., 2014. Quantitative trait loci analysis of fiber quality traits using a random-mated recombinant inbred population in Upland cotton (*Gossypium hirsutum* L.). *BMC Genomics.* 15, 397.

Feng, Z., Zhang, B., Ding, W., Liu, X., Yang, D.L., Wei, P., Cao, F., Zhu, S., Zhang, F., Mao, Y. and Zhu, J.K., 2013. Efficient genome editing in plants using a CRISPR/Cas system. *Cell Res.* 23, 1229–1232.

Feuillet, C., Messmer, M., Schachermayr, G., Keller, B., 1995. Genetic and physical characterization of *LR1* leaf rust resistance locus in wheat (*Triticum aestivum*). *Mol. Gen. Genet.* 248, 553–562.

Flavell, R.B., 1995. Plant biotechnology R & D—the next ten years. *Tibtech.* 13, 313–319.

França, L.T., Carrilho, E. and Kist, T.B., 2002. A review of DNA sequencing techniques. *Quart. Rev Bophys.* 35, 169–200.

Francia, E., Tacconi, G., Crosatti, C., Barabaschi, D., Bulgarelli, D., Dall'Aglio, E., Valè, G., 2005. Marker assisted selection in crop plants. *Plant Cell, Tissue and Organ Culture,* 82, 317–342.

Fulneček, J. and Kovařík, A., 2014. How to interpret methylation sensitive amplified polymorphism (MSAP) profiles?. *BMC Genet.* 15, 1–9.

Fusari, C.M., Lia, V.V., Hopp, H.E., Heinz, R.A., Paniego, N.B., 2008. Identification of single nucleotide polymorphisms and analysis of linkage disequilibrium in sunflower elite inbred lines using the candidate gene approach. *BMC Plant Biol.* 8, 1–14.

Gardiner, J.M., Coe, E.H., Melia-Hancock, S., Hoisington, D.A., Chao, S., 1993. Map in maize using an immortalized F2 population. *Genetics.* 134, 917–930.

Gharizadeh, B., Herman, Z.S., Eason, R.G., Jejelowo, O. and Pourmand, N., 2006. Large-scale pyrosequencing of synthetic DNA: A comparison with results from Sanger dideoxy sequencing. *Electrophoresis.* 27, 3042–3047.

Gorji, A.M., Poczai, P., Polgar, Z., Taller, J., 2011. Efficiency of arbitrarily amplified dominant markers (SCoT, ISSR and RAPD) for diagnostic fingerprinting in tetraploid potato. *American J. Potato Res.* 88, 226–237.

Graner, A., Streng, S., Kellermann, A., Schiemann, A., Bauer, E., Waugh, R., Pellio, B. and Ordon, F., 1999. Molecular mapping and genetic fine-structure of the rym5 locus encoding resistance to different strains of the Barley Yellow Mosaic Virus Complex. *Theor. Appl Genet.* 98, 285.

Graner, A., Streng, S., Drescher, A., Jin, Y., Borovkova, I. and Steffenson, B.J., 2000. Molecular mapping of the leaf rust resistance gene Rph7 in barley. *Plant Breed.* 119, 389–392.

Grimmer, M.K., Trybush, S., Hanley, S., Francis, S.A., Karp, A. and Asher, M.J.C., 2007. An anchored linkage map for sugar beet based on AFLP, SNP and RAPD markers and QTL mapping of a new source of resistance to Beet necrotic yellow vein virus. *Theoret. Appl. Genet.* 114, 1151–1160.

Grimmer, M.K., Kraft, T., Francis, S.A. and Asher, M.J.C., 2008. QTL mapping of BNYVV resistance from the WB258 source in sugar beet. *Plant Breed.* 127, 650–652.

Guo, L. and Lu, Z., 2010. Global expression analysis of miRNA gene cluster and family based on isomiRs from deep sequencing data. *Comput. Biol. Chem.* 34, 165–171.

Gupta, M., Chyi, Y.S., Romero-Severson, J., Owen, J.L., 1994. Amplification of DNA markers from evolutionarily diverse genomes using single primers of simple-sequence repeats. *Theor. Appl. Genet.* 89, 998–1006.

Gupta, P.K., Sonwane, A.A., Singh, N.K., Meshram, C.D., Dahiya, S.S., Pawar, S.S., Gupta, S.P., Chaturvedi, V.K. and Saini, M., 2012. Intracerebral delivery of small interfering RNAs (siRNAs) using adenoviral vector protects mice against lethal peripheral rabies challenge. *Virus Res.* 163, 11–18.

Hailu, G., Asfere, Y., 2020. The role of molecular markers in crop improvement and plant breeding programs: a review. *Agric. J.* 15, 171–175.

Hao, Y.J. and Deng, X.X., 2003. Genetically stable regeneration of apple plants from slow growth. *Plant Cell Tissue Organ Culture.* 72, 253–260.

Heather, J.M. and Chain, B., 2016. The sequence of sequencers: The history of sequencing DNA. *Genomics.* 107, 1–8.

Hemmat, M., Weedon, N.F., Manganaris, A.G., Lawson, D.M., 1994. Molecular marker linkage map for apple. *J. Heredity.* 85, 4–11.

Herzog, E., Frisch, M., 2011. Selection strategies for marker-assisted backcrossing with high-throughput marker systems. *Theor. Appl. Genet.* 123, 251–260.

Hittalmani, S., Parco, A., Mew, T.V., Zeigler, R.S., Huang, N., 2000. Fine mapping and DNA marker-assisted pyramiding of the three major genes for blast resistance in rice. *Theor. Appl. Genet.* 100, 1121–1128.

Hodges, E., Xuan, Z., Balija, V., Kramer, M., Molla, M.N., Smith, S.W., Middle, C.M., Rodesch, M.J., Albert, T.J., Hannon, G.J. and McCombie, W.R., 2007. Genome-wide in situ exon capture for selective resequencing. *Nature Genet.* 39, 1522–1527.

Horst, L., Wenzel, G. 2005. *Molecular Marker Systems in Plant Breeding and Crop Improvement.* Springer Verlag, Germany. pp. 74–76.

Hospital, F., 2009. Challenges for effective marker-assisted selection in plants. *Genetica.* 136, 303–310.

Hsu, P.D., Scott, D.A., Weinstein, J.A., Ran, F.A., Konermann, S., Agarwala, V., Li, Y., Fine, E.J., Wu, X., Shalem, O. and Cradick, T.J., 2013. DNA targeting specificity of RNA-guided Cas9 nucleases. *Nature Biotechnol.* 31, 827–832.

Hu, X., Sullivan-Gilbert, M., Gupta, M. and Thompson, S.A., 2006. Mapping of the loci controlling oleic and linolenic acid contents and development of fad2 and fad3 allele-specific markers in canola (*Brassica napus* L.). *Theoret. Appl. Genet.* 113, 497–507.

Hyman, E.D., 1988. A new method of sequencing DNA. *Anal. Biochem.* 174, 423–436.

Jaccoud, D., Peng, K., Feinstein, D., Kilian, A. 2001. Diversity arrays: a solid state technology for sequence information independent genotyping. *Nucleic Acids Res.* 29, e25.

Jander, G., Norris, S.R., Rounsley, S.D., Bush, D.F., Levin, I.M. and Last, R.L., 2002. Arabidopsis map-based cloning in the post-genome era. *Plant Physiol.* 129, 440–450.

Jarvis, P., Lister, C., Szabo, V. and Dean, C., 1994. Integration of CAPS markers into the RFLP map generated using recombinant inbred lines of *Arabidopsis thaliana. Plant Mol. Biol.* 24, 685–687.

Jeffreys, A., Wilson, V., Thein, S., 1985. Hypervariable "minisatellite" regions in human DNA. Nature 314, 67–73.

Jena, K.K., Mackill, D.J., 2008. Molecular markers and their use in marker-assisted selection in rice. *Crop Sci.* 48, 1266–1276.

Jeon, J.S., Chen, D., Yi, G.H., Wang, G.L. and Ronald, P.C., 2003. Genetic and physical mapping of Pi5 (t), a locus associated with broad-spectrum resistance to rice blast. *Mol. Genet. Genom.* 269, 280–289.

Jiang, G.L., 2013. Molecular markers and marker-assisted breeding in plants. In: Andersen, S.B. (Ed.), *Plant Breeding from Laboratories to Fields.* Intech Open, Rijeka, Croatia, pp. 45–83.

Jones, E.S., Sullivan, H., Bhattramakki, D. and Smith, J.S.C., 2007. A comparison of simple sequence repeat and single nucleotide polymorphism marker technologies for the genotypic analysis of maize (*Zea mays* L.). *Theoret. Appl. Genet.* 115, 361–371.

Kalendar, R., Antonius, K., Smýkal, P., Schulman, A.H., 2010. iPBS: a universal method for DNA fingerprinting and retrotransposon isolation. *Theor. Appl. Genet.* 121, 1419–1430.

Kalendar, R., Flavell, A.J., Ellis, T.H.N., Sjakste, T., Moisy, C., Schulman, A.H., 2011. Analysis of plant diversity with retrotransposon-based molecular markers. *Heredity.* 106, 520–530.

Kalendar, R., Aizharkyn, K.S., Khapilina, O.N., Amenov, A.A., Tagimanova, D.S., 2017. Plant diversity and transcriptional variability assessed by retrotransposon-based molecular markers. *Russ. J. Genet. Appl. Res.* 21, 128–134.

Kalia, R.K., Rai, M.K., Kalia, S., Singh, R., Dhawan, A.K., 2011. Microsatellite markers: an overview of the recent progress in plants. *Euphytica.* 177, 309–334.

Kelly, N.J., Palmer, M.T., Morrow, C.D., 2003. Selection of retroviral reverse transcription primer is coordinated with tRNA biogenesis. *J. Virol.* 77, 8695–8701.

Kiss, G.B., Csanádi, G., Kálmán, K., Kaló, P. and Ökrész, L., 1993. Construction of a basic genetic map for alfalfa using RFLP, RAPD, isozyme and morphological markers. *Mol. Gen. Genet.* 238, 129–137.

Koelling, J., Coles, M.C., Matthews, P., Schwekendiek, A., 2012. Development of new microsatellite markers (SSRs) for *Humulus lupulus. Mol. Breed.* 30, 479–484.

Kuhn, D., 2012, January. Design of an Illumina Infinium 6k SNPchip for genotyping two large avocado mapping populations. In *Proceedings of the 20th Conference on Plant and Animal Genome*, San Diego.

Kumar, S., Rao, M., 2014. Conventional and molecular breeding for bacterial leaf blight and blast resistance in rice. *Res. Rev. J. Ecol.* 3, 1–3.

Kumar, S.J., Susmita, C., Agarwal, D.K., Pal, G., Rai, A.K., Simal-Gandara, J., 2021. Assessment of genetic purity in rice using polymorphic SSR markers and its economic analysis with grow-out-test. *Food Anal. Meth.* 14, 856–864.

Kurowska, M., Daszkowska-Golec, A., Gruszka, D., Marzec, M., Szurman, M., Szarejko, I. and Maluszynski, M., 2011. TILLING—a shortcut in functional genomics. *J. Appl Genet.* 52, 371–390.

Lagudah, E.S., Krattinger, S.G., Herrera-Foessel, S., Singh, R.P., Huerta-Espino, J., Spielmeyer, W., Brown-Guedira, G., Selter, L.L. and Keller, B., 2009. Gene-specific markers for the wheat gene Lr34/Yr18/Pm38 which confers resistance to multiple fungal pathogens. *Theoret. Appl. Genet.* 119, 889–898.

Lai, K., Duran, C., Berkman, P.J., Lorenc, M.T., Stiller, J., Manoli, S., Hayden, M.J., Forrest, K.L., Fleury, D., Baumann, U. and Zander, M., 2012. Single nucleotide polymorphism discovery from wheat next-generation sequence data. *Plant Biotechnol. J.* 10, 743–749.

Lee, S.C., Choi, D.S., Hwang, I.S. and Hwang, B.K., 2010. The pepper oxidoreductase CaOXR1 interacts with the transcription factor CaRAV1 and is required for salt and osmotic stress tolerance. *Plant Mol. Biol.* 73, 409–424.

Lehmensiek, A., Sutherland, M.W. and McNamara, R.B., 2008. The use of high resolution melting (HRM) to map single nucleotide polymorphism markers linked to a covered smut resistance gene in barley. *Theoret. Appl. Genet.* 117, 721–728.

Leitao, J. M., 2011. "Plant mutation breeding and biotechnology," in Chemical mutagenesis. Eds. Shu, Q. Y., Forster, B. P., and Nakagawa, H. (Wallingford: CABI), 135–158.

Leshchiner, I., Alexa, K., Kelsey, P., Adzhubei, I., Austin-Tse, C.A., Cooney, J.D., Anderson, H., King, M.J., Stottmann, R.W., Garnaas, M.K., Ha, S., 2012. Mutation mapping and identification by whole-genome sequencing. *Genome Res.* 22, 1541–1548.

Li, G., Quiros, C.F., 2001. Sequence-related amplified polymorphism (SRAP), a new marker system based on a simple PCR reaction: its application to mapping and gene tagging in *Brassica. Theor. Appl. Genet.* 103, 455–461.

Li, X., Qian, Q., Fu, Z., Wang, Y., Xiong, G., Zeng, D., Wang, X., Liu, X., Teng, S., Hiroshi, F., Yuan, M., 2003. Control of tillering in rice. *Nature.* 422, 618–621.

Lin, X., Kaul, S., Rounsley, S., Shea, T.P., Benito, M.I., Town, C.D., Fujii, C.Y., Mason, T., Bowman, C.L., Barnstead, M. and Feldblyum, T.V., 1999. Sequence and analysis of chromosome 2 of the plant *Arabidopsis thaliana. Nature.* 402, 761–768.

Liu, S. and Anderson, J.A., 2003. Marker assisted evaluation of Fusarium head blight resistant wheat germplasm. *Crop Sci.* 43, 760–766.

Lozano-Juste, J. and Cutler, S.R., 2014. Plant genome engineering in full bloom. *Trends. Pant Sci.* 19, 284–287.

Lu, R., Malcuit, I., Moffett, P., Ruiz, M.T., Peart, J., Wu, A.J., Rathjen, J.P., Bendahmane, A., Day, L. and Baulcombe, D.C., 2003. High throughput virus-induced gene silencing implicates heat shock protein 90 in plant disease resistance. *EMBO J.* 22, 5690–5699.

Lundqvist, U. (2014). Scandinavian mutation research in barley—a historical review. *Hereditas* 151, 123–131.

Lynch, M., Walsh, B., 1998. Genetics and analysis of quantitative traits. Sunderland, MA: *Sinauer.* Vol. 1, pp. 535–557.

Ma, X., Zhang, Q., Zhu, Q., Liu, W., Chen, Y., Qiu, R., Wang, B., Yang, Z., Li, H., Lin, Y. and Xie, Y., 2015. A robust CRISPR/Cas9 system for convenient, high-efficiency multiplex genome editing in monocot and dicot plants. *Mol. Plant.* 8, 1274–1284.

Madhumati, B., 2014. Potential and application of molecular markers techniques for plant genome analysis. *Int. J. Pure App. Biosci.* 2, 169–188.

Maeda, M., Uryu, N., Murayama, N., Ishii, H., Ota, M., Tsuji, K. and Inoko, H., 1990. A simple and rapid method for HLA-DP genotyping by digestion of PCR-amplified DNA with allele-specific restriction endonucleases. *Human Immunol.* 27, 111–121.

Mago, R., Spielmeyer, W., Lawrence, G., Lagudah, E., Ellis, J. and Pryor, A., 2002. Identification and mapping of molecular markers linked to rust resistance genes located on chromosome 1RS of rye using wheat-rye translocation lines. *Theor. Appl Genet.* 104, 1317–1324.

Makarova, K.S., Haft, D.H., Barrangou, R., Brouns, S.J., Charpentier, E., Horvath, P., Moineau, S., Mojica, F.J., Wolf, Y.I., Yakunin, A.F. and Van Der Oost, J., 2011. Evolution and classification of the CRISPR–Cas systems. *Nature Rev. Microbiol.* 9, 467–477.

Mammadov, J.A., Chen, W., Ren, R., Pai, R., Marchione, W., Yalçin, F., Witsenboer, H., Greene, T.W., Thompson, S.A. and Kumpatla, S.P., 2010. Development of highly polymorphic SNP markers from the complexity reduced portion of maize [*Zea mays* L.] genome for use in marker-assisted breeding. *Theoret. Appl. Genet.* 121, 577–588.

Manivannan, A., Choi, S., Jun, T.H., Yang, E.Y., Kim, J.H., Lee, E.S., Lee, H.E., Kim, D.S., Ahn, Y.K., 2021. Genotyping by sequencing-based discovery of SNP markers and construction of linkage map from F5 population of pepper with contrasting powdery mildew resistance trait. *BioMed Res. Int.* 2021, 8.

Mardis, E.R., 2008. Next-generation DNA sequencing methods. *Annu. Rev. Genomics Hum. Genet.* 9, 387–402.

Mardis, E.R., 2008. The impact of next-generation sequencing technology on genetics. *Trends. Genet.* 24, 133–141.

Martin, B., Nienhuis, J., King, G., Schaefer, A., 1989. Restriction fragment length polymorphisms associated with water-use efficiency in tomato. *Science.* 243, 1725–1728.

Martin, G.B., Williams, J.G., Tanksley, S.D., 1991. Rapid identification of markers linked to a Pseudomonas resistance gene in tomato by using random primers and near-isogenic lines. *Proc. Natl. Acad. Sci.* 88, 2336–2340.

Matsuoka, Y., Mitchell, S.E., Kresovich, S., Goodman, M., Doebley, J., 2002. Microsatellites in *Zea*—variability, patterns of mutations, and use for evolutionary studies. Theor. Appl. Genet. 104, 436–450.

Mba, C., 2013. Induced mutations unleash the potentials of plant genetic resources for food and agriculture. *Agronomy* 3, 200.

Mba, C., Afza, R., and Shu, Q. Y., 2011. "Plant mutation breeding and biotechnology," in Mutagenic radiations: X-rays, ionizing particles and ultraviolet. Eds. Shu, Q. Y., Forster, B. P., and Nakagawa, H. (Wallingford: CABI), 83–90.

McCallum, C.M., Comai, L., Greene, E.A. and Henikoff, S., 2000. Targeting Induced Local Lesions in Genomes (TILLING) for plant functional genomics. *Plant Physiol.* 123, 439–442.

Melchinger, A.E., Messmer, M.M., Lee, M., Woodman, W.L. and Lamkey, K.R., 1991. Diversity and relationships among US maize inbreds revealed by restriction fragment length polymorphisms. *Crop Sci.* 31, 669–678.

Metzker, M.L., 2010. Sequencing technologies—the next generation. *Nature Rev. Genet.* 11, 31–46.

Meyers, B.C., Tingey, S.V. and Morgante, M., 2001. Abundance, distribution, and transcriptional activity of repetitive elements in the maize genome. *Genome Res.* 11, 1660–1676.

Michaels, S.D., Amasino, R.M., 1998. A robust method for detecting single-nucleotide changes as polymorphic markers by PCR. *Plant J.* 14, 381–385.

Monden, Y., Fujii, N., Yamaguchi, K., Ikeo, K., Nakazawa, Y., Waki, T., Hirashima, K., Uchimura, Y., Tahara, M., 2014. Efficient screening of long terminal repeat retrotransposons that show high insertion polymorphism via high-throughput sequencing of the primer binding site. *Genome.* 57, 245–252.

Morjane, H., Geistlinger, J., Harrabi, M., Weising, K., Kahl, G., 1994. Oligonucleotide fingerprinting detects genetic diversity among *Ascochyta rabiei* isolates from a single chickpea field in Tunisia. *Curr. Genet.* 26, 191–197.

Morozova, O. and Marra, M.A., 2008. Applications of next-generation sequencing technologies in functional genomics. *Genomics.* 92, 255–264.

Muchero, W., Diop, N.N., Bhat, P.R., Fenton, R.D., Wanamaker, S., Pottorff, M., Hearne, S., Cisse, N., Fatokun, C., Ehlers, J.D. and Roberts, P.A., 2009. A consensus genetic map of cowpea [*Vigna unguiculata* (L) Walp.] and synteny based on EST-derived SNPs. *Proc. Natl. Acad. Sci.* 106, 18159–18164.

Nadeem, M.A., Nawaz, M.A., Shahid, M.Q., Doğan, Y., Comertpay, G., Yıldız, M., Hatipoğlu, R., Ahmad, F., Alsaleh, A., Labhane, N., Ozkan, H., 2018. DNA molecular markers in plant breeding: current status and recent advancements in genomic selection and genome editing. *Biotechnol. Biotechnol. Equip.* 32, 261–285.

Nakamura, Y., Leppert, M., O'Connell, P., Wolff, R., et al., 1987. Variable number of tandem repeats (VNTR) markers for human gene mapping. Science 235, 1616–1622.

Niewoehner, O., Jinek, M., Doudna, J.A., 2013. Evolution of CRISPR RNA recognition and processing by Cas6 endonucleases. *Nucleic Acids Res.* 42, 1341–1353.

Niewohner, J., Salamini, F., Gebhardt, C., 1995. Development of PCR assay diagnostic for RFLP markers closely linked to allele GROI and HI conferring resistance to root cyst nematode *Globodera rostochiensis* potato. *Mol. Breed.* 1, 65–78.

Nilkanta, H., Amom, T., Tikendra, L., Rahaman, H. and Nongdam, P., 2017. ISSR marker based population genetic study of *Melocanna baccifera* (Roxb.) Kurz: a commercially important bamboo of Manipur, *North-East India. Scientifica.* 2017, 26–38.

Nongdam, P., Nirmala, C. 2007. Genetic variability in four species of *Cymbidium* based on isozyme markers. *Physiol. Mol. Biol. Plants.* 13, 65–68.

Novaes, E., Drost, D.R., Farmerie, W.G., Pappas, G.J., Grattapaglia, D., Sederoff, R.R. and Kirst, M., 2008. High-throughput gene and SNP discovery in Eucalyptus grandis, an uncharacterized genome. *BMC Genom.* 9, 1–14.

Okou, D.T., Steinberg, K.M., Middle, C., Cutler, D.J., Albert, T.J. and Zwick, M.E., 2007. Microarray-based genomic selection for high-throughput resequencing. *Nature Methods.* 4, 907–909.

Olson, M., Hood, L., Cantor, C.H., Botstein, D., 1989. A common language for physical mapping of the human genome. *Science.* 24, 1434–1435.

Padmakar, B., Sailaja, D., Aswath, C., 2015. Molecular exploration of guava (*Psidium guajava* L.) genome using SSR and RAPD markers: a step towards establishing linkage map. *J. Hortic. Sci.* 10, 130–135.

Pakniyat, H., Powell, W., Baird, E., Handley, L.L., Robinson, D., Scrimgeour, E.M., Nevo, E., Hackett, C.A., Caligari, P.D.S., Forste, B.P. 1997. AFLP variation in wild barley (*Hordeum spontaneum* C. Koch) with reference to salt tolerance and associated ecogeography. *Genome.* 40, 332–341.

Paran, I., Kesseli, R., Michelmore, R., 1991. Identification of restriction fragment length polymorphism and random amplified polymorphic DNA markers linked to downy mildew resistance genes in lettuce, using near-isogenic lines. *Genome.* 34, 1021–1027.

Parchman, T.L., Geist, K.S., Grahnen, J.A., Benkman, C.W. and Buerkle, C.A., 2010. Transcriptome sequencing in an ecologically important tree species: assembly, annotation, and marker discovery. *BMC Genom.* 11, 1–16.

Pham, A.T., Shannon, J.G. and Bilyeu, K.D., 2012. Combinations of mutant FAD2 and FAD3 genes to produce high oleic acid and low linolenic acid soybean oil. *Theor. Appl. Genet.* 125: 503–515.

Poczai, P., Varga, I., Bell, N.E., Hyvönen, J. 2011. Genetic Diversity Assessment of bittersweet (*Solanum dulcamara*, Solanaceae) germplasm using conserved DNA derived polymorphism and intron-targeting markers. *Annals Appl. Biol.* 159, 141–153.

Poland, J.A. and Rife, T.W., 2012. Genotyping-by-sequencing for plant breeding and genetics. *Plant Genome.* 5, 92–102.

Qi, L.S., Larson, M.H., Gilbert, L.A., Doudna, J.A., Weissman, J.S., Arkin, A.P. and Lim, W.A., 2013. Repurposing CRISPR as an RNA-guided platform for sequence-specific control of gene expression. *Cell.* 152, 1173–1183.

Ramakrishna, W., Chowdari, K.V., Lagu, M.D., Gupta, V.S., Ranjekar, P.K., 1995. DNA fingerprinting to detect genetic variation in rice using hypervariable DNA sequences. *Theor. Appl. Genet.* 90, 1000–1006.

Reddy, B.V., Reddy, C.V.C.M., Sekhar, A.C., Reddy, P.C.O., Rajasekhar, P., Srinivasulu, K., 2021. Role of molecular based markers methods and their applications in crop improvement. Plant Cell Biotechnol. Mol. Biol. 22, 38–54.

Ribaut, J.M., Betran, J., 1999. Single large-scale marker-assisted selection (SLS-MAS). *Molecular Breeding.* 5, 531–541.

Ronaghi, M., 2001. Pyrosequencing sheds light on DNA sequencing. *Genome Res.* 11, 3–11.

Ronaghi, M., Karamohamed, S., Pettersson, B., Uhlén, M. and Nyrén, P., 1996. Real-time DNA sequencing using detection of pyrophosphate release. *Anal. Biochem.* 242, 84–89.

Ronaghi, M., Uhlén, M. and Nyrén, P., 1998. A sequencing method based on real-time pyrophosphate. *Science.* 281, 363–365.

Rothberg, J.M., Hinz, W., Rearick, T.M., Schultz, J., Mileski, W., Davey, M., Leamon, J.H., Johnson, K., Milgrew, M.J., Edwards, M. and Hoon, J., 2011. An integrated semiconductor device enabling non-optical genome sequencing. *Nature.* 475, 348–352.

Russell, J.R., Bayer, M., Booth, C., Cardle, L., Hackett, C.A., Hedley, P.E., Jorgensen, L., Morris, J.A. and Brennan, R.M., 2011. Identification, utilisation and mapping of novel transcriptome-based markers from blackcurrant (*Ribes nigrum*). *BMC Plant Biol.* 11, 1–11.

Sanger, F. and Coulson, A.R., 1975. A rapid method for determining sequences in DNA by primed synthesis with DNA polymerase. *J. Mol.* Biol. 94, 441–448.

Sanger, F., Nicklen, S. and Coulson, A.R., 1977. DNA sequencing with chain-terminating inhibitors. *Proc. Natl. Acad. Sci.* 74, 5463–5467.

Sant, V.J., Patankar, A.G., Sarode, N.D., Mhase, L.B., Sainani, M.N., Deshmukh, R.B., Ranjekar, P.K., Gupta, V.S., 1999. Potential of DNA markers in detecting divergence and in analysing heterosis in Indian elite chickpea cultivars. *Theor. Appl. Genet.* 98, 1217–1225.

Sardesai, N., Kumar, A., Rajyashri, K., Nair, S. and Mohan, M., 2002. Identification and mapping of an AFLP marker linked to Gm7, a gall midge resistance gene and its conversion to a SCAR marker for its utility in marker aided selection in rice. *Theor. Appl Genet.* 105, 691–698.

Sarfatti, M., Katan, J., Fluhr, R., Zamir, D., 1989. An RFLP marker in tomato linked to the *Fusarium oxysporum* resistance gene *I2*. *Theor. Appl. Genet.* 78, 755–759.

Schachermayr, G., Messmer, M.M., Feuillet, C., Winzeler, H., Keller, B., 1995. Identification of molecular markers linked to the *Agropyron elongatum*-derived leaf rust resistance gene *lr24* in wheat. *Theor. Appl. Genet.* 90, 982–990.

Schmierer, D.A., Kandemir, N., Kudrna, D.A., Jones, B.L., Ullrich, S.E. and Kleinhofs, A., 2004. Molecular marker-assisted selection for enhanced yield in malting barley. *Mol. Breed.* 14, 63–473.

Schulz, B., Eckstein, R.L. and Durka, W., 2013. Scoring and analysis of methylation-sensitive amplification polymorphisms for epigenetic population studies. *Mol. Ecol. Resour.* 13, 642–653.

Semagn, K., Bjørnstad, Å. and Ndjiondjop, M.N., 2006. An overview of molecular marker methods for plants. *African J. Biotechnol.* 5, 24–38.

Seyfarth, R., Feuillet, C., Schachermayr, G., Winzeler, M. and Keller, B., 1999. Development of a molecular marker for the adult plant leaf rust resistance gene Lr35 in wheat. *Theor. Appl Genet.* 99, 554–560.

Shan, Q., Wang, Y., Li, J., Zhang, Y., Chen, K., Liang, Z., Zhang, K., Liu, J., Xi, J.J., Qiu, J.L. and Gao, C., 2013. Targeted genome modification of crop plants using a CRISPR-Cas system. *Nature Biotechnol.* 31, 686–688.

Shattuck-Eidens, D., Bell, R.N., Mitchell, J.T., McWorther, V.C., 1991. Rapid detection of maize DNA sequence variation. *GATA* 8, 240–245.

Shen, L., Courtois, B., McNally, K.L., Robin, S. and Li, Z., 2001. Evaluation of near-isogenic lines of rice introgressed with QTLs for root depth through marker-aided selection. *Theor. Appl.* Genet. 103, 75–83.

Shendure, J. and Ji, H., 2008. Next-generation DNA sequencing. *Nat. Biotechnol.* 26, 1135–1145.

Shi, A., Chen, P., Li, D., Zheng, C., Zhang, B. and Hou, A., 2009. Pyramiding multiple genes for resistance to soybean mosaic virus in soybean using molecular markers. *Mol. Breed.* 23, 113–124.

Slatkin, M., 1987. Gene flow and population structure of natural populations. *Science.* 263, 787–792.

Spaniolas, S., May, S.T., Bennett, M.J., Tucker, G.A., 2006. Authentication of coffee by means of PCR-RFLP analysis and lab-on-a-chip capillary electrophoresis. *J. Agric. Food Chem.* 54, 7466–7470.

Springer, N.M., Ying, K., Fu, Y., Ji, T., Yeh, C.T., Jia, Y., Wu, W., Richmond, T., Kitzman, J., Rosenbaum, H. and Iniguez, A.L., 2009. Maize inbreds exhibit high levels of copy number variation (CNV) and presence/absence variation (PAV) in genome content. *PLoS Genet.* 5, e1000734.

Stadler, L. J., 1928. Genetic effects of X-Rays in maize. *Proc. Natl. Acad. Sci.* 14, 69–75.

Steele, K.A., Price, A.H., Shashidhar, H.E., Witcombe, J.R., 2006. Marker-assisted selection to introgress rice QTLs controlling root traits into an Indian upland rice variety. *Theor. Appl. Genet.* 112, 208–221.

Stuber, C.W., Lincoln, S.E., Wolff, D.W., Helentjaris, T. and Lander, E., 1992. Identification of genetic factors contributing to heterosis in a hybrid from two elite maize inbred lines using molecular markers. *Genetics.* 132, 823–839.

Suenaga, K., Singh, R.P., Huerta-Espino, J. and William, H.M., 2003. Microsatellite markers for genes Lr34/Yr18 and other quantitative trait loci for leaf rust and stripe rust resistance in bread wheat. *Phytopathology.* 93, 881–890.

Svitashev, S., Young, J.K., Schwartz, C., Gao, H., Falco, S.C. and Cigan, A.M., 2015. Targeted mutagenesis, precise gene editing, and site-specific gene insertion in maize using Cas9 and guide RNA. *Plant Physiol.* 169, 931–945.

Szarejko, I., Szurman-Zubrzycka, M., Nawrot, M., Marzec, M., Gruszka, D., Kurowska, M., 2017. "Biotechnologies for Plant Mutation Breeding: Protocols," in *Creation of a TILLING Population in Barley After Chemical Mutagenesis with Sodium Azide and* MNU. Eds. Jankowicz-Cieslak, J., Tai, T. H., Kumlehn, J., and Till., B. J. Cham: Springer International Publishing, pp. 91–111.

Tamura, Y., Hattori, M., Yoshioka, H., Yoshioka, M., Takahashi, A., Wu, J., Sentoku, N., Yasui, H., 2014. Map-based cloning and characterization of a brown planthopper resistance gene BPH26 from *Oryza sativa* L. *ssp. indica cultivar ADR52. Scientific Reports.* 4, 1–8.

Tanksley, S.D., 1993. Mapping polygenes. *Ann. Rev. Genet.* 27, 205–233.

Tanksley, S.D., Ganal, M.W., Martin, G.B., 1995. Chromosome landing: A paradigm for map based gene cloning in plants with large genomes. *Trends Genet.* 11, 63–68.

Thudi, M., Li, Y., Jackson, S.A., May, G.D. and Varshney, R.K., 2012. Current state-of-art of sequencing technologies for plant genomics research. *Brief. Funct. Genomics.* 11, 3–11.

Toojinda, T., Tragoonrung, S., Vanavichit, A., Siangliw, J.L., Pa-In, N., Jantaboon, J., Siangliw, M. and Fukai, S., 2005. Molecular breeding for rainfed lowland rice in the Mekong region. *Plant Prod. Sci.* 8, 330–333.

Torres, A.M., Weeden, N.F. and Martin, A., 1993. Linkage among isozyme, RFLP and RAPD markers in *Vicia faba. Theor. Appl. Genet.* 85, 937–945.

Trick, M., Long, Y., Meng, J. and Bancroft, I., 2009. Single nucleotide polymorphism (SNP) discovery in the polyploid Brassica napus using Solexa transcriptome sequencing. *Plant Biotechnol* J. 7, 334–346.

Uzun, B., Cagirgan, M.I., 2009. Identification of molecular markers linked to determinate growth habit in *Sesame. Euphytica.* 166, 379–384.

van Berloo, R., Aalbers, H., Werkman, A. and Niks, R.E., 2001. Resistance QTL confirmed through development of QTL-NILs for barley leaf rust resistance. *Mol. Breed.* 8, 187–195.

van Orsouw, N.J., Hogers, R.C., Janssen, A., Yalcin, F., Snoeijers, S., Verstege, E., Schneiders, H., van der Poel, H., van Oeveren, J., Verstegen, H. and van Eijk, M.J., 2007. Complexity reduction of polymorphic sequences (CRoPS™): a novel approach for large-scale polymorphism discovery in complex genomes. *PLoS One.* 2, e1172.

Varshney, R.K., 2009. Gene-based marker systems in plants: high throughput approaches for marker discovery and genotyping. In: Molecular Techniques in Crop Improvement: 2nd Edition. Springer, Netherlands, pp.119–142. ISBN 978-90-481-2966-9

Virk, P.S., Newbury, H.J., Jackson, M.T., Ford-Loyd, B.V., 1997. In: Callow, J.A., Newbury, H.J. (eds), *Biotechnology and Plant Genetic Resources: Conservation and Use*. CABI Publishing. pp. 21–23.

Vos, P., Hogers, R., Bleeker, M., Reijans, M., Lee, T.V.D., Hornes, M., Friters, A., Pot, J., Paleman, J., Kuiper, M. and Zabeau, M., 1995. AFLP: a new technique for DNA fingerprinting. *Nucleic Acids Res.* 23, 4407–4414.

Watanabe, S., Hideshima, R., Xia, Z., Tsubokura, Y., Sato, S., Nakamoto, Y., Yamanaka, N., Takahashi, R., Ishimoto, M., Anai, T., Tabata, S., 2009. Map-based cloning of the gene associated with the soybean maturity locus E3. *Genetics.* 182, 1251–1262.

Welsh, J., McClelland, M. 1990. Fingerprinting genomes using PCR with arbitrary primers. *Nucleic Acids Res.* 18, 7213–7218.

Willcox, M.C., Khairallah, M.M., Bergvinson, D., Crossa, J., Deutsch, J.A., Edmeades, G.O., Gonzalez-de-Leon, D., Jiang, C., Jewell, D.C., Mihm, J.A. and Williams, W.P., 2002. Selection for resistance to Southwestern corn borer using marker-assisted and conventional backcrossing. *Crop Sci.* 42, 1516–1528.

Williams, J.G., Kubelik, A.R., Livak, K.J., Rafalski, J.A. and Tingey, S.V., 1990. DNA polymorphisms amplified by arbitrary primers are useful as genetic markers. *Nucleic Acids Res.* 18, 6531–6535.

Williams, K.J., 2003. The molecular genetics of disease resistance in barley. *Australian J. Agric. Res.* 54, 1065–1079.

Williamson, V.M., Ho, J.Y., Wu, F.F., Miller, N., Kaloshian, I., 1994. A PCR-based marker tightly linked to the nematode resistance gene, Mi, in tomato. *Theor. Appl. Genet.* 87, 757–763.

Winter, P., Kahl, G., 1995. Molecular marker technologies for plant improvement. *World J. Microbiol. Biotechnol.* 11, 438–448.

Wu, J., 2017. Construction and application of identification and analysis process of full-length LTR-retrotransposons (D). *Anhui Agric. Univ,* 56.

Xiao, J., Li, J., Yuan, L., Tanksley, S.D., 1995. Dominance is the major genetic basis of heterosis in rice as revealed by QTL analysis using molecular markers. *Genetics.* 140, 745–54.

Xiao, J., Li, J., Grandillo, S., Ahn, S.N., Yuan, L., Tanksley, S.D., McCouch, S.R., 1998. Identification of trait-improving quantitative trait loci alleles from a wild rice relative, *Oryza rufipogon*. *Genetics* 150, 899–909.

Xiao, N., Wang, H., Yao, W., Zhang, M., Ming, R., Zhang, J., 2020. Development and evaluation of SSR markers based on large scale full-length transcriptome sequencing in sugarcane. *Trop. Plant Biol.* 13, 343–352.

Xu, Y. and Crouch, J.H., 2008. Marker-assisted selection in plant breeding: From publications to practice. *Crop Sci.* 48, 391–407.

Yan, Z.Z., Yan, R., Fan, W., Kai, L., Zhang, D.Y., Qi, Y., Zhang, Y.F., Zhao, Y.F., Zhang, J.Y., 2017. Cross-species transferability of EST-SSR markers developed from the transcriptome of Melilotus and their application to population genetics research. Sci. Rep. 2017;7. doi: 10.1038/s41598-017-18049-8.

Yang, G.P., Saghai Maroof, M.A., Xu, C.G., Zhang, Q., Biyashev, R.M., 1994. Comparative analysis of microsatellite DNA polymorphism in landraces and cultivars of rice. *Mol. Gen. Genet.* 245, 187–194.

Young, D.N., 1999. A cautiously optimistic vision for marker-assisted breeding. *Mol. Breed.* 5, 505–510.

Yuan, Y., SanMiguel, P.J. and Bennetzen, J.L., 2003. High-Cot sequence analysis of the maize genome. *Plant J.* 34, 249–255.

Zhang, Q., Gao, Y.J., Maroof, M.S., Yang, S.H. and Li, J.X., 1995. Molecular divergence and hybrid performance in rice. *Mol. Breed.* 1, 133–142.

Zhao, X., Kochert, G., 1992. Characterisation and genetic mapping of short, highly repeated interspersed DNA sequence from rice (*Oryza sativa* L). *Mol. Gen. Genet.* 2311, 353–359.

Zhou, R., Zhu, Z., Kong, X., Huo, N., Tian, Q., Li, P., Jin, C., Dong, Y. and Jia, J., 2005. Development of wheat near-isogenic lines for powdery mildew resistance. *Theor. Appl. Genet.* 110, 640–648.

Zietkiewicz, E., Rafalski, A., Labuda, D., 1994. Genome fingerprinting by simple sequence repeat (SSR)-anchored polymerase chain reaction amplification. *Genomics.* 20, 176–183.

7 Cryptic Genetic Variation in Evolution and Crop Improvement

Gurbachan S. Miglani,[1] *Wajhat Un Nisa,*[2] *and Surinder K. Sandhu*[2]
[1]School of Agricultural Biotechnology, Punjab Agricultural University, Ludhiana, Punjab, India
[2]Department of Plant Breeding and Genetics, Punjab Agricultural University, Ludhiana, Punjab, India
Corresponding Email: miglanigs1945@gmail.com

7.1 INTRODUCTION

Cryptic genetic variation (CGV) is a kind of genetic variation which doesn't have any profound influence on the standard phenotype observed in the population; rather it can modify the phenotype which appears after environmental disturbances or the introduction of new alleles (Gibson et al., 2004). Even in this "genome age" many things about genes are still shrouded in mystery. This is especially true for "cryptic mutations" because the genes for these mutations are hidden and have unexpected effects on traits.

CGV, accidentally incurred by genetic or environmental interference, is believed to change the penetrance of various expressions. The transition to a new environment is related to the phenotypic diversification of many taxa. It is believed that when a single ancestral lineage faces new pressures, spectacular adaptive changes happen. Examples of Charles Darwin's finches, Hawaiian silver sword plant (*Argyroxiphium sandwicense* subspecies *macrocephalum*) and East African Rift Valley Cichlid have been documented (Petrin et al., 2005). The transformation of the environment is also a characteristic of human disturbance: the introduction of species is usually the result of human transportation, and the destruction of habitat leads to changes in species interactions and the way in which resources are obtained. In a new or stressful environment, whenever encountered by a population, the additional genetic variation (phenotypic variation at the population level caused by the genetic variation) actually increases. More precisely, we can say that under typical environments, the influence of individual genetic variation on traits will increase in nominal terms (Hoffman and Merila, 1999). Moreover, understanding on what basis the populations adjust to ever rapidly changing conditions paves the way to suggest that the prevailing environment itself affects the phenotypic expression of genetic variation, which undoubtedly

 DOI: 10.1201/9781003392897-7

provides the basic and ground material for adjusting in changing environment (Gibson and Dworkin, 2004).

Observations on phenotypic differences between populations under common environmental conditions can reveal that these differences have potential genetic basis. However, exposure to new and opposite environments may trigger the release of CGV that would otherwise be unobservable (Schlichting, 2008). The foremost issue in plant breeding is the kind of genetic variation that is to be considered for improving the phenotype. Although all forms of genomic changes can lead to yield variation and heterosis, the part of epigenetic variation in crop improvement is widespread and varied and little is known about it (Springer and Schmitz, 2017).

7.2 HISTORICAL SIGNIFICANCE

CGV can fuel evolutionary alternate. This was first confirmed via C.H. Waddington's classic experiment which dates back to 1942. He used ether vapor in the emerging embryos of *Drosophila* to set off a severe reply (Waddington, 1942). In answer, a share of embryos developed a change in their phenotype, a second thorax. He made selections in *Drosophila* with bithorax character for over 20 years and eventually resulted in fixation of the trait. Similarly, Waddington (1953) exposed *Drosophila* pupae to heat shock, resulting in the development of a character known as *crossveinless* (lacking a characteristic crossvein in adults). Thus, both phenomena laid the initial foundation of the notion referred to as genetic assimilation (Waddington, 1953) that environmentally triggered modifications ought to grow to be fixed in a population. In 1940, Waddington coined the term canalization, which refers to the property of various developmental pathways responsible for producing the phenotypes regardless of any genetic and environmental change which could otherwise hamper the normal development of organism (Hall, 1992).

Genetic variability inside the genotype is built up during the course of time, without that variation being phenotypically expressed. And this concealed variability can be drawn into the attention and exposed to selection via genetic assimilation. However, this principle became controversial after Theodosius Dobzhansky and Ernst Mayr sought that Waddington supported Lamarckian inheritance (Pigliucci et al., 2006). They did not accept the view that the inheritance of acquired characteristics had occurred and proclaimed that Waddington had performed selection of the variants that were already present in the population under study (Gilbert, 2013).

The different phenotypes are phenocopies of phenotypes produced through mutations. This statement was used as major proof by Waddington to argue that assimilated phenotypes have genetic foundation rather than any environmental cause. The movement of time when environmental agents lead to assimilation and frequently coincides with the time of action of the mutant gene that is responsible for the expression of corresponding phenotype, thereby providing an indication which interprets that both phenotypes were formed by activation of comparable developmental processes (Hadorn, 1961). Genetically assimilated phenotypes result from a polygenic basis as the genes are located on numerous chromosomes to which Darwin agreed later (Waddington, 1957). Phenocopies and genetic assimilation, in conjunction with canalization, inherit a massive concealed genetic variability that may be aroused both via a mutation or selection when exposed to an environmental provocation, which means that genes are present, however, beneath the brink required for their activation. The environmental inducement modifies the brink of activation of these genes and therefore permit their expression under a sudden change.

Copious proof of CGV in plants has come from within-population variation pertaining to phenotype whenever any novel environments is imposed (Helgadottir and Snaydon, 1986; Schlichting, 2008; Clausen et al., 1940). Classic work of Cooper (1954) specified that whenever manmade selection was forced on unseen dissimilarities for the time of flowering which were exposed due to any kind environmental change, it can probably be the cause of rapid genetic divergence of populations.

7.3 HOW IS CGV STORED?

There has always been a dilemma about how CGV is stored. Hermisson and Wagner (2004) precisely pointed out that hidden mutations do not require canalization as a prerequisite, as any genetic system with epistatic relationships between loci can store mutations (Schlichting, 2008). Moreover, higher the quantity of genes contributing to a characteristic, the more will be the alternative phenotypes of that trait (Sultan, 1987; Goldstein and Holsinger, 1992). In the following decades, the idea of hiding mutations in this way has been observed in various situations (Rice, 2000; True and Haag, 2001). Every living organism consists of greater number of genes which are engaged in complex ways and responsible for different cascades of pathways, but still there is no question to the fact that mathematical evolutionary notion is subjugated with a gene-by-gene standpoint which assumes that irrespective of the genetic background every gene has the same effect. Any sort of interaction between the genes (epistasis) tends to occupy a position in different tendencies, viz., progression of recombination, canalization, and reproductive isolation, but does not fall in the standard accounts of phenotypic adaptation (Thomson et al., 2013). However, the variety of epistatic communications, which include canalization, capacitance, and genetic redundancy, are responsible for upkeeping a specific trait even though fair amount of allelic variation is present (Schlichting, 2008).

As a consequence of the current polyploidization, record genes in tetraploid and hexaploid wheat species occur in multiple functional copies, mentioned as homeologs. These duplicated genes shield the rapid natural modifications going on in the giant and dynamic wheat genomes (Dubcovsky, 2007). These duplicate genes alleviate the rapid natural changes generated by the great genomes of wheat. This variation remains concealed to nature and human choice, since changes in loss of functions in a single wheat homeolog are often disguised by joblessness of other homologs. This becomes an advantage in the way that redundancy confers tolerance to mutations, thereby creating mutant populations (Tsai et al., 2013). This property was exploited in wheat to sequence the mutants for speedy identification of mutations in the dissimilar copies of the wheat genes. These mutations can be further united to expose previously concealed variation, thereby conferring lack of spring types related to *vrn-2* which has a low adaptive worth, as in maximum accessions of cultivated diploid wheat *Triticum monococcum* (Yan et al., 2004) and many of the diploid barley varieties (Karsai et al., 2005; Szucs et al., 2007) have a spring growth habit linked with recessive *vrn-2* mutations. Based on these indications, it was hypothesized that these recessive mutations at the *VRN2* locus in wheat had remained concealed from selection for about 8000 years by the virtue of idleness of multiple homeologs (Krasileva et al., 2016). However, buffering action of genetic redundancy has already been explored (Wagner, 1999; Hartman et al., 2001; Frank, 2003; Gu et al., 2003; Wagner, 2005a; Ihmels et al., 2007).

Differential regulation of genes with respect to different environments and at developmental stages has been documented by various researchers (Chen et al., 2002; Hui et al., 2003; Makova and Li, 2003; Michaut et al., 2003; Rabbani et al., 2003; Scutt et al., 2003; Seki et al., 2003; Kluger et al., 2004; Wellmer et al., 2004; Schmidt and Baldwin, 2006; Brisson et al., 2007) which accounts for gene-environment interaction and thrusts on the fact that innate variation will not be subjected to selection (Schlichting, 2008). Whenever we speak of local adaptation, it is the conflicting natural selection that promotes it; however, at the same time what leads to this local adaptation in light of genetics is not clearly understood (Jill et al., 2012).

Antagonistic pleiotropy leads to local adaptation where only certain alleles are preferred, that too in selected habitats, and the polymorphism is sustained by selection. On the other hand, conditional neutrality favoritisms few alleles in one condition but remains neutral in other environments because of segregating neutrally in the conditions which are quite nonadaptive. However, they are available for selection only with the right conditions as prerequisite. This happens because if the alleles show uniform penetrance under all environmental conditions, it leads to ever-present exposure to natural selection, and most probably the alleles which couldn't fit in a particular environment will be lost pertaining to negative selection (Paaby and Rockman, 2014). Hermisson and Wagner (2004) proposed an equation: $\Delta G \approx \alpha\Delta(0+\rho\alpha s\langle v0\rangle Ne$, where Δ_G acts as the coefficient of concealed variation, and is the measure of the quantity of recently released genetic variation which is comparative to the genetic variation that was phenotypically expressed under the longstanding conditions or which can also imply the relative change in evolvability. $\alpha\Delta_0$ is the released variation for an even dispersal and N_e is the population size. The equation explains that Δ_G increases due to the expression of conditionally neutral loci which are actually the subset of alleles at a particular locus that have not been subjected to selection under the preceding environmental conditions. One scenario holds the thought that in large populations these neutral loci can pertain to enormous hidden variations, but what if the environment had existed earlier and had become "rare" lately? There would be the past memory which implies that relaxation to the balance between the mutation and selection in the past environment was not complete at the time change occurred in environment. As the variation around these types of loci was abridged as a result of selection in earlier generations, which holds good for the fact that conditionally neutral loci are not yet in balance of mutation and drift. Qualitatively, it was concluded that the frequency of rare environments needs to be less for more contribution of conditionally neutral loci to Δ_G.

7.4 WHAT CAN BE THE POSSIBLE CAUSE?

It is an avid question of how the trait values get changed by altering the environment with genetic background that rarely produces any effect. In terms of quantitative genetics, gene by environment (G × E) and gene by gene (G × G) or epistatic communications need to be studied (Falconer, 1996). Usually what happens is that the trait value is changed by changing the niche, or sometimes no effect is conceived which definitely qualifies for G × E but does not necessarily result in expression of CGV (Gibson et al., 2004). Similar is the case for epistasis which doesn't have any contribution to standing genetic variation. But one class of G × E interaction that is thought to be responsible is decanalization. These mutations tend to have negligible input to the reasonable array of phenotypes when observed in a population. However, these changes play a significant part in modification of a particular phenotype which ensues after a certain amount of perturbation in environment has occurred or because of new alleles (Gibson and Dworkin, 2004). Therefore, the possible genetic mechanisms can change because of the changes in genetic pathways or altered threshold required for expression of a particular phenotype.

7.5 OBSERVATIONS ON CRYPTIC VARIATION IN PLANTS

Usually, the question of how discrete states of morphological traits evolve stays unanswered. One condition is that single-gene modifications inspire the evolution of recent discrete character states and that evolution depends on the incidence of recent single-gene mutations. Another cause can be that modifications in certain number of genes are prerequisite for raising a character or population overhead a certain edge, important for expressing the alternate or novel state of character. To apprehend the assertion, cryptic genetic variants were recognized in teosinte for characters which can be invariant in teosinte (Figure 7.1), thereby emphasizing that such CGV can make a contribution to the evolution of novelty while restructured to surpass the edge essential for phenotypic expression or by stabilizing the changing effects of essential mutations (Lauter and Doebley, 2002). Polyploidy is a principal spectacle in plant evolution. Many crop species have experienced recent genome duplication procedures as well; therefore, mutant seed stocks and sequence data serve as public collection for speedy identity of mutations within side the unique copies of the wheat genes, which may be united to find formerly hidden variation (Krasileva et al., 2016).

All existing life forms are the result of three influences—mutation as cause of heritable variation, environmental aspects influencing the selection of those mutations, and time during which the genotype and environment continuously interrelate—after which an evolutionary alteration is observed (Nuvak and Bruner, 1992). Induced mutations by different technologies and induced stresses have been used to generate the hidden and more specifically novel versions of variation in different plant species (Table 7.1) so that the selection could act upon these variations and further can be utilized for breeding purposes. Combining of the traditional methods with the novel skills of high-throughput sequencing has resulted in the development of effective reverse genetic strategies in polyploidy crops. Sequencing genomes or exomes of huge mutant populations can produce full-size databases of mutations for most genes.

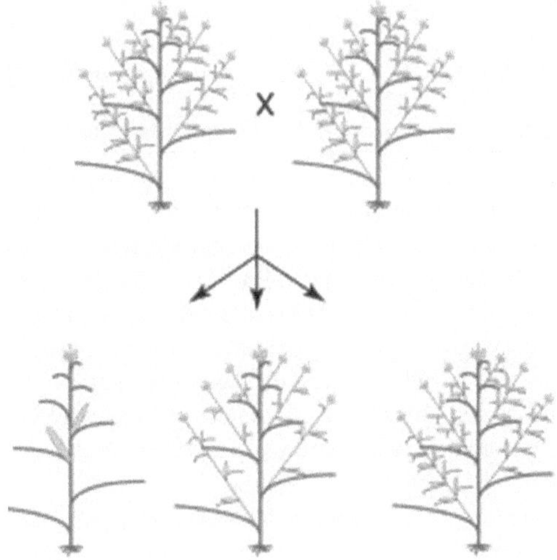

FIGURE 7.1 Two lines of teosinte that are phenotypically very similar were crossed, and their progeny were then test-crossed to an inbred line of maize. Each cross gave rise to a range of phenotypes ranging from maize-like (bottom left) to teosinte-like (bottom right), indicating the expression of cryptic genetic variation hidden in the parents.

Source: Reproduced, with permission, from Gibson and Reed, 2008.

Thereafter, this mutant collection, combined with genome editing, can be cast off in polyploid species to mix mutations in all copies of a gene (homeologs) and to reveal phenotypic versions that became formerly concealed with the aid of using functional redundancy among homeologs in case of wheat (Uauy et al., 2017). CGV has additionally been pronounced for its function in the paternalistic regulation of seed development from *Arabidopsis* which carries a pool of hidden variation (Pires et al., 2016). There is always a struggle between motherly and fatherly alleles over resource apportionment during seed development, where a motherly buffer acts on the fatherly effects and was revealed by a new technique for mapping of parent-of-origin influences.

Epigenomics is a fast-rising area of research which is increasing our capability to provide an explanation for detected phenotypic difference through documentation of various cellular products comprising RNAs, protein-DNA interactions, chromatin modifications, and chromatin accessibility (Lane et al., 2014). DNA modification is the most deeply examined epigenetic change responsible for various phenotypes and cellular mechanisms. Silencing of various genes in the genome occurs due to DNA methylation, and after identification of such methylation patterns, phenotypic variation can be generated as a consequence of expression of cryptic alleles. As established in *Arabidopsis thaliana* for epigenetic recombinant inbred lines (epiRILs), substantial concealed variations exist in the genome which can be exposed by changing various DNA methylation designs on genome-wide basis (Johannes et al., 2009;

TABLE 7.1
Different Strategies Followed to Decipher Changes in Phenotype

Plant	Salient observation with various techniques used	Reference
Antirrhinum majus L. (Snapdragon)	Flower asymmetry was used to determine the extent to which the underlying regulatory genes varied in expression among closely related species for genes *CYCLOIDEA* (CYC) and *RADIALIS* (RAD); mutant alleles were used to demonstrate that the species lie on a plateau in gene expression-morphology space so that the variation has no detectable phenotypic effect. However, phenotypic differences can be revealed by shifting genotypes off the plateau through genetic crosses.	Rosas et al. (2010)
Pinus sylvestris L. (Scots pine)	During the flood treatment, saplings from drier regions were exposed to water-logging conditions, and under threshold they expressed their tolerance against flooding which was way more than saplings originating from western sites where precipitation is substantially greater, signifying the impact of natural selection.	Donnely et al. (2018)
Triticum aestivum L. (Bread wheat)	Wheat has the fair ability to tolerate high densities of induced mutations which was exploited to sequence and catalogue around 10 million mutations in the protein-coding regions of 2735 mutant lines of tetraploid and hexaploid wheat.	Krasileva et al. (2016)
	Independent variation in methylation, genotype, and transposon copy number was found using sodium bisulfite treatment and targeted gene enrichment so far unexploited; sources of variation have had a significant impact on the wheat genome, and ancestral methylation states become preferentially "hard coded" as single-nucleotide polymorphisms via 5-methylcytosine deamination. These mechanisms also drive local adaption, impacting important traits such as heading date and salt tolerance.	Gardiner et al. (2018)
Zea mays L. (Maize)	Ozone decreasing the photosynthetic gain was detected via Face technology which revealed that selection under elevated ozone is needed to identify deleterious alleles in the world's largest commodity crop.	Choquette et al. (2019)

Reinders et al., 2009; Roux et al., 2011). But the following problem is faced for major cereals: creating similar epigenomic populations to incorporate recognized epimutations into existing breeding programs remains unknown (Lexing et al., 2015). This is largely due to the richness of transposon sequences in crop genomes, especially in rice and maize (Hu et al., 2014; Li et al., 2014).

7.6 EVIDENCE SUPPORTING CGV

CGV is certainly unique and mysterious, as the variation pertains to effects based on conditions and has different forms which consist of gene-by-gene (G × G) interactions, together with dominance and epistasis and gene-by-environment (G × E) interactions (Paaby et al., 2014). The most exclusive characteristic of CGV is the state which tends to induce any kind of allelic effects that were not present in the population before. However, the rarity seizes the opportunity for any selection to act on it, thus allowing it to accumulate. CGV encompasses both molecular genetic aspects, which pertains to the fact that these cryptic variants are basically polymorphic loci which don't have any significant no effect on phenotype unless perturbations due to rare environmental conditions have occurred. The perspective of quantitative genetics deals with this concept: whenever a population faces unusual or rare environmental conditions there is an upsurge in additive genetic variance (additive: V_A), which is only part of variation that responds to selection or, we can say, is heritable.

The experimental investigation of cryptic variation can be pursued via various mapping approaches like quantitative trait loci (QTL) and association studies and other strategies like mutagenesis and introgressions (Gibson et al., 2004). Different studies carried out on determining bristle and crossvein in *D. melanogaster* employed the screening of conditional phenotypes. If we consider the example of penetrance for *scute* and *crossveinless* mutations which was temperature dependent (Waddington, 1953; Rendel, 1959; Dykhuizen and Hartl, 1980), experiments have proven that certain regions/loci dealing with metabolic pathways are expressed under nutrient-limiting conditions. These studies supported the fact that such kind of allelic variation can be maintained and upon some stress will contribute toward survival in a respective population. In the domesticated species, selection of novel phenotypes along with their stabilization is due to cryptic variation as stated by Lauter and Doebley (2002). *Zea mays* ssp. *mexicana* and *parviglumis*, as two subspecies of teosinte, have difference in inflorescence architecture (Figure 7.1), but when both were crossed to the modern cultivated maize, difference in segregating generation up to a large effect was observed, which could possibly be a fine strategy for emphasizing the expression of CGV. Then, phenotypic differences can be contrasted, followed by mapping the regions governing cryptic phenotypes especially using QTL-mapping procedures. A study has already been done for *Ubx* mutation (Gibson et al., 1999) in which a locus having major effect was considered the cause for around 70% of variation in enlargement of the halteres.

7.7 CRYPTIC VARIATION IN ADAPTATION AND EVOLUTION

7.7.1 Adaptive Significance of Cryptic Genetic Variation

Sixteen populations of *D. melanogaster* were collected from different sites in Mexico by Sarah Bedichek Pipkin of Howard University, Washington, D.C. One of us (G.S.M. as a Ph.D. scholar [1976–1980], under the guidance of Sarah Pipkin and Franklin R. Ampy) used these natural populations for the detection and analysis of cryptic variation for alcohol dehydrogenase (ADH). For this purpose, isochromosomal lines were constructed using genetic markers so that these lines were homozygous for chromosome 2, which carries the *alcohol dehydrogenase* (*Adh*) locus (Miglani, 1980; Miglani, 2018). These isochromosomal lines were homozygous either for Adh^{I} or Adh^{II} allele. Crude extract was instantly used for measuring ADH activity spectrophotometrically. Statistical analysis exposed cryptic variation in the activity of alcohol dehydrogenase (ADHI and ADHII) in 12 Adh^{I}/Adh^{I} and 40 Adh^{II}/Adh^{II} isochromosomal lines.

The presence of additional genetic variation in ADH activity was demonstrated in the 52 isochromosomal lines following treatment with three denaturants, namely, guanidine hydrochloride (GuHCl), urea, and heat (Miglani and Ampy, 1981a). A substantial upsurge in the coefficient of variation (CV) values observed subsequent to treatment with these denaturants provided evidence of cryptic genetic variation in isochromosomal lines originally showing statistically the same native ADH activity. Mean alcohol dehydrogenase (ADH) activity levels within ADHI and ADHII isochromosomal lines were significantly different among the Mexican populations of *Drosophila melanogaster* used. Also, the comparative amount of genetic variation differed among and within the ADHI and ADHII strains studied. The quantitative landscape of the differences in ADH activity reinforced the perception of the additive gene action of the *Adh* locus (Miglani and Ampy, 1981c).

In total, 35 cryptic variant lines that revealed statistically significant differences in mean ADH activity levels following treatment with GuHCl, urea, or heat were used to study regulatory properties of cryptic variants and decipher the mechanisms implicated in adjustment of alcohol metabolism in *Drosophila* (Miglani and Ampy, 1981b). For this purpose, studies were performed at three developmental stages (i.e., late third-instar larval, preemergence pupal, and adult) of *Drosophila* grown in different environments (i.e., at 18°C and 28°C). Spectrophotometric evaluations of native and residual ADH activities subsequent to treatment with a denaturant were performed. The three denaturants affected various developmental stages differently at the two temperatures. Differences in response of cryptic variants after treatment with the denaturants during different developmental stages suggested that this variation might have an adaptive significance by compensating in metabolism in response to changes in the environmental conditions.

7.7.2 Evolutionary Role of Cryptic Genetic Variation

The accumulation of CGV occurs under weakened selection, which can serve as an ultimate source of evolutionary innovations. However, this kind of selection could have been probably caused by buildup of strongly deleterious or lethal alleles which can swamp the effect of possibly adaptive alleles when they are exposed. Revealing and assimilating CGV might be an influential force in evolution as adaptive combinations of mutations can appear more often if the weakened selection acts on such variation. Therefore, these combinations of mutations may be relevant to the evolution of new adaptations. A model was proposed by Masel (2006) who assumed that deleterious mutations might be raw material for particular environment. In a population, Hayden and coworkers (2011) studied group I RNA enzymes (ribozymes)

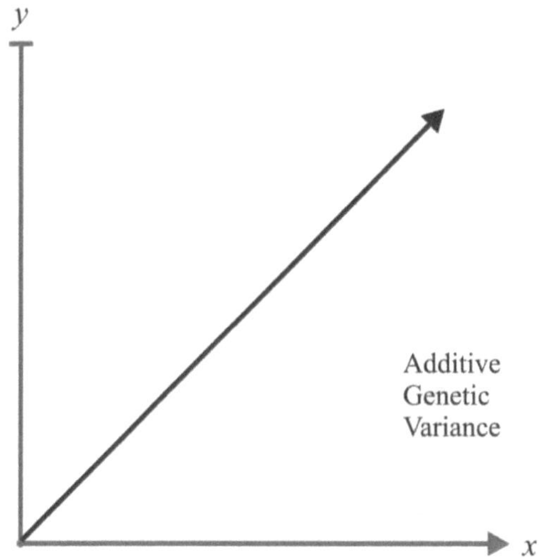

FIGURE 7.2 Increase in additive genetic variance usually occurs for the traits (polygenic) after the change in environment or any kind of interaction with genetic element or environment is generic in nature. This predicted increase can be considerable in particular large populations and only if there are conditionally neutral alleles at the loci underlying the trait.

which resulted from the *Azoarcus* pre-tRNA. In the existence of high concentrations of denaturants (7.5M urea) or at high temperatures (80°C), this ribozyme is able to retain higher activity which signifies that once the cryptic variation has accumulated, it can rapidly adapt to any novel substrate than a population without cryptic variation.

Undoubtedly, various studies have confirmed that hidden variation has evolutionary role (Hansen et al., 2000; Rutherford and Lindquist, 1998). Chaperone Hsp90 which depicts mutational buffering by acting as a capacitor is considered a classic example of the above statement. The mutations of any kind were responsible for revealing the hidden variation which Hermisson and Wagner (2004) clearly demonstrated in an analytical framework, whereby it was concluded that this phenomenon has generic base and can be due to differential causes which do not need specific molecular mechanisms (Figure 7.2).

7.8 INTEGRATING CGV INTO OUR PLANT BREEDING PROGRAMS

In crop plants, different threshold is required to express the new state of character as not all the alleles are universally penetrant; some show dominance while others reveal epistasis or G × E relationship. Living systems are crafted in a robust way which makes the trait persistent under perturbations through canalization; rather we can say that accumulation of variation at genetic level occurs without having influence on phenotype. Currently, world faces the siege of global warming and consequent climate changes crippling our agriculture. Prolonged droughts due to rise in temperature, excessive flooding, increase in the pest and disease incidence, and deterioration of grain quality are the major crises. Genetic vulnerability affects standing genetic variation due to long-term directional selection. Therefore, to adapt to novel environments, CGV can play a significant role and contribute to adaptive evolution. This in turn will open up an avenue for standing genetic variation, either with beneficial or deleterious effects to respond to selection which will lead to two forms of phenotypes in the population: one part will be preadapted to new conditions and another will carry the burden of deleterious alleles known as maladaptive phenotypes (Paaby and Rockman, 2014).

7.8.1 Evolutionary Plant Breeding

As it is believed that natural selection promotes the fittest, Suneson (1956) put forward a concept of evolutionary plant breeding. The system implied use of evolutionary populations developed after crossing F_2 in all possible combinations and subjecting them to decentralized selection, i.e., distributing to farmers of target environment where nature can play its game and release the variations. Later on, this system was discarded due to heterogeneity and massive size. The crucial question was about existence of any kind of novel variations which could have implication on novel traits. This was apprehended through a study in late 1990s where the scald resistance was not found in the original parental population of barley but after 45 rounds of selection at the infested location, resistant alleles were detected (Maroof Saghai et al., 1983). Yet, not many studies have been conducted on crops with the perspective of evolutionary plant breeding.

7.8.2 Feralization Promoting Cryptic Variation

Feralization term is defined as the phenomenon of conversion of cultivated species back to wild form. The food we have comes from the lap of domesticated species that are a consequence of rounds of nonnatural selection performed on wild relatives during the course of evolution under the process known as domestication. Similarly, in recent years, a process called dedomestication or feralization was confirmed in various crops like wheat, rice, and barley along with livestock (Wu et al., 2021). In a more simplified way, it is not a complete conversion, but rather acquisition of phenotype similar to the wild under loosened artificial selection or by omitting the intensive farming practices (Gering et al., 2019). Various pathways are held responsible for this process, but major question is how it unlocks cryptic variation. Usually our concept befalls the fact that domestication is responsible for a number of genetic bottlenecks which dramatically reduces genetic variation (Doebley et al., 2006) and has the least potential for evolving under changing conditions due to genetic homogeneity. But feralization has turned this fact upside down by confirming that cultivated crops can undergo adaptive evolution. A breakthrough study was carried out in weedy rice *Oryza sativa* f. *spontanea* which is a con-specific weed of rice and the best example of dedomestication. It lacks primary dormancy like cultivated rice, but to a great surprise, it gained a novel strategy to sense the optimum thermal changes or cues required for germination at different regimes, thereby promoting rapid evolution (Xia et al., 2011). This clearly established that despite low genetic variation, cryptic beneficial alleles were released upon selection pressure in different environments which preexisted in cultivated progenitor populations. The feral plants live in the wild but originate from domesticated individuals. Feral plants thus offer a proofreading concept for studying convergent evolution along with the above-discussed concept of conditional neutrality.

These dedomesticates can leverage the needs of modern plant breeding. The novel variations can be exploited to develop or improve our elite varieties. Considering the example of weedy rice, dedomestication not only develops response to temperature but also serves as a source of resistance to various biotic stress like blast and abiotic stresses like cold, drought, and salinity along with herbicide tolerance which is quite crucial (Moretto et al., 2016; Wang et al., 2018). Hence, it can be envisaged that rapid adaption has diversified their genetic pools by imparting novel traits which can be a boon during the siege of climate change. All that matters in this case is the identification of these lately expressed valuable variations which completely depends on advances in the genomics and skill of the breeder.

7.8.3 Plasticity: An Evolutionary Advantage

Land plants are sessile in nature, unlike animals. In order to counter the negative effects of environment, they tend to change at morphological level, which is called plasticity. A generalized definition of plasticity is the capability of a given genotype to produce different phenotypes in varied environments. During their life cycle, plants encompass plasticity at developmental as well as physiological level by changing the morphology, structure, and number of organs formed via multiple meristematic growing points that can retort independently to environmental indications (Gilroy and Trewavas, 2001; Trewavas, 2015) also by fine-tuning to changing conditions at the cell and tissue level. Now if we drive down the alley of difference between evolution and plasticity, it is said evolution occurs between the generations while plasticity is characteristic a feature attained within the generation by keeping up with rapid and novel environmental change spatially and temporally. This plasticity can be adaptive, though; this might be because of some hidden perturbations as already discussed above, making it ubiquitous. Practically, its usage will be to predict evolutionary response under different anthropogenic environmental conditions.

Plasticity of developmental programs in plants is extensively studied but the underpinning mechanisms are not clear. Plants have the explicit property of adjusting the architecture of their root system which is the driver of energy in rejoinder to different edaphic indicators such as mineral nutrient concentrations and water gradients (Kellermeier et al., 2014). Shahzad et al. (2020) reported that cryptic variation maintains a balance during the perturbation of low potassium by supporting the lateral root developed in *Arabidopsis*. This property is envisaged by CLASSY chromatin remodeling factor (CLSY1) as a constituent of RNA-directed DNA methylation (RdDM) by silencing *IAA27* when auxin-driven path is inhibited under low K. Thus, unraveling the response of different parts can help to select a phenotype under minimal inputs for sustainable growth.

7.8.4 Crop Wild Relatives

The quest is still going on for digging the genes that could impart resistance against various climatic and manmade stresses. In the current climate volatility, widening the genetic base is the need of the hour. Tanksley and McCouch (1997) state that crop wild relatives (CWRs) carry the treasure of alien alleles which is useful in many aspects and CGV. This introgressed variation can be expressed in cultivated gene pool only through a systematic breeding scheme. The reduction in genetic gains occurs when the breeding value stops responding to selection. Usually, it becomes a bit easier to use diversity from cultivated rather than exotic plants. However, we can't forego the fact that during the vagaries of environment wild relatives have evolved through selection pressures with distinct degrees. But for mainstream breeding, this cryptic variation and linkage drag may pose a paucity for commercial productions with respect to maladaptiveness. Therefore, this variation needs to be introgressed into our elite lines. Advanced Backcross QTL (AB-QTL) (Tanksley and Nelson, 1996) analysis is an effective scheme to exploit wild species by concurrent detection and transfer of valuable QTLs from unadapted germplasm into the performing best breeding lines. Other prebreeding tools are chromosome segment substitution lines (CSSLs) which comprise complete library of introgression lines with chromosomal segments of typically a distant genotype in an adapted background and are treasured genetic resources for identifying new genomic regions and QTL hot spots. Stepped aligned inbred recombinant strain (STAIRS) approach can be used to conduct comparisons of the phenotypes of pairs or sets of lines in which only the genetic differences are located in short defined regions of one selected chromosome so that one can identify candidate genes. A strategized flowchart (Figure 7.3) explains how we can use the abovementioned sources of hidden variation in the breeding program.

7.9 CONCLUSIONS

In today's context, studying regular quantitative traits is a bit harder. The study of CGV will require novel experimental powers. Besides, most quantitative and evolutionary genetic theory adopts additive gene effects, and this paradigm is challenged by CGV. Whenever the plants hit environmental perturbation, the part of diversity because of CGV is incidentally preadapted to novel environments, which can aid the evolutionary adaptation of any population and it is quite intriguing that this genotypic diversity is a sign of widespread epistasis as well. Most individual mutations do not deliver a large fitness advantage alone; rather it can be done in amalgamation, which signifies the importance of epistasis. Therefore, power to discover such mutations cryptically is central, peculiarly for those genotypes where interaction of mutations forms the base high fitness. So, all the observations discussed here back theoretical work demonstrating that the unleashing of cryptic variation following any environmental disturbance is an asset of every existing genetic system, and understanding how it manages to influence physiology and evolution is at the initial stages of exploration. Moreover, cutting-edge genomic skills will help to accelerate the investigations of CGVs, by aligning the impact of genotype onto phenotype.

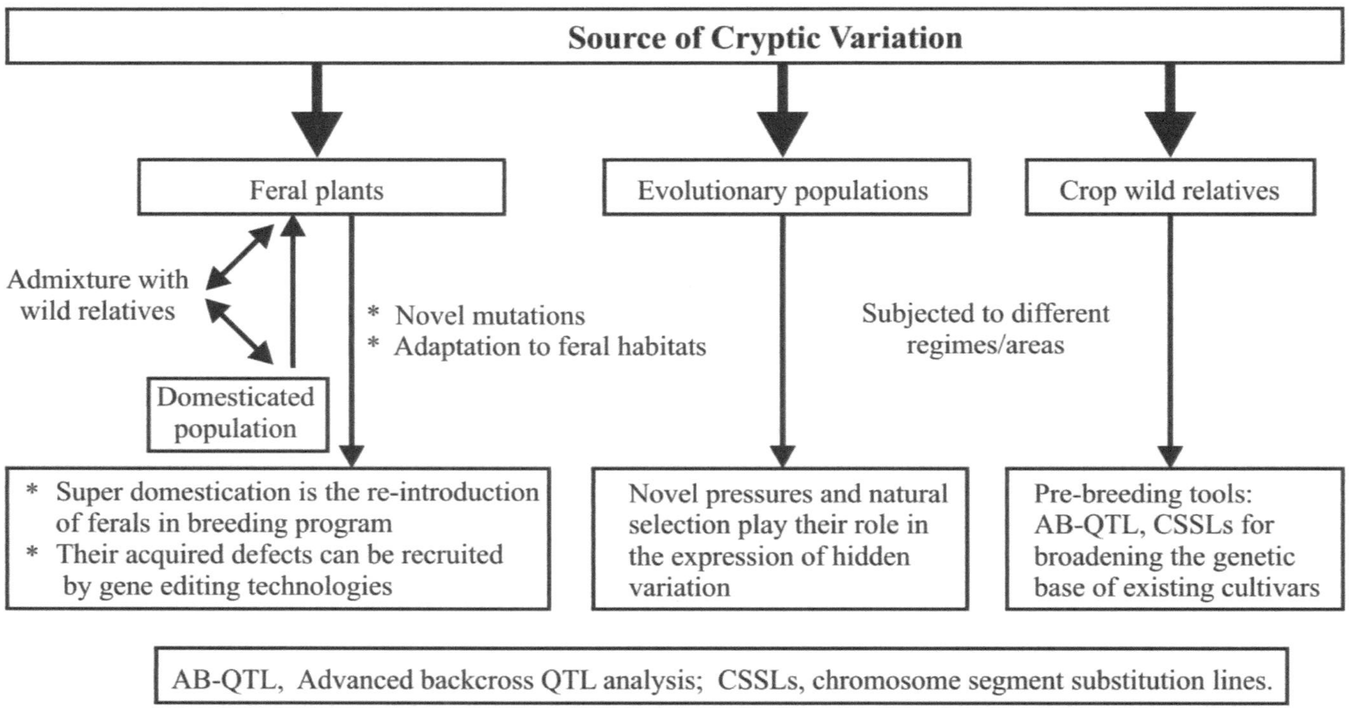

FIGURE 7.3 A strategy for using CGV in plant breeding. The simply represents an overview of exploiting this variation for commercial breeding by including crop wild relatives, evolutionary populations, and other sources which have undergone fair amount of selection pressure. Various modern technologies can be put to use to tag the variation.

REFERENCES

Baldwin, B., Sanderson, G., Michael, J., 1998. Age and rate of diversification of the Hawaiian silversword alliance (Compositae). *Proc. Natl. Acad. Sci U.S.A.* 95(16), 9402–9406.

Choquette, N.E., Ogut, F., Wertin, T.M.; Montes, C.M., Sorgini1, C.A., et al., 2019. Uncovering hidden genetic variation in photosynthesis of field-grown maize under ozone pollution. *Global Change Biol.* 25, 1–12.

Clausen, J., Keck, D.D., Hiesey, W.M., 1940. *Experimental studies on nature of species.* Washington, Carnegie Institute.

Cooper, J.P., 1954. Studies on growth and development in *Lolium.* IV. Genetic control of heading responses in local populations. *J. Ecol.* 42, 521–556.

Doebley, J., Gaut, B.S., Smith B.D., 2006. The molecular genetics of crop domestication. *Cell* 127, 1309–1321.

Donnelly, K., Cavers, S., Cottrell, J.E., Ennos, R.A., 2018. Cryptic genetic variation and adaptation to waterlogging in Caledonian Scots pine, *Pinus sylvestris* L. Ecol. *Evo.* 8, 8665–8675.

Dubcovsky, J., Dvorak, J., 2007. Genome plasticity a key factor in the success of polyploidy wheat under domestication. *Science* 316(5833), 1862–1866.

Dykhuizen, D., Hartl, D.L., 1980. Selective neutrality of 6 PGD allozymes in *E. coli* and the effects of genetic background. *Genetics* 96, 801–817.

Gardiner, L.J., Joynson, R., Omony, J., Pilcher, R.R., Olohan, L., et al., 2018. Hidden variation in polyploidy wheat drives local adaptation. *Genome Res.* 28(1), 319–1332.

Gering, E., Incorvaia, D., Henriksen, R., Conner, J., Getty, T., Wright, D., 2019. Getting back to nature: feralization in animals and plants. *Trends Ecol. Evol.* 34, 1137–1151.

Gibson, G., Dworkin, I., 2004. Uncovering cryptic genetic variation. *Nat. Rev. Genet.* 5, 681–690.

Gibson, G., Reed, L.K., 2008. Cryptic genetic variation. *Curr. Biol.* 18(21), R989–R990.

Gibson, G., Wemple, M., van Helden, S., 1999. Potential variance affecting homeotic *Ultrabithorax* and *Antennapedia* phenotypes in *Drosophila melanogaster. Genetics* 151, 1081–1091.

Gilbert, S., 2013. *A conceptual history of modern embryology.* New York: Springer USA, pp. 280.

Gilroy, S., A. Trewavas., 2001. Signal processing and transduction in plant cells: The end of the beginning? *Nat. Rev. Mol. Cell Biol.* 2, 307–314.

Goldstein, D.B., Holsinger, K.E., 1992. Maintenance of polygenic variation in spatially structured populations roles for local mating and genetic redundancy. *Evolution* 46, 412–429.

Gupta, S., Singh, B.B., 2009. Utilization of potential germplasm and species spectrum in improvement of pulse crops. In: Ali, M., Gupta S., Basu, P.S., Naimuddin, K. (eds). *Legumes for ecological sustainability.* Kanpur, India: Indian Institute of Pulse Research, pp. 332–341.

Hadorn, E., 1961. *Developmental physiology and lethal factors.* New York: John Wiley & Sons, pp. xviii + 355.

Hall, B.K., 1992. Waddington's legacy in development and evolution. *Am. Zool.* 32, 113–122.

Hansen, T.F., Carter, A.J., Chiu, C.H., 2000. Gene conversion may aid adaptive peak shifts. *J. Theor. Biol.* 207(4), 495–511.

Hayden, E., Ferrada, E., Wagner, A., 2011. Cryptic genetic variation promotes rapid evolutionary adaptation in an RNA enzyme. *Nature* 474, 92–95.

Helgadottir, A., Snaydon, R.W., 1986. Patterns of genetic variation among populations of *Poa pratensis* L. and *Agrostis capillaris* L. From Britain and Iceland. *J. Appl. Ecol.* 23, 703–719.

Hermisson, J., Wagner, G.P., 2004. The population genetic theory of hidden variation and genetic robustness. *Genetics* 168, 2271–2284.

Hoffman, A.A., Merila, J., 1999. Heritable variation and evolution under favourable and unfavourable conditions. *Trends Ecol. Evol.*14, 96–101.

Hu, L., Li, N., Xu, C., Zhong, S., Lin, X., et al., 2014. Mutation of a major CG methylase in rice causes genome-wide hypomethylation, dysregulated genome expression, and seedling lethality. *Proc. Natl. Acad. Sci. U.S.A.* 111, 10642–10647.

Ihmels, J., Collins, S.R., Schuldiner. M., Krogan, N.J., Weissman, J.S., 2007. Backup without redundancy: Genetic interactions reveal the cost of duplicate gene loss. *Mol. Sys. Biol.* 3, 86. Doi: 10.1038/msb4100127

Ji, L., Neumann, D.A., Schmitz, R.J., 2015. Crop epigenomics: Identifying, unlocking, and harnessing cryptic variation in crop genomes. *Mol. Plant* 8, 860–870.

Jill, T.A., Lee, C.R., Rushworth, C.A., 2012. Genetic trade-offs and conditional neutrality contribute to local adaptation. *Mol. Ecol.* 22(3), 699–708.

Johannes, F., Porcher, E., Teixeira, F.K., Saliba-Colombani, V., Simon, M., et al., 2009. Assessing the impact of transgenerational epigenetic variation on complex traits. *PLoS Genet.* 5, e1000530. doi:10.1371/journal.pgen.1000530

Kellermeier, F., Armengaud, P., Seditas, T.J., Danku, J., Salt, D.E., Amtmann, A., 2014. Analysis of the root system architecture of *Arabidopsis* provides a quantitative readout of crosstalk between nutritional signals. *Plant Cell* 26(4), 1480–1496.

Krasileva, V.K., Hans, A., Vasquez, G., Howella, T., Paul, B., Francine, P., et al., 2016. Uncovering hidden variation in polyploidy wheat. *Proc. Natl. Acad. Sci. U.S.A.* 114, 1–9.

Lane, A. K., Niederhuth, C.E., Ji, L., Schmitz, R.J., 2014. PENCODE: a plant encyclopedia of DNA elements. *Annu Rev. Genet.* 48, 49–70.

Lauter, N., Doebley, J., 2002. Genetic variation for phenotypically invariant traits detected in teosinte: implications for the evolution of novel forms. *Genetics* 160, 333–342.

Li, Q., Eichten, S.R., Hermanson, P.J., Zaunbrecher, V.M., Song, J., et al., 2014. Genetic perturbation of the maize methylome. *Plant Cell* 26, 4602–4616.

Maroof Saghai, M.A., Webster, R.K., Allard, R.W., 1983 Evolution of resistance to scald, powdery mildew, and net blotch in barley composite cross II populations. *Theor. Appl. Genet.* 66, 279–283.

Masel, J., 2004. Genetic assimilation can occur in the absence of selection for the assimilating phenotype, suggesting a role for the canalization heuristic. *J. Evo. Biol.* 17(5), 1106–1110.

Masel, J., 2006. Cryptic genetic variation is enriched for potential adaptations. *Genetics* 172(3), 1985–1991.

McGuigan, K., Sgro, C.M., 2009. Evolutionary consequences of cryptic genetic variation. *Trends Ecol. Evol.* 24, 305–311.

Merotto, A. Jr., Goulart, I.C., Nunes, A.L., Kalsing, A., Markus, C., et al., 2016. Evolutionary and social consequences of introgression of nontransgenic herbicide resistance from rice to weedy rice in Brazil. *Evol. Appl.* 9, 837–846.

Miglani, G.S., 1980. Regulatory properties of cryptic variants at the *alcohol dehydrogenase* (*Adh*) locus in *Drosophila melanogaster*. Ph.D. thesis. Howard University, Washington, D.C., p. 145.

Miglani, G.S., 2018. *Cryptic variation at Adh locus in Drosophila melanogaster.* Lambert Academic Publishing, Mauritius, p. 103.

Miglani, G.S., Ampy, F.R., 1981a. Changes in native alcohol dehydrogenase activity in *Drosophila* upon treatment with guanidine hydrochloride, urea and heat. *J. Biosci.* 3(3), 275–284.

Miglani, G.S., Ampy, F.R., 1981b. *Drosophila* alcohol dehydrogenase: Developmental studies on cryptic variant lines. *Biochem. Genet.* 19, 947–954.

Miglani, G.S., Ampy, F.R., 1981c. Statistical analysis of cryptic variation in *Drosophila* alcohol dehydrogenase. *Proc. Ind. Acad. Sci. (Anim. Sci.)* 90, 545–552.

Nuvak, F.J., Brunner, H., 1992. Plant breeding: Induced mutation technology for crop improvement. *IAEA Bulletin* 4, 25–33.

Olsen, S., 2004. *Evolution in Hawaii: A supplement to teaching about evolution and the nature of science.* The National Academies Press, Washington, D.C.

Paaby, A.B., Rockman, M.V., 2014. Cryptic genetic variation, evolution's hidden substrate. *Nat. Rev. Genet.* 15(2), 247–258.

Petren, K., Grant, P.R., Grant, B.R., Keller, L.F., 2005. Comparative landscape genetics and the adaptive radiation of Darwin's finches: the role of peripheral isolation. *Mol. Ecol.* 14(10), 2943–2957.

Pigliucci, M., 2006. Phenotypic plasticity and evolution by genetic assimilation. *J. Exptl. Biol.* 209(12), 2362–2367.

Pires, N.D., Bemer, M., Müller, L.M., Baroux, C., Spillane, C., et al., 2016. Quantitative genetics identifies cryptic genetic variation involved in the paternal regulation of seed development. *PLoS Genet.* 12(1), e1005806. Doi:10.1371/journal.Pgen.1005806

Pratap, A., Gupta, S.K., 2009. Biotechnological interventions in host plant resistance, eds Peshin, R. and Dhawan, A.K. in *Integrated Pest Management: Innovation, Dissemination and Impact.* Dordrecht: Springer Publishers, pp. 183–207.

Reinders, J., Wulff, B.B., Mirouze, M., Mari-Ordonez, A., Dapp, M., et al., 2009. Compromised stability of DNA methylation and transposon immobilization in mosaic Arabidopsis epigenomes. *Genes Dev.* 23, 939–950.

Rendel, J.M., 1959. Canalization of the scute phenotype of *Drosophila. Evolution* 13, 425–439.

Rice, S.H., Wolf, J.B., Brodie, E.D., Wade, M.J., 2000. The evolution of developmental interactions. In *Epistasis and the evolutionary process.* New York: Academic Press, pp. 92–98.

Rosas, U., Nick, H.B; Copsey, L., Reuille, P.B.D., Coen, E., 2010. Cryptic variation between species and the basis of hybrid performance. *PLoS Biol.* 8(7), e1000429. Doi:10.1371/journal.pbio.1000429

Roux, F., Colome-Tatche, M., Edelist, C., Wardenaar, R., Guerche, P., et al., 2011. Genome-wide epigenetic perturbation jump-starts patterns of heritable variation found in nature. *Genetics* 188, 1015–1017.

Rutherford, S.L., Lindquist, S., 1998. Hsp90 as a capacitor for morphological evolution. *Nature* 396(6709), 336–342. Doi:10.1038/24550

Schlichting, C.D., 2008. Hidden reaction norms, cryptic genetic variation, and evolvability. *Ann. New York Acad. Sci.* 1133, 187–203.

Schluter, D., 2000. *The ecology of adaptive radiation.* New York, United States: Oxford University Press, pp. 300.

Schmidt, D.D., Baldwin, I.T., 2006. Transcriptional responses of *Solanum nigrum* to methyl jasmonate and competition: a glasshouse and field study. *Func. Ecol.* 20, 500–508.

Shahzad, Z., Eaglesfield, R., Carr, C., et al., 2020. Cryptic variation in RNA-directed DNA-methylation controls lateral root development when auxin signalling is perturbed. *Nat. Commun.* 11, 218. Doi:10.1038/s41467-019-13927-3

Springer, N.M., Schmitz, R.J., 2017. Exploiting induced and natural epigenetic variation for crop improvement. *Nat. Rev. Genet.* 18(9), 563–575. Doi:10.1038/nrg.2017.45

Sultan, S.E., 1987. Evolutionary implications of phenotypic plasticity in plants. *Evol. Biol.* 21, 127–178.

Suneson, C.A., 1956. An evolutionary plant breeding method. *Agron. J.* 48(4), 188–191.

Tanksley, S.D., McCouch, S.R., 1997. Seed banks and molecular maps: unlocking genetic potential from the wild. *Science* 277, 1063–1066.

Tanksley, S.D., Nelson, J.C., 1996. Advanced backcross QTL analysis: a method for the simultaneous discovery and transfer of valuable QTLs from unadapted germplasm into elite breeding lines. *Theor. Appl. Genet.* 92, 191–203. Doi:10.1007/BF00223376

Thompson, J., Charpentier, A., Bouguet, G., Charmasson, F., Roset, S., et al., 2013. Evolution of a genetic polymorphism with climate change in a Mediterranean landscape. *Proc. Natl. Acad. Sci. U.S.A.* 110, 2893–2897. Doi:10.1073/pnas.1215833110

Trewavas, A.J., 2015. *Plant behaviour and intelligence.* London: Oxford University Press, pp. 291.

True, J.R., Haag, E.S., 2001. Developmental system drift and flexibility in evolutionary trajectories. *Evol. Dev.* 3, 109–119.

Tsai, H., Missirian, V., Ngo, K.J., Tran, R.K., Chan, S.R., et al., 2013. Production of a high-efficiency TILLING population through polyploidization. *Plant Physiol.* 161, 1604–1614.

Uauy, C., Wulff, B.B.H., Dubcovsk, J., 2017. Combining traditional mutagenesis with new high-throughput sequencing and genome editing to reveal hidden variation in polyploid wheat. *Annu. Rev. Genet.* 51 , 435–435.

Waddington, C.H., 1942. Canalization of development and the inheritance of acquired characters. *Nature* 150(3811), 563–565.

Waddington, C.H., 1953. Genetic assimilation of an acquired character. *Evolution* 7(2), 118–126.

Waddington, C.H., 1956. Genetic assimilation of the *bithorax* phenotype. *Evolution* 10 , 1–13.

Waddington, C.H., 1957. *The strategy of the genes.* London: Allen and Unwin.

Wagner, A. 1999. Redundant gene functions and natural selection. *J. Evo. Biol.* 12, 1–16.

Wang W.S., Mauleon, R., Hu, Z., Chebotarov, D., Tai, S.S., Wu, Z.C., Li, M., Zheng, T.Q., Fuentes, R.R., Zhang, F., et al., 2018. Genomic variation in 3,010 diverse accessions of Asian cultivated rice. *Nature* 557, 43–49.

Wu, D., Lao, S., Fan, L., 2021. De-domestication: An extension of crop evolution. *Trends Pl. Sci.* 26(6), 560–574.

Xia, H.B., Xia, H., Ellstrand, N.C., Yang, C., Lu, B.R., 2011. Rapid evolutionary divergence and ecotypic diversification of germination behavior in weedy rice populations. *New Phytol.* 191 , 1119–1127.

Yan, L., Helguera, M., Kato, K., Fukuyama, S., Sherman, J., et al., 2004. Allelic variation at the *VRN-1* promoter region in polyploid wheat. *Theor. Appl. Genet.* 109, 1677–1686.

8 Mutagenesis in Plant Tissue Culture

Gurwinder Kaur, Gautam Chhabra, Umesh Preethi Praba, Rajveer Singh, Shabneek Kaur, Jessica Kaur, and Yogesh Vikal
School of Agricultural Biotechnology, PAU, Ludhiana, Punjab, India
*Corresponding Email: yvikal-soab@pau.edu

8.1 INTRODUCTION

The primary aim of plant breeding is the development of plant varieties for sustainable food production through the exploitation of natural and induced genetic diversity. Plant breeders have successfully recombined the desired genes from the existing available gene pool and the related plant species by sexual hybridization and successfully developed new cultivars with desirable traits such as high yield and abiotic and biotic stress resistance. In recent years, it has become more challenging due to the nonexistence or lack of availability of desired genotypes. However, mutagenesis is one of the major breakthroughs in plant breeding for inducing genetic variability. Compared to traditional cross-breeding methods, mutagenesis can modify only a few characters in an otherwise promising cultivar without significantly altering the remaining and often unique genetic background. The frequency rate of spontaneous mutations is much lower and, therefore, difficult to exploit by the plant breeders. Induced mutations can be caused by physical mutagens (gamma rays, X-rays, ion beams, etc.) or chemical mutagens. Site-directed mutagenesis can be defined as a process of generating a mutation at a defined site of a DNA molecule. However, insertional mutagenesis is caused by DNA insertions, either through T-DNA insertion or genetic transformation or transposable elements activation. The mutagen treatment breaks the nuclear DNA, and during the DNA repair mechanism, new heritable mutations are induced randomly. The induced mutations can cause changes in either chromosomal or genomic regions or cytoplasmic organelles, and these genetic variations enable the plant breeders to select useful mutants such as flower color, flower shape, disease resistance, and early flowering types (Jain and Maluszynski, 2004). The induced genetic variability in most of the economically important plant species results in the 3218 mutant varieties developed worldwide (Venugopalan and Suprasanna, 2022). Mutation induction is a flexible, workable, unregulated, nonhazardous, and low-cost alternative to genetically modified organisms (GMOs).

Mutagenesis can be applied to different types of plant materials such as seeds, stems, leaves, buds, tuber, rhizomes, bulbs, pollen, anther, embryos, microspores, callus, and cell cultures (Suprasanna et al., 2014; Bado et al., 2015). Seeds are easy

 DOI: 10.1201/9781003392897-8

to maintain in large amounts, easy to store and transport, especially posttreatment, and also the material of choice in sexually propagated species (Bado et al., 2015). However, conventional mutagenesis demands the screening of a large mutagenized population as well as proper selection methods after mutagenesis. Therefore, *in vitro* mutagenesis offers some advantages over *in vivo* mutagenesis such as high mutation frequency, uniform treatment of mutagen, use of selective agent for single cell system as well as homogenous cell population, and fast screening of large populations in a relatively small space. Despite these advantages, chimerism is one of the major problems in random mutagenesis of plant tissue, where same plant has a mixture of two or more genetically different types of cells (Mba, 2013). However, this major limitation can be overcome through *in vitro* random mutagenesis in which single mutagenic cell gives rise to whole plant. This approach of chimerism elimination is used in different crops such as Chrysanthemum and banana (Datta and Chakrabarty, 2009; Christov et al., 2014; Jankowicz-Cieslak and Till, 2016). Another major problem in random mutagenes is heterozygosity of mutated loci (Mba, 2013). Out of two alleles, only one copy of the allele is mutated in diploid organisms. This could be challenging when the phenotype is revealed by the homozygous mutation. However, it can be addressed through selfing for many generations as well as double haploids induction, especially in sexually propagated plants (Mba, 2013). However, double haploids are restricted to those plant species in which anther/ovule cultures have been established (Maluszynski et al., 2009). Haploid mutagenesis has been developed in different plant species including cereals (wheat, triticale, maize), liverwort, *Brassica napus*, and tobacco (Shen et al., 2015). *In vitro* mutagenesis may enable the screening of agronomic traits such as herbicide, salts and metal tolerance, waterlogging, drought, and cold or the selection of the embryo-gametophyte-lethal mutations in asexually reproducible plants (Suprasanna et al., 2014).

8.2 MUTAGENIC AGENTS

Agents that cause artificial mutations are termed mutagens. These are classified into two major groups, viz. chemical mutagens and physical mutagens. Point mutations are commonly caused by chemical agents; however, chromosomal aberrations and reorganizations are mostly induced by physical mutagens. The occurrence rate and mutation types depend upon the dosage of mutagen, its rate of exposure, and administration of the mutagen rather than its type. However, selection of mutagen will depend upon a particular researcher's circumstances, such as its ease of use and availability at low price, safety concerns during usage, types of tissue, and also effectiveness in inducing certain genetic variations.

Physical Mutagens

The physical agents that induce DNA damage among living organisms are referred to as physical mutagens. Physical mutagens (e.g., X-rays, gamma rays [γ-rays], cosmic rays, and ionizing radiation) cause hereditary aberrations, and more than 50% of mutant varieties were developed through physical mutagens (Table 8.1). Radiations are divided into ionizing and nonionizing radiations on the basis of their capacity to produce ions. Nonionizing radiations such as UV lights are weak mutagens that

TABLE 8.1
Types of Commonly Used Physical Mutagens for Plant-induced Mutagenesis

Type of radiation	Source	Characteristics	Energy	Hazard
X-rays	X-ray machine	Electromagnetic radiation; penetrates tissues from a few mm to many cm	50–300 keV	Dangerous, penetrating
γ-rays	Radioisotopes and nuclear reaction	Electromagnetic radiations similar to X-rays; penetrating through whole parts; sources are ^{60}Co (Cobalt-60) and ^{137}Cs (Caesium-137)	Up to several MeV	Dangerous, very penetrating
Neutron (fast, slow and thermal)	Nuclear reactors or accelerators	Uncharged particle, slightly heavier than proton, observable only through interaction with nuclei penetrate tissues to many centimeters; source is 235U	From less than 1 eV to several MeV	Very hazardous
α-particles	Radioactive isotopes or accelerators	High ionising, heavier than other radiations, penetrate tissues up to few centimetres	2–9 MeV	May be dangerous
β-particles, fast electrons, or cathode rays	Radioisotopes or particle accelerators	An electron (– or +) ionizing much	Up to several MeV	Very dangerous
Protons or deuterons	Nuclear reactors or accelerators	Produced in nuclear reactors and accelerators; derived from hydrogen nucleus; penetrate tissues up to several centimeters	Up to several GeV	Very dangerous
Low-energy ion beams	Particle accelerators		Dozens of keV	Dangerous
High-energy ion beams	Particle accelerators	Produced positively charged ions are accelerated at a high speed (around 20%–80% of the speed of light) and deposit high energy on a target	Up to GeV	Dangerous

cannot affect the structure of atoms; however, ionizing radiations are strong enough to affect the atomic structure in mutagenized material, including living plants as well as animal tissues. When ionizing radiation passes through tissue, it can dislocate an electron from its orbit to produce ions. The ionizing radiations are discharged from radioactive isotopes or can be produced artificially from reactors and further deployed through accelerators.

Although discoveries of X-rays (Röntgen, 1895), radioactivity (Becquerel, 1896), and radioactive elements (Curie and Curie, 1898) during the 19th century provided sound foundations for mutagenesis, it came to reality in the late 1920s after the discovery of Muller and Stadler in which the mutagenic effects of X-rays on fruit flies and maize, respectively, were studied. Apart from ionizing radiations, other mutagens such as alpha (α) and beta (β) particles and neutrons are most commonly used. Physical mutagens have the ability to produce reactive ions, which in turn cause abnormalities at the DNA level (deletion, addition, and/or substitution of nucleotides by breaking the chemical bonds of DNA molecules) to gross chromosomal breakages and rearrangements. The mutations induced by γ-rays are mostly large deletions (up to 6 Mb) that are not transferred to advanced generation as revealed by Naito et al., 2005 in *Arabidopsis thaliana*. Similarly, γ-rays induced nucleotide substitutions and small deletions (2–16 bp) have been demonstrated in rice (Morita et al., 2009; Sato et al., 2006) and *Arabidopsis* (Yoshihara et al., 2010). Single base substitutions induced by fast neutrons are more prevalent, and G:C to A:T transitions were also focused at pyrimidine dinucleotide sites, signifying that fast neutrons induce mutational covalent linkages between adjacent pyrimidine residues. Γ-rays are effective for inducing point mutations and small deletions, whereas fast neutron bombardment promotes translocations, chromosome losses, and large deletions (Leung et al., 2001; Wu et al., 2005, Liu et al., 2009). Cosmic radiations are also mutagenic to crops; these are high-energy particles (mostly protons) released from outer space (Mei et al., 1998; Ren et al., 2010). Strong cosmic radiations have been used to induce useful mutations in different crops such as rice, wheat, cotton, sweet pepper, tomato, sesame, and maize (Mei et al., 1998, Tanaka et al., 2010).

Ion beam radiations are different in terms of energy transfer, mass deposition, and charged exchange from other physical mutagens. These beams promote linear energy transfer (LET) that can induce lethality as well as chromosomal aberration in comparison to other most commonly used physical agents (Watanabe, 2001). Ion beam–induced double-strand DNA breaks are less repairable as compared to those caused by γ-radiation. Similarly, large DNA alterations (deletions) are more frequently caused by carbon ions than electrons. Despite the ion beams of high energy, low-energy (10–103 keV) beams have been used for mutation breeding, gene transfer, etc. This method has many advantages such as a low damage rate, higher mutation rate, and wider mutation spectrum. Yu et al. (1991) first time used low-energy nitrogen ion irradiation for the mutagenic treatment of rice seeds, and since then, new lines of rice mutants with a higher yield, broader disease resistance, shorter growing period, and high grain quality were grown in China. Low-energy ion beam–induced mutations in Thai jasmine rice of Thailand revealed phenotypic variation including short stature, red/purple color of the leaf sheath, collar, auricles, ligules, and dark brown stripes on the leaf blade, dark brown seed coat, and pericarp (Anuntalabhochai et al., 2004;

Phanchaisri et al., 2007). The ion beam exposure of plant materials for few minutes triggered numerous genetic changes. Callus is more sensitive to ion beam radiations as compared to other plant materials. Ion beam mutagenesis includes many advantages such as high survival rates with low dose, high mutagenesis rates, and wide variations. The ion beam radiations usually alter a single characteristic. Therefore, a new plant variety can be developed through mutant selection for desired trait while retaining the existing valuable ones. The time period for the development of a new variety is only 3 years as compared to conventional methods. This approach has been particularly successful in flower breeding programs. The sterile Vervena mutant, "Temari Bright Pink," was the first flower developed and marketed using ion beam mutagenesis in 2002. The effective use of ion beam radiations is also demonstrated in Dahlia, Verbena, Petunia, Dianthus, Cherry blossoms, and Delosperma. The development of semidwarf buckwheat and salt-tolerant rice are successful examples of ion beam mutagenesis (Morishita et al., 2003; Hayashi et al., 2007). In addition, three mutants of sweet pepper have been developed through ion beam irradiation: two dwarf plants and a yellow pepper (Honda et al., 2006). These mutant phenotypes are reported to be recessive and single gene mutations. Other mutational studies did not demonstrate the monogenic homozygous recessive mutant identification in the M_1 generation. However, homozygous recessive mutations induced in nuclear gene through ion beam radiations can be easily detected in the M_1 generation (Abe et al., 2000; Imanishi et al., 2007). Therefore, the time required to develop a new variety could be significantly reduced. This technique also induces high-rate mutations at relatively low doses without severely affecting growth. Therefore, ion beam is an efficient tool for crop improvement through mutation breeding.

Ultraviolet radiation was first described by Johann Wilhelm Ritter in 1801. It is light energy emitted between the wavelengths of 100 and 400 nm, i.e., between the electromagnetic spectra of X-ray and visible light. The mutagenic effect of UV produces pyrimidine dimers in interaction with DNA. The lesions due to pyrimidine dimers further affect DNA replication and transcription, leading to chromosomal rearrangements and lethal mutations. However, due to the weak penetration, their utilization has been limited only to materials and equipment irradiation used in tissue culture. Similar to UV rays, α-particles have low penetrability as compared to X-rays and γ-rays. The energy of α-particles varies from 3 to 7 MeV which corresponds to the longest and shortest half-lives of α-emitting nuclides. The α-particles are easily absorbed due to their charge and large mass, and after absorption, travel only to a few centimeters in air. However, these are strongly ionizing, and their high doses can cause all symptoms of radiation poisoning. The chromosomal damage from α-particles is about 100 times greater than any other radiation. Due to low penetrability, α-particles are rarely used for mutation induction in crops. However, β-particles are as effective as X- or γ-rays except for their low penetrability, similar to α-particles. To overcome the low penetrability of β-particles, various methods have been developed for plant mutagenesis. One method of β-particles administration is to submerge plant materials (e.g., seeds or propagules) in 32^P solution which is then integrated into germline cells of M_1 plants to cause mutations. This method has been successfully used for the development of rice and cotton mutant varieties. However, the use of radioactive isotopes becomes a safety concern in growing these materials.

TABLE 8.2
Chemical Mutagens Commonly Used in Induced Mutagenesis

Chemical mutagens	Mode of action
Alkylating agents (EMS, DES, NMU and ethylene amine)	Alkylate phosphate groups and bases. Mispairing and loss of bases
Acridines (Ethidium bromide, acridine orange and proflavine)	Intercalate between bases and results in deletion and addition of bases
Base analogs (2-Amino purine, hypoxanthine)	Base pair substitution
Nitrous acid (Hydroxyl amine, sodium azide)	Replacement of amino group with a hydroxyl group and conversion of cytosine to a modified base

Source: Toker et al. (2007); Raina et al. (2021).

8.2.1 Chemical Mutagens

In contrast to physical agents, chemical mutagens are easy to use, do not involve any specialized instrument, and also cause high mutation frequency. Chemical mutagens can cause point mutations, single base-pair changes, or single-nucleotide polymorphisms (SNPs). These are mostly applied on seeds, seedlings, and different plant propagules such as bulbs, tubers, corms and rhizomes, vegetative cuttings, scions, or *in vitro*–cultured tissues (Suprasanna et al., 2012; Xu et al., 2012). There are various chemicals that can induce mutations in prokaryotic as well as eukaryotic cells (Table 8.2). The chemical mutagens can be further classified into the following ways.

8.2.1.1 Alkylating agents

Ethyl methanesulfonate (EMS), methyl methanesulfonate (MMS), ethyl ethanesulfonate (EES), ethylene imines, diethyl sulfate (DES), and nitroso compounds are most commonly used alkylating agents. These are broadly classified into monofunctional, bifunctional, or polyfunctionals on the basis of number of alkyl groups present in the compound (Natarajan, 2005). The mutagen-induced alkylation can be defined as the replacement of an alkyl group (e.g., C_2H_5 of EMS; CH_3 of MMS) for hydrogen in the nitrogenous bases (Sharma and Chopra, 1994). Alkylating agents induce every kind of point mutation, transition, transversion, deletion, and frameshift and a comparatively low frequency of chromosome breakages that cause several chromosomal rearrangements (Maluszynski et al., 2003). Ashburner (1990) and Sharma and Chopra 1994 have explained the effects of alkylation of DNA in the following ways.

8.2.1.1.1 Alkylation of DNA's phosphate groups

Alkylation of phosphate groups leads to the formation of phosphate triesters which are very unstable and release the alkyl group. However, if adequate alkyl groups remain attached, then they interfere with DNA replication process. Then, phosphate triester bond between the sugar and the phosphate group gets hydrolyzed and results in the breakage of the DNA strand.

8.2.1.1.2 Alkylation of bases

Nitrogen (N_7) and oxygen (O_6) in the guanine are favored site for alkylation. O_6 alkylguanine pairs with thymine instead of cytosine and therefore leads to AT→GC transition.

8.2.1.1.3 Depurination

The alkylated guanine is released from the deoxyribose and therefore leaves behind a depurination site. This site can be further occupied through random base during DNA replication and results in transition or transversions (Wani et al., 2014).

8.2.1.2 Base analogs

5-Bromouracil and 5-bromodeoxyuridine have been shown to induce mutagenic effects due to their close resemblance to some of the naturally occurring bases in the DNA molecule. For example, thymine (T) and 5-methyl uracil have a structural analog, 5-bromouracil (5-BU). Therefore, 5-BU can substitute for T during DNA duplication, leading to transition type of mutation. Other mutagens comprise azides (sodium azide), antibiotics, acridines, nitrous acid, and hydroxylamine.

Despite the numerous mutagenic agents, only a few mutagens have been established in plants. Out of these mutagens, alkylating agents have been used in experimental mutagenesis and mutation breeding in plants. Approximately 80% of mutant plant varieties registered in the IAEA database (http://mvgs.iaea.org/Search.aspx) were developed through alkylating agents. These are electrophilic compounds that bind covalently to nucleophilic centers in organic macromolecules. However, these compounds form covalently linked alkyl adducts to the bases and to the phosphodiesters in DNA. Ethyl methanesulfonate (EMS), 1-methyl-1-nitrosourea (MNU), and 1-ethyl-1-nitrosourea (ENU) contribute 64% of total varieties developed by alkylating agents. Both these compounds promote single nucleotide changes through specific nucleotides alkylation (Sega, 1984) that further leads to high-density mutations which are randomly distributed (Greene et al., 2003). EMS treatment mostly induces random G/C to A/T transitions and reported approximately 5% frequency of truncated mutations (Greene et al., 2003). A study by Sato et al. (2006) reported that mutagenesis rate caused by γ rays is 1 per 6.19 Mb of genome, which is relatively lower than chemical mutagens such as EMS. Till et al. (2007) mutagenized rice cv. Nipponbare with 1.5% EMS or sequentially soaked in 1mM sodium azide and observed comparable density of putative mutations as estimated by TILLING (~1/300kb).

8.3 MOLECULAR METHODS FOR MUTATION DETECTION

A number of mutant detection methods have been used to discriminate mutants from wild type on the basis of mutagen types and expected DNA changes. Earlier "forward genetics" was used in which mutant phenotype was identified based on visual assessment of different traits such as chlorophyll deficiency, leaf and stem abnormalities, and seed size and shape. Then, molecular techniques employed DNA fingerprinting and mapping mutations with PCR-based markers. Simple sequence repeats (SSRs) are highly polymorphic and codominant markers. These are not only restricted to noncoding genomic regions but are also located in coding genome sequence, therefore permitting the development of EST (expressed sequence tag)–based SSR markers (Varshney et al., 2005). Multiple genomic regions can be simultaneously targeted with random primers which do not necessitate the prior knowledge of sequence information. The RAPD technique is also fairly simple and does not require restriction digestion, labeled probes, and the use of hazardous chemicals for detection. Unlike SSR markers, these are dominant and their reproducibility is not so high. The AFLP technique is highly versatile and typically permits the analysis of dozens of DNA markers concurrently. Nowadays, SNPs (single-nucleotide polymorphisms) are progressively used as DNA markers because of the availability of genome sequence information of major crops. In comparison to other DNA markers, SNP marker density can be measured on a kilobase scale instead of a megabase scale. Because of their high density, these markers are widely used for high-resolution gene mapping, assessing genetic diversity, and marker-assisted selection. Other molecular markers, such as CAPS, STS, SCAR, ISSR, SRAP, VNTR, and DArT, have been developed and are being utilized in mutation breeding programs. The major limitations of these marker technologies are that they are an indirect approach, done without knowing the genomic sequence, and also costly due to requirement of more time, resources, and technical expertise.

Till the 19th century, phenotypic selection was a fast and easy method for mutant identification from large populations under controlled environmental conditions. Advances in phenomics approaches increased both accuracy and throughput of this approach (Van-Dijak et al., 2020). The frequency of induced mutation modifying gene function will be rare. Therefore, a typical mutagenesis experiment might involve raising the thousands of M_2 families to select a mutant phenotype. With the availability of whole genome sequence information, genotypic screening is an effective way to screen mutant populations. The development of effective tools for determining rare, induced mutations has driven advances in both forward and reverse genetics (Shelake et al., 2019; Abe et al., 2012; Jankowicz-Cieslak et al., 2015; Schneeberger et al., 2009). Technologies have emerged from PCR-based methods to array hybridization to sequence genome for discovering large genomic variations such as insertions, deletions, and rearrangements (Campbell et al., 2006).

Southern blotting is used to detect DNA fragments after restriction enzyme digestion. After digestion, DNA fragments that differ in size and molecular weight on the basis of recognition sites of enzymes are called restriction fragment length polymorphism. Variations in the size of DNA fragments on the basis of modifications of restriction sites are due to different types of mutation that yield RFLP (Maniatis et al., 1982). A modification of the RFLP procedure is PCR/RFLP (size analysis of PCR products), which involves DNA stretch amplification through PCR using labeled nucleotides followed by enzymatic digestion with restriction enzymes followed by gel electrophoresis; it could make the mutant recognition very simple, especially in heterozygous form (Mitra et al., 2003).

Another PCR type, Amplification Refractory mutation system (ARMS) analysis, was investigated for target discovery of common mutations. This procedure involves two complementary reactions and three primers. Out of three primers, one is constant as well as complementary to template in both reactions, and the other two primers differ at their 3′ termini, therefore specific to either wild or mutant allele. The amplification will occur only in one tube in case of homozygous mutant or homozygous wild; however, in heterozygous, amplification will occur in both tubes (Newton et al., 1989).

Heteroduplex migration analysis is the differential mobility of DNA heteroduplexes on polyacrylamide gel electrophoresis as compared to homoduplexes. Heteroduplex from two mismatched strands typically shows an abnormal mobility. PCR products from a suspected carrier are denatured and then permitted to hybridize into a normal PCR product to form heteroduplexes that can be easily detected on gel (Williams and Goodeve, 2002).

Single-strand conformational polymorphism (SSCP) analysis detects the presence or absence of mutation in an amplified exon. Single DNA strands fold to form a complex three-dimensional structure, i.e., maintained by intramolecular bonds. However, it was reported that 80% of induced mutations modify the folding pattern and subsequently alter the migration of DNA on gel. The differential mobility on gel recommends a mutation in an amplified exon that can further be sequenced to know the exact genetic change (Young, 2010; Gupta et al., 2005).

Denaturing gradient gel electrophoresis (DGGE) has higher sensitivity (over 90%) as compared to single-strand conformational polymorphism and heteroduplex migration analyses. PCR products are resolved on a specific denaturing gel that involves high concentrations of denaturing agents such as urea. This method depends on the basic principle that wild and mutant sequences will display dissimilar banding patterns on a gel. Double-strand DNA denatures to form a single strand that terminates at different locations in case of heterozygotes. The appropriate exons will then be sequenced to detect specific underlying mutations (Young, 2010; Tomaiuolo et al., 2003). Denaturing high-performance liquid chromatography (DHPLC) is same as DGGE which identifies heteroduplexes having differential separation after reannealing of the normal and mutant strands. PCR products are injected in a column comprising an increased gradient of mobile phase (acetonitrile) essential to elute homo and heteroduplexes. It is an efficient method for detecting small insertions, deletions, and base substitutions (Turnpenny and Ellard, 2007). However, conformation-sensitive capillary electrophoresis (CSCE) has higher throughput for the detection of heteroduplexes than DHPLC because of the use of fluorescent technology.

Oligonucleotide ligation assay (OLA) uses a pair of oligonucleotides that are designed to anneal to adjacent sequences within a PCR product. After perfect hybridization, these are joined by DNA ligase. Oligonucleotides complementary to normal and mutant sequences are differentially labeled and products are identified by computer software (Turnpenny and Ellard, 2007).

High-resolution melt curve analysis (HRM) uses fluorescent dyes that intercalate with double-strand DNA. This dye is incorporated into the PCR reaction, and then PCR products are denatured to separate the two DNA strands. Fluorescence level decreases upon DNA strands dissociation and the melting profile relies upon size and sequence of PCR product. HRM technique is very sensitive and its detection ability is largely determined by fragment length, sequence composition, mutation identity, PCR quality, and analytical equipment (Milbury et al., 2009).

A reverse genetic approach has been used that adds the high-frequency point mutations caused by EMS treatments, with heteroduplex detection among wild-type and mutant DNA fragments, originally using DHPLC, which has been termed as TILLING. It was first established in *Arabidopsis* and since has been utilized in different crops such as barley, wheat, maize, and sorghum using primarily enzymatic mismatch cleavage methods for mutation detection (McCallum et al., 2000). TILLING is a regularly used method for study of EMS-mutated populations and has also been employed to detect 2–4 bp deletions in rice that were induced by γ-radiation. This approach is termed EcoTILLING, where it has been used to detect natural mutations in a population (Cordeiro et al., 2006). Alternatively, capillary electrophoresis (CE) is an effective technique that has been used for detection of DNA polymorphism in natural and mutated populations. CE has better efficiency, sensitivity, and throughput. This technique has been reported to be highly accurate in differentiating between two SNPs of the EAAC-1 gene that corresponds to three haplotypes, which were then validated by cycle sequencing. More recently, next-generation sequencing approaches have been utilized for mutation discovery and TILLING (Szantai et al., 2005; Lee et al., 2014).

8.4 APPLICATIONS OF *IN VITRO* MUTAGENESIS

8.4.1 *In Vitro* Mutagenesis for Biotic Stress

A variety of biotic stresses can reduce crop production, including fungal diseases, insects, nematodes, viruses, bacteria, and parasitic weeds. Many crops also experience problems due to weeds (Atkinson and Urwin, 2012; Rasmussen et al., 2013). The development of strategies for resistance to diseases and pests in plants is seen as a powerful way to support the sustainability of agriculture due to the reduction in chemical inputs into the environment (Gimenez et al., 2018). Induced *in vitro* mutagenesis for resistance to plant diseases has been proposed as an alternative or complementary strategy for crop protection (Table 8.3). In a study, combined induced microspore mutation, regeneration, pathogen, and herbicide selection to produce *Brassica napus* mutants resistant to *Alternaria brassicicola* and chlorsulfuron (Ahmad et al., 1991). Using the same protocol, mutants with *Sclerotinia sclerotiorum* tolerance were observed in rapeseed (Liu et al., 1997), and the mutants that were resistant to *Pectobacterium carotovorum* subsp. *carotovorum* (soft rot disease) were selected in *Brassica campestris* ssp. *Pekinensis* (Zhang and Takahata, 1999). Mangal and Sharma (2002) *in vitro* mutagenized callus of *Brassica oleracea* var. botrytis (cauliflower) with EMS and γ-rays and screened them for resistance against the *Xanthomonas campestris* pv. *campestris* (black rot pathogen). The survival of calli is reduced with enhanced concentration of culture filtrate (CF). At 30% level of CF calli survived, regenerated to complete plant, and showed resistance towards black rot.

TABLE 8.3
Applications of *in vitro* Mutagenesis for Biotic Stress Resistance

Plant species	Common name	Explant	Mutagen/ mutant	Selection agent and level used	Disease resistance	References
Brassica napus	Rapeseed	Microspore culture	UV-irradiation; LD_{50} 20s	Spore suspension containing 10^6 spores/ml of *A. brassicicola*	*Alternaria brassicicola*	Ahmad et al. (1991)
				30 µg/ml Glean (active ingredient chlorsulfuron)	Chlorsulfuron	
Malus domestica cv. Greensleeves	Apple	Axenic shoot culture	Somaclone R20 63, R46 3, R47 11 and R78 3	Inoculated with 10^8 cfu/ml of *E. amylovora* strain T	*Erwinia amylovora* (fire blight pathogen)	Donovan et al. (1994)
Musa spp., AAA Group cv. Highgate	Banana	Shoot culture derived from corms	γ-Rays	Conidial suspension (3 × 10^4 spores/ml) of *F. oxysporum* f. sp. *cubense*	*Fusarium oxysporum* f. sp. *cubense* (fusarial wilt or Panama disease)	Bhagwat and Duncan (1998)
Brassica campestris L. ssp. *Pekinensis*	Chinese cabbage	Microspore culture	UV-irradiation; LD_{50} 12s	Culture filtrate of *P. carotovorum* subsp. *carotovorum*	*Pectobacterium carotovorum* subsp. *Carotovorum* (soft rot disease)	Zhang and Takahata (1999)
Prunus persica (L.) Batsch	Peach	Callus derived from immature embryos	Somaclone 122-1	Detached-leaf bioassay	*Xanthomonas campestris* pv. *pruni* (bacterial leaf spot) and *Pseudomonas syringae* pv. *syringae* (bacterial canker)	Hammerschlag (2000)

Brassica oleracea var. *botrytis* L.	Cauliflower	Callus derived from seeds	EMS 3% and γ-rays (77.28 rads/min)	Different percent concentrations of toxic culture filtrate (0, 5, 10, 15, 20, 25, 30, 35)	*Xanthomonas campestris* pv. *campestris* (black rot pathogen)	Mangal and Sharma (2002)
Saccharum spp.	Sugarcane	Calli from the susceptible cultivar B6749	γ-rays 2 krad	Inoculation with strain B of *Sugarcane Mosaic Virus*	*Sugarcane Mosaic Virus*	Zambrano et al. (2003)
Brassica napus	Rapeseed	Microspore culture	EMS 0.15%	Oxalic acid (3 mmol/l; half lethal dose was 3.1 mmol/l)	*Sclerotinia sclerotiorum* (stem rot)	Liu et al. (2005)
Solanum surattense Burm. f.	Kantakari	Cotyledons	EMS 0.1% and γ-rays 10 Gy	Selective medium supplemented with 750 mg/l streptomycin sulfate	Chloroplast-encoded streptomycin resistance	Swamy et al. (2005)
Saccharum officinarum	Sugarcane	Callus cultures derived from spindles and buds	NaN_3 solution (0.003%) and γ-rays (10 and 20 Gy)	Spraying 3 ml of uredospore suspension of *P. melanocephala* (4–5 × 10^5 uredospores/mL)	*Puccinia melanocephala* (brown rust)	Oloriz et al. (2011)

In vitro mutagenesis and somaclonal variation are two significant tissue culture–based approaches that enable improvement in one or a few genotype-based features without affecting the rest of the genome (Predieri, 2001). Moreover, somaclonal variation has a similar genetic variation spectrum to induced mutation because both result in a qualitatively similar range of DNA changes (Brar and Jain, 1998; Cassells et al., 1998). Hammerschlag (2000) performed detached-leaf bioassay in *Prunus persica* (L.) Batsch (peach) derived somaclone 122-1, which showed dynamic resistance toward *Xanthomonas campestris* pv. *pruni* (bacterial leaf spot) and *Pseudomonas syringae* pv. *syringae* (bacterial canker). The somaclone 122-1 is produced from callus culture of peach cultivar Redhaven. Donovan et al. (1994) differentiated *in vitro* and *in vivo* methods to screen out the resistant apple somaclones (R20 63, R46 3, R47 11 and R78 3) for fire blight pathogen (*Erwinia amylovora*) and observed that *in vitro* screening had a higher selection pressure. Bhagwat and Duncan (1998) were exposed banana (*Musa* spp., AAA Group cv. Highgate) *in vitro*–grown shoot apices cultures to various doses of γ-rays to determine the efficacy of inducing mutations. Their aim was to develop mutants that were resistant to deadly fungus *Fusarium oxysporum* f. sp. *cubense* that caused fusarium wilt or Panama disease in banana. The plants that were considered tolerant to the fungus were multiplied *ex vitro* for field screening analysis. Liu et al. (2005) established a protocol for the development of doubled haploid in *Brassica napus* plants with enhanced resistance to *Sclerotinia sclerotiorum*. Haploid seedlings regenerated from microspore culture were used to generate haploid calli for *in vitro* mutagenesis selection. Mutation was induced by EMS, and screening for resistant mutants was performed on media supplemented with oxalic acid.

In higher plants, chloroplast-encoded antibiotic resistance plays a vital role in plant breeding experiments. The Streptomycin-resistant plantlets showing chloroplast-encoded mutants in *Solanum surattense* were generated through mutagenic treatment (EMS and γ-rays). The irradiated cotyledon explants showed a high frequency, i.e., 16.5% of resistant shoots as compared to EMS-treated cotyledon explants (Swamy et al., 2005).

8.4.2 *In Vitro* Mutagenesis for Abiotic Stress

Abiotic stress typically interferes with a number of cellular processes as well as intricate metabolic pathways. Drought, waterlogging, salinity, temperature, and heavy metals are among the abiotic stressors that have a significant impact on crop yield. Also, it is anticipated that UV exposure would significantly impact crops and farming systems as the ozone layer thins. The abiotic stress-tolerance mechanism can be better understood by screening several cultivars for their tolerance to abiotic stress (Mittler, 2006; Sergeeva et al., 2006; Godoy et al., 2021). To choose the plants that are resilient to abiotic stressors, *in vitro* mutagenesis has been widely used (Table 8.4).

Rapeseed herbicide resistance has recently been a key concern for many breeders. The microspore and embryo culture approach has been used in herbicide-resistant breeding to select embryos that are resistant to herbicides at the embryonic stage. Swanson et al. (1989) mutagenized *Brassica napus* cv. Topas, microspore

TABLE 8.4
Applications of *in vitro* Mutagenesis for Abiotic Stress Resistance

Plant species	Common name	Explant	Mutagen	Selection agent and level used	Abiotic stress	References
Brassica napus cv. Topas	Canola	Microspore culture	20 μM ethyl nitrosourea (ENU)	Pursuit (0, 50, 100, 300, and 500 g/ha), Assert (0, 400, 600, 800, 1000, and 1200 g/ha), and Scepter (0, 50, 100, 300, and 500 g/ha)	Imidazolinones tolerance	Swanson et al. (1989)
Solanum tuberosum L. cv. Kufri Jyoti and Kufri Chandramukhi	Potato	Stem-cuttings	γ-rays (20 and 40 Gy)	Irradiated plantlets were screened for microtuberization at 28 ± 1°C	Heat tolerance	Das et al. (2000)
Hordeum vulgare L.	Barley	Callus culture derived from immature embryos	EMS 10 mg/l, NaN_3 0.5 mg/l and γ-rays 5000 rad	30 μM Al for 24 h	Al tolerance	Zhu et al. (2003)
Capsicum annuum	pepper	Cotyledons from EMS-treated seedlings	0.1% EMS	10 mg/l and 20 mg/l atrazine	Atrazine resistance	Venkataiah et al. (2005)
Chrysanthemum morifolium Ramat. cv. Juhua	Chrysanthemum	Leaf segments	0.025% EMS	250 mM NaCl	Salinity	Hossain et al. (2006)
Ipomoea batatas L.	Sweet potato	Callus culture derived from leaf segments	0.5% EMS	200 mM NaCl	Salinity	Luan et al. (2007)
Arachis hypogaea L. cv. Huayu 22	Peanut	Embryonic leaflets	Pingyangmycin 4 mg/l	15 and 20 g/l NaCl	Salinity	Zhao et al. (2013)

culture with 20 μM ethyl nitrosourea and the mutant developed showed tolerance against diverse imidazolinones, i.e., Pursuit, Assert, and Scepter. Venkataiah et al. (2005) investigated atrazine-resistant plants through *in vitro* mutagenesis in pepper (*Capsicum annuum* L.) by applying low dosage of 0.1% EMS to the cotyledon culture. In media containing selected amounts of sucrose (0.5%) and atrazine (20 mg/l), mutagenized cotyledon explants generated herbicide-resistant plants. Various morphogenetic responses were noticed when the concentration of sucrose (0.5%–5%) changed. Maximum shoot regeneration was seen at 2% sucrose, and as sucrose content increased (to 3%–5%), the ability to regenerate became less effective. However, the explants were completely bleached when the sucrose concentration was reduced from 2% to 0.5%, which made it possible to choose plants that were herbicide resistant. After regenerating green shoots on rooting medium containing 10 mg/l atrazine, 1 mg/l indole acetic acid, and 0.5% sucrose, complete atrazine-resistant plantlets were developed.

Chemical mutagen EMS (0.025%) was administered to the explant (leaf segments) of chrysanthemum (*Chrysanthemum morifolium* Ramat. cv. Juhua) in order to create salt-tolerant strains. This method led to the stable development of one NaCl-tolerant chrysanthemum variant (E2). The ability of the plant to sustain both bloom quality and yield under stress circumstances was used to assess salt tolerance (Hossain et al., 2006). Similarly to this, Luan et al. (2007) employed EMS to cause mutation and *in vitro* testing for salinity tolerance and plant regeneration of sweet potato (*Ipomoea batatas* L.). After being treated with 0.5% EMS for 0, 1, 1.5, 2, 2.5, and 3 h, calli made from leaf explants were rinsed four times with sterile distilled water. For the purpose of selecting mutant cell lines, salt-tolerant calli were subcultured on media supplemented with 200 mM NaCl. This procedure was done five times (20 days each). After being propagated, it was discovered that in these mutants, salt tolerance was greater than that of control plants.

In two commercial potato cultivars, Kufri Jyoti and Kufri Chandramukhi, Das et al. (2000) produced heat-tolerant mutants through *in vitro* mutagenesis of *in vitro* grown plantlets. Shoots that had received γ-rays (20 and 40 Gy) were micropropagated for three cycles. At 28°C, a significant portion of the micropropagated shoots produced microtubers. High-temperature-induced microtubers had deformed shapes, but their field germination was normal. High-temperature stress increased the frequency of chlorophyll variations in the material obtained from γ-rays; however, only around 40% of the plants had normal leaf tissue, whereas control plants had entirely destroyed leaves.

Zhu et al. (2003) created 12 aluminum-tolerant barley (*Hordeum vulgare* L.) cell lines using EMS, NaN_3, and γ-rays. For analysis of the mechanisms underlying Al tolerance, four mutant cell lines with Al-tolerance were chosen. The relative growth of the four mutant cell lines (M_3, M_{10}, M_{11}, and M_{25}) in a modified MS medium was at least 26% greater on day 10 than the original line when the four cell lines were treated with 30 μM Al for 24 hours. Al accumulation was much less in the three mutant cell lines (M_3, M_{10}, and M_{25}) than in the control cell line. Analysis of organic acids showed that M_3, M_{10}, and M_{25} have increased citrate and/or malate production in response to Al, which is highly linked with the greater Al tolerance of these three mutant cell lines. This shows that similar to other grass species, the Al exclusion mechanism by organic acids is active in barley cell lines.

8.4.3 *In Vitro* Mutagenesis for Nutritional Quality

Crop plant's ability to absorb nutrients and withstand stress may be greatly improved if suitable mutants affecting the metabolism or catabolism of important amino acids could be discovered, identified, and developed into new varieties. Crop plants' nutritional value is significantly improved by inducing mutations (Table 8.5). The successful introduction of multiple mutant genes has significantly increased the nutritional content of commercial crop cultivars (Jain and Suprasanna, 2011). The benefits of biochemical selection pressure given by *in vitro* procedures can help with the recovery of certain metabolic mutants (Suprasanna and Rao, 1997; Suprasanna et al., 2012). There are often higher concentrations of the equivalent natural free amino acids in specific plant cells that are resistant to amino acid analogues or specific amino acid combinations. For selection in cultivated plant cells and recovery of resistant cell lines or plants, many amino acid analogues have been used.

Rapeseed is presently, after palm and soybean, the third-most significant source of edible vegetable oil worldwide. Consequently, enhancing quality by altering the mix of fatty acids is currently a key goal in breeding this crop. However, radiation would prevent embryonic development and microspore embryogenesis in *Brassica* species. Choosing an optimum irradiation content is therefore essential for mutation. By immediately isolating *Brassica carinata* microspores and exposing them to UV light, as well as by immersing the roots in a solution containing 0.05% colchicine for up to 3 hours, causing chromosomal doubling, and the resulting mutants were used to analyze the composition of glucosinolate and erucic acid. Using this method, three groups of doubled haploid lines with varying levels of glucosinolate and erucic acid content were found from a population of 270 doubled haploid lines (Barro et al., 2003). Similar to this, *B. carinata* microspores were subjected to chemical mutagenesis, i.e., EMS (0%–0.5%), in order to discover lines with changed erucic acid content. Nine lines had been found from a population of over 400 recovered doubled haploid plants that show promising beneficial alterations in erucic acid concentration in the seed oil. Erucic acid concentrations in three lines were below 25%, with a minimum of 17.1%, while in six lines, they were above 52% as compared to control plants (Barro et al., 2001).

Cassava variety PRC 60a was subjected to *in vitro* mutagenesis using the cyclic somatic embryogenic system developed by Joseph et al. (2004). They claimed that the ideal dose for causing mutations to choose the embryos as good experimental materials is 50 Gy of γ-rays. They noticed that more than 50% of the mutant lines that had been regenerated had different morphologies from plants of the wild type. This method of introducing genetic variability led to the development of novel cassava cultivars with significant phenotypic variations.

Explants from the potato cultivars Shepody and Atlantic that were propagated *in vitro* were given five doses of γ-rays, i.e., 0, 2, 4, 6, and 8 Gy to test the stimulating effects of low irradiation on the development and quality of microtubers. Both

TABLE 8.5
Applications of *in vitro* Mutagenesis for Nutritional Quality

Plant species	Common name	Explant	Mutagen	Trait improvement	References
Brassica carinata	Ethiopian rape	Microspore culture	EMS (0%–0.5%)	Erucic acid	Barro et al. (2001)
Pyrus communis L.	Pear	*In vitro*–grown shoots	γ-rays (3.5 Gy)	Early bearing and consistent productivity	Predieri and Zimmerman (2001)
Brassica carinata Braun	Ethiopian mustard	Microspore culture	UV-irradiation; LD_{50} 8 min	Glucosinolate reduced and erucic acid improved	Barro et al. (2003)
Chrysanthemum morifolium Ramat cv. Lalima	Chrysanthemum	Ray florets	γ-rays (0.5 Gy and 1 Gy)	Flower color and shape	Misra et al. (2003)
Manihot esculenta Crantz var. PRC	Cassava	Cotyledonary segments derived somatic embryos	γ-rays 50 Gy	Altered starch yield and composition	Joseph et al. (2004)
Dendranthema grandiflora Tzvelev	Chrysanthemum	Immature floral pedicels	0.77% EMS	Petal color	Latado et al. (2004)
Solanum tuberosum L. var. Shepody and Atlantic	Potato	Nodal segment	γ-rays (0, 2, 4, 6, and 8 Gy)	Microtuberization period prolonged, microtuber number and fresh mass, starch content, ascorbic acid, protein content and reducing sugar contents	Li et al. (2005)

cultivars started tuberization 5 days earlier than control. However, radiation treatment for cv. Atlantic showed that the entire microtuberization phase was extended by 10–15 days with 4, 6, and 8 Gy doses of γ-rays. Both the quantity of microtubers (up by 116.7% and 34.5% over the control in Shepody and Atlantic, respectively) and the fresh mass (up by 77.6% and 23.2% in Shepody and Atlantic, respectively) was significantly increased. icrotubers starch content was raised by 2–4 Gy of low-dose radiation. Large dosages (6–8 Gy) increased the amount of ascorbic acid and reduced sugar. The protein content of microtubers was efficiently raised by 4–6 Gy treatments as well (Li et al., 2005).

Mutagenesis is a vital technique for growing fruit and ornamental plants. According to Predieri and Zimmerman (2001) six cultivars of pear (*Pyrus communis* L.) had their *in vitro* shoots exposed to γ-rays (3.5 Gy). Only 2 of the 97 variations that were chosen exhibited chimerical behavior. Microcuttings from the irradiated shoots and extra nonirradiated microcuttings were rooted after three subcultures in order to establish plants for the survey orchard shoots. Improved production-related traits, such as early bearing and consistent productivity, were considered when screening orchard trees. Misra et al. (2003) used γ-rays (0.5 and 1 Gy) to alter flower color and shape of *Chrysanthemum morifolium* Ramat cv. Lalima. Ray florets were bombarded with γ-radiation after being inoculated on MS medium supplemented with 1.07 μM α-naphthaleneacetic acid (NAA) and 8.87 μM benzyladenine (BAP) (0.5 and 1 Gy). The γ-irradiated plants (0.5 Gy) produced two mutants that were vegetatively propagated and provided true-to-type flowers. Latado et al. (2004) created new cultivars of chrysanthemum (*Dendranthema grandiflora* Tzvelev) using 0.77% EMS in immature flower pedicels. In total, 48 mutants were developed with a change in petal color out of the 910 plants obtained from the pedicels treated with EMS (light-pink, yellow, white, bronze, pink-salmon, and salmon color). They were mostly uniform in terms of phenotype.

8.5 MUTANT INFORMATION RESOURCES

Rapid and unrestricted access to mutants is required to evaluate their value and potential for use in environments other than that in which they were developed. Crop mutants are also becoming increasingly important in functional genomics and proteomics. Hence, it is crucial to gather, preserve, and ease the distribution of mutants of significant crops so that plant breeders, geneticists, and biotechnologists around the world can access them without restriction. Moreover, collections of these mutants would preserve the genetic diversity conserved in regional variants (Oladosu et al., 2016).

The Mutant Variety Database (MVD), which collects information on officially approved mutant agricultural varieties of both seed and vegetative crops, is maintained by the Joint FAO/IAEA Centre of Nuclear Techniques in Food and Agriculture (https://mvd.iaea.org). The database illustrates the role of mutation breeding in ensuring global food security and crop adaptability to climate change. It provides data on superior mutant varieties that have been spread globally over more than a hundred distinct crop species. The type of mutagen employed, the dosage, the traits enhanced, and the variety's released agronomic data are all listed for each variety

TABLE 8.6
List of Mutant Varieties Generated through *in vitro* Mutagenesis in Cereals

Plant species	Mutant variety ID	Variety name	Year of registration	Explant	Mutagen/mutant	Trait improvement details	Country name
Triticum aestivum L.	757	Hezu 8	1992	Callus culture	γ-rays 10 Gy	High yield, early maturity and scab resistance	China
	2647	H6765	2004	Pollen culture	γ-rays 1.5 Gy	High yield, tolerance to drought and salinity	China
	3428	Hangmai901	2011	Doubled haploid	Physical mutagen	High tolerance to drought	China
Oryza sativa L.	1169	Aya	1992	Anther culture	γ-rays	Low amylose content and high eating quality	Japan
	2367	Ganzaoxiang 56	2004	Anther culture	γ-rays 300 Gy	High grain yield and good eating quality	China
	2377	Teyou 37	2001	*In vitro* culture	γ-rays 30 Gy	High grain yield and good quality	China
	2391	DT36	2002	*In vitro* culture	γ-rays 50 Gy	High yield, short duration and photoperiod insensitivity, tolerance to various biotic and abiotic stresses	Vietnam
	2446	Hatsu-akane	1990	Protoplast culture	Somaclonal mutant	Short stem	Japan
	2447	Hatsu-yume	1990	Protoplast culture	Somaclonal mutant	Shorter stem	Japan
	2448	Sumi-takara	1991	Anther culture	Somaclonal mutant	Hairless leaves	Japan
	2449	Yume-kaori	1993	Protoplast culture	Somaclonal mutant	Short stem and lower 1000-kernel weight	Japan
	2452	Hareyaka	1995	Protoplast culture	Somaclonal mutant	High tolerance to lodging	Japan

Oryza sativa L.	2454	Yume-gokochi	1995	Protoplast culture	Somaclonal mutant	Reduced amylose content	Japan
	2457	Ohita 3 Gou	1997	Protoplast culture	Somaclonal mutant	Shorter stem and slightly smaller grain	Japan
	2460	Yume-ekubo	2000	Protoplast culture	Somaclonal mutant	Short stem and late maturity	Japan
	2462	Iwata 15 Gou	2001	Callus culture	Somaclonal mutant	Low amylose content	Japan
	2671	Nanhua 11	1987	Callus culture	Carbon dioxide laser irradiation	High yield, resistance to bacterial diseases and wide adaptability	China
	2673	Zhongzu 1	2000	*In vitro* culture	γ-rays	Early maturity, high yield, high protein content and disease resistance	China
	2896	GINES	2007	*In vitro* culture	Protons 20 Gy	Tolerance to salinity and good quality	Cuba
	3077	Yume-ippai	2003	*In vitro* culture	Somaclonal mutant	Short culm	Japan
	3092	Yume-no-hana	2004	*In vitro* culture	Somaclonal mutant	Low amylose content and good taste	Japan
	3128	Akineiro	2005	Anther culture	Chemical mutagen	Medium maturity, low amylose content, high lodging resistance	Japan
	3232	Wonhaebyeo	2007	Callus culture	γ-rays 50 Gy	Tolerance to salinity	Republic of Korea
	3444	Goldami 1ho	2011	Anther culture	γ-rays	High protein and amino acids	Republic of Korea
	4862	INCA LP-7	1998	*in vitro* culture	Chemical mutagen	Resistant to, *Steneotarsonemus spinki*, drought and salinity tolerance	Cuba

Source: FAO/IAEA Mutant Variety Database 2022, https://mvd.iaea.org

in the database. A novel plant variety that has been developed via the following processes is referred to as a mutant variety in the MVD: (a) Direct application of physical or chemical mutagenesis and somaclonal variation and (b) indirect usage of mutant lines that are used as parental lines in crossbreeding programs. Using this database, the mutant varieties of cereals that could only be created through tissue culture techniques were analyzed and ranked in accordance with the different types of mutagens and explant used. The internationally registered varieties were recognized by their unique mutant variety ID. All of these specifications along with improved traits are described in Table 8.6.

Several genomic consortia are working to develop databases and other resources for functional genomics. Moreover, databases for functional genomics are available, including OryGenesDB (http://orygenesdb.cirad.fr), RiceGE (http://signal.salk.edu/cgi-bin/RiceGE), and Gramene (www.gramene.org), where information about the flanking sequence tags has been compiled and mutants for important genes can be discovered. These databases establish connections between rice genes and other grass genes, directing functional searches to the resources for rice mutants. Along with other national initiatives, the International Rice Functional Genomics Collaboration (IRFGC) set out to create mutant resources in order to identify the function of every rice gene, mostly through reverse genetics techniques (Hirochika et al., 2004).

REFERENCES

Abe, A., Kosugi, S., Yoshida, K., Natsume, S., Takagi, H., Kanzaki, H., Matsumura, H., Yoshida, K., Mitsuoka, C., Tamiru, M., Innan, H., 2012. Genome sequencing reveals agronomically important loci in rice using MutMap. *Nat. Biotechnol.* 30(2), 174–178.

Abe, T., Bae, C.H., Ozaki, T., Wang, J.M., Yoshida, S., 2000. Stress induced by heavy ion beams. *Gamma Field Symp.* 39, 45–56.

Ahmad, I., Day, J.P., Macdonald, M.V., Ingram, D.S., 1991. Haploid culture and UV mutagenesis in rapid-cycling *Brassica napus* for the generation of resistance to chlorsulfuron and *Alternaria brassicicola*. *Ann. Bot.* 67(6), 521–525.

Anuntalabhochai, S., Chundet, R., Phanchaisri, B., Yu, L.D., Promthep, S., Jamjod, S., Vilaithong, T., 2004. Mutation induction in Thai purple rice by low-energy ion beam. *Proceedings of the 9th Asia Pacific Physics Conference* (9th APPC), Hanoi, Vietnam. 1–6.

Ashburner M, 1990 *Drosophila: A Laboratory Handbook.* Cold Spring Harbor Laboratory Press, Cold Spring Harbor, NY.

Atkinson, N.J., Urwin, P.E., 2012. The interaction of plant biotic and abiotic stresses: from genes to the field. *J. Exp. Bot.* 63, 3523–3543.

Bado, S., Forster, B. P., Nielen, S., Ali, A. M., Lagoda, P. J., Till, B. J., Laimer, M., 2015. Plant mutation breeding: current progress and future assessment. *Plant Breed. Rev.* 39, 23–88.

Barro, F., Fernandez-Escobar, J., De La Vega, M., Martin, A., 2001. Doubled haploid lines of *Brassica carinata* with modified erucic acid content through mutagenesis by EMS treatment of isolated microspores. *Plant Breed.* 120(3), 262–264.

Barro, F., Fernandez-Escobar, J., De la Vega, M., Martin, A., 2003. Modification of glucosinolate and erucic acid contents in doubled haploid lines of *Brassica carinata* by UV treatment of isolated microspores. *Euphytica.* 129(1), 1–6.

Becquerel, H., 1896. Sur les radiations émises par phosphorescence. *Comptes Rendus.* 122, 420–421.

Bhagwat, B., Duncan, E.J., 1998. Mutation breeding of Highgate (*Musa acuminata*, AAA) for tolerance to *Fusarium oxysporum* f. *sp. cubense using gamma irradiation. Euphytica.* 101 , 143–150.

Brar, D.S., Jain, S.M., 1998. Somaclonal variation: mechanism and applications in crop improvement. In: *Somaclonal variation and induced mutations in crop improvement* (eds) Jain S.M., Brar D.S., Ahloowalia B.S., Kluwer Academic Publishers, Dordrecht. 15–37.

Campbell, B.W., Hofstad, A.N., Sreekanta, S., Fu, F., Kono, T.J., O'Rourke, J.A., Vance, C.P., Muehlbauer, G.J. Stupar, R.M., 2016. Fast neutron-induced structural rearrangements at a soybean NAP1 locus result in gnarled trichomes. *Theor. Appl. Genet.* 129, 1725–1738.

Cassells, A.C., Deadman, M.L., Brown, C.A., Griffin, E., 1998. Field resistance to late blight (*Pytophtora infectans* (Mont.) De Bary in potato (*Solanum tuberosum* L.) somaclones associated with instability and pleiotropic effects. *Euphytica.* 57, 157–167.

Christov, N.K., Ilchovska, D., Hristov, K.N., 2014. Chemical mutagenesis, mutation breeding and quantitative genetic analyses of maize mutants: from theory to practice. In*: Mutagenesis: exploring genetic diversity of crops* (eds) Tomlekova N.B., Kozgar M.I., Wani M.R., Wageningen, Academic Publishers. 427–430.

Cordeiro, G., Eliott, F.G., Henry, R.J., 2006. An optimized ecotilling protocol for polyploids or pooled samples using a capillary electrophoresis system. *Anal. Biochem.* 355, 145–147.

Curie, P., Curie, M., 1898. Sur une substance nouvelle radio-active, contenue dans la pechblende (On a new radioactive substance contained in pitchblende), C. R. hebd. Séanc. *Acad. Sci. Paris.* 127, 175–178.

Das, A., Gosal, S.S., Sidhu, J.S., Dhaliwal, H.S., 2000. Induction of mutations for heat tolerance in potato by using *in vitro* culture and radiation. *Euphytica.* 114, 205–209.

Datta, S.K., Chakrabarty, D., 2009. Management of chimera and in vitro mutagenesis for development of new flower colour/shape and chlorophyll variegated mutants in chrysanthemum. In: *Induced plant mutations in the genomics era* (ed) Shu Q.Y., *Food and Agriculture Organization of the United Nations*, Rome. 303–305.

Donovan, A.M., Morgan, R., Valombra-Piagnani, C., Ridout, M.S., James, D.J., Garrett, C.M.E., 1994. Assessment of somaclonal variation in apple. I. Resistance to the fire blight pathogen, *Erwinia amylovora. J. Hortic. Sci.* 69(1), 105–113.

Ge, H., Li, Y., Fu, H., Long, G., Luo, L., Li, R., Deng, Z., 2015. Production of sweet orange somaclones tolerance to citrus canker disease by *in vitro* mutagenesis with EMS. *Plant Cell Tiss. Organ Cult.* 123, 29–38.

Gimenez, E., Salinas, M., Manzano-Agugliaro, F., 2018. Worldwide research on plant defense against biotic stresses as improvement for sustainable agriculture. *Sustainability.* 10(2), 391. doi:10.3390/su10020391

Godoy, F., Olivos-Hernández, K., Stange, C., Handford, M., 2021. Abiotic stress in crop species: improving tolerance by applying plant metabolites. *Plants.* 10(2), 186. doi:10.3390/plants10020186

Greene, E.A., Codomo, C.A., Taylor, N.E., Henikoff, J.G., Till, B.J., Reynolds, S.H., Enns, L.C., Burtner, C., Johnson, J.E., Odden, A.R., Comai, L., Henikoff, S., 2003. Spectrum of chemically induced mutations from a large-scale reverse-genetic screen in Arabidopsis. *Genetics.* 164(2), 731–740.

Gupta, V., Arora, R., Ranjan, A., Bairwa, N.K., Malhotra, D.K., Udhayasuriyan, P.T., Saha, A., Bamezai, R., 2005. Gel-based nonradioactive single-strand conformational polymorphism and mutation detection: limitations and solutions. *Methods Mol. Biol.* 291, 247–261.

Hammerschlag, F.A., 2000. Resistant responses of peach somaclone 122-1 to *Xanthomonas campestris* pv. *pruni* and to *Pseudomonas syringae* pv. *syringae*. *Hort. Science.* 35(1), 141–143.

Hayashi, Y., Takehisa, H., Kazama, Y., Ichida, H., Ryuto, H., Fukunishi, N., Abe, T., Kamba, C., Sato, T., 2007. Effects of ion beam irradiation on mutation induction in rice. In: *Cyclotrons and their applications, Eighteenth International Conference*. Tohoku University, Miyagi, Japan, 237–239.

Hirochika, H., Guiderdoni, E., An, G., Hsing, Y.I., Eun, M.Y., Han, C., Upadhyaya, N., Ramachandran, S., Zhang, Q., Pereira, A., Sundaresan, V., Leung, H., 2004. Rice mutant resources for gene discovery. *Plant Mol. Biol.* 54, 325–334.

Honda, I., Kikuchi, K., Matsuo, S., Fukuda, M., & Saito, H., & Ryoto, H., & Fukunishi, N., Abe, T., 2006. Heavy-ion-induced mutants in sweet pepper isolated by M1 plant selection. *Euphytica.* 152, 61–66.

Hossain, Z., Mandal, A.K.A., Datta, S.K., Biswas, A.K., 2006. Development of NaCl-tolerant strain in *Chrysanthemum morifolium* Ramat. through *in vitro* mutagenesis. *Plant Biol.* 8(4), 450–461.

Imanishi, S., Noguchi, A., Nagata, M., Abe, T. and Honda, I. 2007. Construction of libraries of 'Micro-Tom' tomato mutations induced by heavy-ion bombardment. *Acta Hortic.* (745) 485–489. DOI: https://doi.org/10.17660/ActaHortic.2007.745.34

Jain, S.M., Maluszynski, M., 2004. Induced mutations and biotechnology on improving crops. In: *In vitro applications in crop improvement: recent progress* (eds) Mujib A., Cho M., Predieri S., Banerjee, S., IBH, Oxford, India. 169–202.

Jain, S.M., Suprasanna, P., 2011. Induced mutations for enhancing nutrition and food production. *GeneConserve.* 40, 201–215.

Jankowicz-Cieslak, J., Till, B.J., 2016. Forward and Reverse Genetics in Crop Breeding. In: *Advances in Plant Breeding Strategies: Breeding, Biotechnology and Molecular Tools* (eds) Al-Khayri J., Jain S., Johnson D., Springer, Cham. doi:10.1007/978-3-319-22521-0_8

Joseph, R., Yeoh, H.H., Loh, C.S., 2004. Induced mutations in cassava using somatic embryos and the identification of mutant plants with altered starch yield and composition. *Plant Cell Rep.* 23, 91–98.

Latado, R.R., Adames, A.H., Neto, A.T., 2004. *In vitro* mutation of chrysanthemum (*Dendranthema grandiflora* Tzvelev) with ethylmethanesulphonate (EMS) in immature floral pedicels. *Plant Cell Tiss. Organ Cult.* 77(1), 103–106.

Lee, L.S., Till, B.J., Hill, H., Huynh, O.A., Jankowicz-Cieslak, J., 2014. Mutation and mutation screening. In: *Cereal genomics: methods and protocols* (eds) Henry R.J., Furtado A., Springer. 77–95.

Leung, H., Wu, C., Baraoidan, M., Bordeos, A., Ramos, M., Madamba, S., Cabauatan, P., Vera Cruz, C., Portugal, A., Reves, G., Bruskiewich, R., McLaren, G., Gregorio, G., Bennett, J., Brar, D., Khush, G., Schnable, P., Wang, G. and Leach, J. 2001. Deletion mutants for functional genomics: progress in phenotyping, sequence assignment, and database development. In: *Rice Genetics IV* (eds) Khush G., Brar D., Hardy B., New Delhi Science Publishers. 239–251.

Li, H.Z., Zhou, W.J., Zhang, Z.J., Gu, H.H., Takeuchi, Y., Yoneyama, K., 2005. Effect of gamma radiation on development, yield and quality of microtubers *in vitro* in *Solanum tuberosum* L. *Biol. Plant.* 49(4), 625–628.

Liu, G., Aspinall, M.D., Ma, X., et al., 2009. An investigation of the digital discrimination of neutrons and gamma-rays with organic scintillation detectors using an artificial neural network. In: *Nuclear Instruments and Methods in Physics Research Section A: Accelerators, Spectrometers, Detectors and Associated Equipment*. 607(3), 620–628.

Liu, S., Wang, H., Zhang, J., Fitt, B.D.L., Xu, Z., Evans, N., Liu, Y., Yang, W., Guo, X., 2005. *In vitro* mutation and selection of doubled haploid *Brassica napus* lines with improved resistance to *Sclerotinia sclerotiorum*. *Plant Cell Rep.* 24, 133–144.

Liu, Y., Liu, H.Y., Zheng, Z.Y., 1997. The application of microspore culture in screening disease resistance to *Sclerotinia sclerotiorum* in oilseed rape. *Southwest China J. Agric. Sci.* 10, 108–112.

Luan, Y.S., Zhang, J., Gao, X.R., An, L.J., 2007. Mutation induced by ethylmethanesulphonate (EMS), *in vitro* screening for salt tolerance and plant regeneration of sweet potato (*Ipomoea batatas* L.). *Plant Cell Tiss. Organ Cult.* 88, 77–81.

Maluszynski, M., Szarejko, I., Bhatia, R., Nichterlein, K., Lagoda, P.J.L., 2009. Mutation techniques. In: *Plant breeding and farmer participation* (eds) Ceccarelli S., Guimaraes E.P., Weltzien E., Food and Agriculture Organization of the United Nations. 159–194.

Maluszynski, M., Szarejko, I., Maluszynska, J., 2003. Mutation techniques. In: *Encyclopedia of applied plant sciences* (eds) Thomas B., Murphy D.J., Murray B.G., Elsevier Academic Press, San Diego. 186–201.

Mangal, M., Sharma, D.R., 2002. *In vitro* mutagenesis and cell selection for the induction of black rot resistance in cauliflower. *J. Hortic. Sci. Biotechnol.* 77(3), 268–272.

Maniatis, T., Fritsch, E., Sambrook, J. 1982. *Molecular cloning: a laboratory manual Textbook.* Cold Spring Harbor Laboratory.

Mba, C., 2013. Induced mutations unleash the potentials of plant genetic resources for food and agriculture. *Agronomy.* 3, 200–231.

McCallum, C.M., Comai, L., Greene, E.A., Henikoff, S., 2000. Targeting induced local lesions IN genomes (TILLING) for plant functional genomics. *Plant Physiol.* 123(2), 439–442.

Mei, M., Qin, Y., Sun, Y., 1998. Morphological and molecular changes of maize plants after seeds been flown on recoverable satellite. *Adv. Space Res.* 22, 1691–1697.

Milbury, C.A., Li, J., Makrigiorgos, G.M., 2009. COLD-PCR-Enhanced high resolution melting enables rapid and selective identification of low-level unknown mutations. *Clin. Chem.* 55, 2130–2143.

Misra, P., Datta, S.K., Chakrabarty, D., 2003. Mutation in flower colour and shape of *Chrysanthemum morifolium* induced by gamma-radiation. *Biol. Plant.* 47(1), 153–156.

Mitra, R.D., Butty, V.L., Shendure, J., Williams, B.R., Housman, D.E., Church, G.M., 2003. Digital genotyping and haplotyping with polymerase colonies. *Proc. Natl. Acad. Sci. USA* 100(10), 5926–5931.

Mittler, R., 2006. Abiotic stress, the field environment and stress combination. *Trends Plant Sci.* 11, 15–19.

Morishita, T., Yamaguchi, H., Degi, K., Shikazono, N., Hase, Y., Tanaka, A., Abe, T., 2003. Dose response and mutation induction by ion beam irradiation in buckwheat. *Nucl. Instrum. Meth. Phys. Res. B.* 206, 565–569.

Morita, R., Kusaba, M., Iida, S., Yamaguchi, H., Nishio, T., Nishimura, M., 2009. Molecular characterization of mutations induced by gamma irradiation in rice. *Genes Genet. Syst.* 84(5), 361–370.

Naito, K., Kusaba, M., Shikazono, N., Takano, T., Tanaka, A., Tanisaka, T., Nishimura, M., 2005. Transmissible and nontransmissible mutations induced by irradiating *Arabidopsis thaliana* pollen with gamma-rays and carbon ions. *Genetics.* 169(2), 881–889.

Natarajan, A.T., 2005. Chemical mutagenesis: from plants to human. *Curr. Sci.* 89(2), 312–317.

Newton, C.R., Graham, A., Heptinstall, L.E., Powell, S.J., Summers, C., Kalsheker, N., Smith, J.C., Markham, A.F., 1989. Analysis of any point mutation in DNA. *The amplification refractory mutation system (ARMS). Nucleic Acids Res.* 17(7), 2503–2516.

Oladosu, Y., Rafii, M.Y., Abdullah, N., Hussin, G., Ramli, A., Rahim, H.A., Miah, G., Usman, M., 2016. Principle and application of plant mutagenesis in crop improvement: A review. *Biotechnol. Biotechnol. Equip.* 30(1), 1–16.

Oloriz, M.I., Gil, V., Rojas, L., Veitia, N., Höfte, M., Jiménez, E., 2011. Selection and characterisation of sugarcane mutants with improved resistance to brown rust obtained by induced mutation. *Crop Pasture Sci.* 62, 1037–1044.

Phanchaisri, B., Chandet, R., Yu, L.D., Vilaithong, T., Jamjod, S., Anuntalabhochai, S., 2007. Low-energy ion beam-induced mutation in Thai jasmine rice (*Oryza sativa* L. cv. KDML 105). *Surf. Coat. Technol.* 201(19–20), 8024–8028.

Predieri, S., 2001. Mutation induction and tissue culture in improving fruits. *Plant Cell Tiss. Organ Cult.* 64, 185–210.

Predieri, S., Zimmerman, R.H., 2001. Pear mutagenesis: *in vitro* treatment with gamma-rays and field selection for productivity and fruit traits. *Euphytica.* 117(3), 217–227.

Raina, A., Sahu, P.K., Laskar, R.A., Rajora, N., Sao, R., Khan, S., Ganai, R.A. 2021. Mechanisms of genome maintenance in plants: playing it safe with breaks and bumps. *Front. Genet.* 12, 675686. doi:10.3389/fgene.2021.675686

Rasmussen, S., Barah, P., Suarez-Rodriguez, M.C., Bressendorff, S., Friis, P., Costantino, P., Bones, A.M., Nielsen, H.B., Mundy, J., 2013. Transcriptome responses to combinations of stresses in Arabidopsis. *Plant Physiol.* 161, 1783–1794.

Ren, W., Zhang, Y., Deng, B., Guo, H., Cheng, L., Liu, Y. 2010. Effect of space flight factors on alfalfa seeds. *Afr. J. Biotechnol.* 9, 7273–7279.

Röntgen, W.C., 1895. Über eine neue Art von Strahlen (Vorläufige Mittheilung). Sitzungsberichte der Physikalisch-medizinischen Gesellschaft zu Würzburg. 29, 132–141.

Sato, Y., Shirasawa, K., Takahashi, Y., Nishimura, M., Nishio, T., 2006. Mutant selection from progeny of gamma-ray-irradiated rice by DNA heteroduplex cleavage using *Brassica* petiole extract. *Breed. Sci.* 56, 179–183.

Schneeberger, K., Ossowski, S., Lanz, C., Juul, T., Petersen, A.H., Nielsen, K.L., Jørgensen, J.E., Weigel, D., Andersen, S.U., 2009. SHOREmap: simultaneous mapping and mutation identification by deep sequencing. *Nat. Methods* 6(8), 550–551.

Sega, G.A., 1984. A review of the genetic effects of ethyl methanesulfonate. *Mutat. Res. Genet. Toxicol.* 134(2–3), 113–142.

Sergeeva, L.I., Keurentjes, J.J., Bentsink, L., Vonk, J., van der Plas, L.H., Koornneef, M., Vreugdenhil, D., 2006. Vacuolar invertase regulates elongation of *Arabidopsis thaliana* roots as revealed by QTL and mutant analysis. *Proc. Natl. Acad. Sci. U.S.A.* 103, 2994–2999.

Sharma, R.P., Chopra, V.L., 1994. Mutation breeding. In: *Plant breeding: theory and practice* (eds) Chopra V.L., Oxford and IBH Pub. Co, New Delhi, India. 409–466.

Shelake, R.M., Pramanik, D., Kim, J.Y., 2019. Evolution of plant mutagenesis tools: a shifting paradigm from random to targeted genome editing. *Plant Biotechnol. Rep.* 13, 423–445.

Shen, Y., Pan, G., Lubberstedt, T., 2015. Haploid strategies for functional validation of plant genes. *Trends Biotechnol.* 33, 611–620.

Suprasanna, P., Rao, P.S., 1997. Selection of mutants using plant cell and tissue culture techniques. In: *Advances in Plant Physiology* (ed) Hemantaranjan, A., Scientific Publishers, Jodhpur, India. 1, 103–122.

Suprasanna, P., Jain, S.M., Ochatt, S.J., Kulkarni, V.M., Predieri, S 2012. Applications of *in vitro* techniques in mutation breeding of vegetatively propagated crops. In: *Plant mutation breeding and biotechnology* (eds) Shu Q.Y., Forster B.P., Nakagawa H., CABI Publishing, Wallingford. 371–385.

Suprasanna, P., Mirajkar, S.J., Patade, V.Y., Jain, S.M. 2014. Induced mutagenesis for improving plant abiotic stress tolerance. In: *Mutagenesis: exploring genetic diversity of crops* (eds) Tomlekova N.B., Kozgar, M.I., Wani, M.R., Wageningen, Academic Publishers, The Netherlands. 345–376.

Swamy, N.R., Ugandhar, T., Praveen, M., Rambabu, M., Upender, M., 2005. Induction of streptomycin resistant plantlets in *Solanum surattense* through *in vitro* mutagenesis. *Plant Cell Tiss. Organ Cult.* 80, 201–207.

Swanson, E.B., Herrgesell, M.J., Arnoldo, M., Sippell, D.W., Wong, R.S.C., 1989. Microspore mutagenesis and selection: Canola plants with field tolerance to the imidazolinones. *Theor. Appl. Genet.* 78, 525–530.

Szantai, E., Ronai, Z., Szilagyi, A., Sasvari-Szekely, M., Guttman, A., 2005. Haplotyping by capillary electrophoresis. *J. Chromatogr. A.* 1079(1–2), 41–49.

Tanaka, A., Shikazono, N., Hase, Y., 2010. Studies on biological effects of ion beams on lethality, molecular nature of mutation, mutation rate, and spectrum of mutation phenotype for mutation breeding in higher plants. *J. Radiat. Res.* 51, 223–233.

Till, B.J., Cooper, J., Tai, T. H., Colowit, P., Greene, E.A., Henikoff, S., Comai, L. 2007. Discovery of chemically induced mutations in rice by TILLING. *BMC Plant Biol.* 7, 1–12.

Toker, C., Yadav, S., Solanki, I.S., 2007. Mutation breeding. *In: Lentil: An ancient crop for modern times* (eds) Yadav, S.S., McNeil, D.L., Stevenson, P.C., Springer, Dordrecht. 209–224.

Tomaiuolo, R., Spina, M., Castaldo, G., 2003. Molecular diagnosis of cystic fibrosis: comparison of four analytical procedures. *Clin. Chem. Lab. Med.* 41(1), 26–32.

Turnpenny, P., Ellard, S., 2007. DNA technology and applications. In: *Emery's Elements of Medical Genetics* (13th ed.), Churchill Livingstone Elsevier's Press, Philadelphia.

Van-Dijk, A.D.J., Kootstra, G., Kruijer, W., de-Ridder, D., 2020. Machine learning in plant science and plant breeding. *iScience.* 24(1), 101890. doi: 10.1016/j.isci.2020.101890

Varshney, R.K., Sigmund, R., Börner, A., Korzun, V., Stein, N., Sorrells, M.E., Langridge, P., Graner, A., 2005. Interspecific transferability and comparative mapping of barley EST-SSR markers in wheat, rye and rice. *Plant Sci.* 168(1), 195–202.

Venkataiah, P., Christopher, T., Karampuri, S., 2005. Selection of atrazine-resistant plants by *in vitro* mutagenesis in pepper (*Capsicum annuum*). *Plant Cell Tiss. Organ Cult.* 83, 75–82.

Venugopalan, V.P., Suprasanna, P., 2022. Use of radiation in food and agriculture. *Curr. Sci.* 123(3), 370–376.

Wani, M.R., Kozgar, M.I., Tomlekova, N.B., Khan, S., 2014. Selection for polygenic variability in early mutant generations of mungbean (*Vigna radiata* (L.) Wilczek). In*: Mutagenesis: Exploring Genetic Diversity of Crops* (eds) Tomlekova N.B., Kozgar M.I., Wani M.R., Wageningen Academic Publishers, The Netherlands. 213–232.

Watanabe, H., 2001. Significance and expectations of ion beam breeding. In: *Gamma Field Symposia*, No. 40, Institute of Breeding, NIAS, Ibaraki, Japan. 15–19.

Williams, I.J., Goodeve, A.C. (2002). Conformation-Sensitive Gel Electrophoresis. *In: PCR Mutation Detection Protocols* (eds) Theophilus B.D.M., Rapley R., Humana Press. 187, 137–150.

Wu, J.L., Wu, C., Lei, C., Baraoidan, M., Bordeos, A., Madamba, M.R., Ramos-Pamplona, M., Mauleon, R., Portugal, A., Ulat, V.J., Bruskiewich, R., Wang, G., Leach, J., Khush, G., Leung, H., 2005. Chemical- and irradiation-induced mutants of indica rice IR64 for forward and reverse genetics. *Plant Mol Biol.* 59(1), 85, 97.

Xu, M., Rao, Z., Dou, W., Yang, J., Jin, J., Xu, Z., 2012. Site-directed mutagenesis and feedback-resistant N-acetyl-L-glutamate kinase (NAGK) increase Corynebacterium crenatum L-arginine production. *Amino Acids.* 43, 255–266.

Yoshihara, R., Nozawa, S., Hase, Y., Narumi, I., Hidema, J., Sakamoto, A.N., 2013. Mutational effects of γ-rays and carbon ion beams on Arabidopsis seedlings. *J. Radiat. Res.* 54(6), 1050–1056.

Young I (2010) *Gene structure andfunction. In: Medical Genetics.* (2nd ed.), Oxford University Press, New York.

Yu, Z., Deng, J., He, J., Huo, Y., Wu, Y., Wang, X., Lui, G., 1991. Mutation breeding by ion implantation. *Nucl. Instrum. Methods Phys. Res. B.* 59, 705–708.

Zambrano, A.Y., Demey, J.R., Fuchs, M., González, V., Rea, R., De Sousa, O., Gutiérrez, Z., 2003. Selection of sugarcane plants resistant to SCMV. *Plant Sci.* 165(1), 221–225.

Zhang, F.L., Takahata, Y., 1999. Microspore mutagenesis and *in vitro* selection for resistance to soft rot disease in Chinese cabbage (*Brassica campestris* L. *ssp. pekinensis*). *Breed Sci.* 49(3), 161–166.

Zhao, M.X., Sun, H.Y., Ji, R.R., Hu, X.H., Sui, J.M., Qiao, L.X., Chen, J., Wang, J.S., 2013. *In vitro* mutagenesis and directed screening for salt-tolerant mutants in peanut. *Euphytica.* 193 , 89–99.

Zhu, M.Y., Pan, J., Wang, L., Gu, Q., Huang, C., 2003. Mutation induced enhancement of Al tolerance in barley cell lines. *Plant Sci.* 164(1), 17–23.

9 Genetic Improvement Using Induced Mutagenesis with Special Reference to Pulses

A Review

Vinod Janardan Dhole, Souframanien Jegadeesan, and Dhanasekar Punniyamoorthy*
Nuclear Agriculture and Biotechnology Division,
Bhabha Atomic Research Centre, Trombay, Mumbai,
Maharashtra, India
*Corresponding Email: vjdhole@barc.gov.in

9.1 INTRODUCTION

Pulses play a very important role in nutritional security by providing a cheap protein source to the vegetarian population of developing and under developed countries in Asia and Africa. Among the pulse-growing countries, India can be regarded as the home for most of the pulse crops like chickpea (*Cicer arietinum*), pigeon pea (*Cajanus cajan*), mungbean (*Vigna radiata*), black gram (*Vigna mungo*), lentils (*Lens esculenta*) and some arid legume pulses like cowpea (*Vigna unguiculata*), moth bean (*Vigna aconitifolia*), and horse gram (*Macrotyloma uniflorum*). Farmers have been cultivating various pulse crops since their domestication and references to various pulses were found in the ancient Indian Sanskrit literature like the *Vedas*, viz., *khalva* (chickpea), *adhaki* (pigeon pea), *mudga* (mungbean), *masha* (black gram), and *masura* (lentils) (Nene, 2006). Being legumes, pulse crops play a very important role in sustainable agriculture by improving soil fertility through crop rotation and symbiotic nitrogen fixation. All the pulses are rich in protein along with other nutrients, minerals, and vitamins. Among the pulses, mungbean possesses the most easily digestible protein, followed by moth bean and black gram. Pulses are also rich in carbohydrates, vitamin C, folic acid, thiamine, iron, zinc, potassium, magnesium, copper, manganese, and phosphorous (Mubarak, 2005, Taunk et al., 2012, Dahiya et al., 2013).

Pulses are more popular in countries with a predominantly vegetarian population like India as it is the cheapest protein source. On average, an adult man would require about 60 g and an adult woman would require 50 g of protein per day, while

DOI: 10.1201/9781003392897-9

a lactating woman would require 75 g per day. To fulfill the daily protein requirement through a balanced vegetarian diet, a minimum of 50 g of pulses per capita per day should be included in the daily diet. The pulse crop production could not cope with the ever-growing population of India, which resulted in declined per capita availability of pulses from 60.7 g in 1951 to 40 g per day per capita in 1998 (Anonyms, 2000). In the years 2020–2021, the situation slightly improved in India, with 45 g per capita availability of pulses (Anonyms, 2022). In India, the area under pulses slightly increased from 263.9 lakh ha in 2010–2011 to 287.83 lakh ha in 2020–2021, with 182.32 and 254.63 lakh tons of production, respectively (Anonyms, 2022). During the last decade, the average productivity of pulse crops also slightly increased from 691 to 885 kg/ha. However, the productivity of pulses in India is very low as compared to global productivity (994 kg/ha). Among the pulses, field pea has the highest productivity of 1440 kg/ha followed by chickpea (1192 kg/ha), lentil (1017 kg/ha), pigeon pea (914 kg/ha), mungbean (601 kg/ha), and black gram (538 kg/ha). The low productivity of pulses in India is attributed to the type of pulses grown, growing conditions like rain-fed and marginal land with minimum inputs, susceptibility to various biotic and abiotic stresses, inherent physiological reasons like excessive flower drop, poor translocation of food material from source to sink and low genetic variability, and cultivars with a narrow genetic base having a low buffering capacity to changing environment. There is an urgent need to increase production and productivity to cope with the ever-growing demand for pulses. In India, 26.99 lakh tons of pulses were imported in the year 2021–2022, which was 9.44% more than imports during the previous year, 2020–2021 (24.66 lakh tons) (Anonyms, 2022).

The increased pulse production was mainly attributed to improved varieties, plant protection measures, agronomical practices, and irrigation facilities. Various plant breeding methods were applied for the genetic improvement of pulse crops, like selection, hybridization, including interspecific hybridization, and mutation breeding to create new genetic variability. Mutation breeding is the most underutilized method of plant breeding due to a lack of expertise and knowledge for mutagenic treatments, mutant screening, low frequency of desirable mutants in a small population, and the random nature of mutations. The demonstration of induced mutation in maize and barley by Stadler in 1928 gave birth to mutation breeding in food crops. Since the birth of mutation breeding, a large number of mutant varieties (3402) have been developed through induced mutation in various crops, including pulses (MVD, IAEA, 2023). However, the share of the pulse varieties (311) developed through induced mutations is very small as compared to cereals (1622). Overall, research on crop improvement and genomics has been neglected in pulse crops, which are often regarded as orphan legumes (Varshney et al., 2009). In the near future, mutation breeding can play an important role in the improvement of pulse crops by creating new genetic variability and novel traits. The novel mutants will help to understand gene expression and biochemical pathways at the molecular level.

9.1.1 Why Use Mutation Breeding in Pulse Crops?

Most of the pulse crops are self-pollinated (except pigeon pea) with small, delicate flowers accompanied by excessive flower drop and poor source-to-sink translocation, which makes hybridization difficult. Some of the crops, like black gram, moth bean, and horse gram mostly cultivated in India and the Indian subcontinent, have very little genetic variability. Lack of natural genetic variability in the pulse crops limits the genetic improvement for yield and biotic and abiotic stress tolerance. The narrow genetic base in modern varieties of pulses leads to the low buffering capacity to various biotic and abiotic stresses in the changing environmental conditions. Pulses are not very rich in wild germplasm, e.g., mungbean has two wild relatives (*Vigna radiata* var. *sublobata* and *Vigna radiata* var. *etulose*) while there is only one in black gram (*Vigna mungo* var. *silvestris*). Similarly, other crops also have one or two wild species in their primary gene pool, e.g., two each in chickpea and lentil, and one each in pigeon pea, black gram, common bean, and lathyrus (Kumar et al., 2004). This offers very limited options for transferring the alien genes from wild relatives to cultivated varieties in pulse crops. In the process of domestication, only a few species were selected, which led to limited genetic variability (Ladizinsky, 1985). Very few wild relatives, breeding behavior, and the nature of domestication events may limit the gene flow between the new cultivars and their wild relatives (Cooper et al., 2001). Due to lack of genetic variability in the primary gene pool, very few genotypes were used to breed new varieties, which resulted in a narrow genetic base with less buffering capacity to different stresses. In the genetic base of 138 chickpea varieties developed through hybridization, 50% is contributed by 28 parents, while 3 parents, Rabat, Pb 7, and JG 62, have contributed over 10% of the genes (Srivastava et al., 2017). Similarly, 57 genotypes were used to develop 47 pigeon pea varieties, while 26 genotypes were repeatedly used to develop 70 black gram varieties (Kumar et al., 2004). In the development of 49 mungbean varieties, 71 parents were used, and the top 10 ancestors contributed 79% of the genetic base. The T1 parent was involved in the development of 35% of varieties (Kumar et al., 2004). The examples above revealed "how broad the genetic base of modern varieties of pulse crops was". The most commonly used breeding method, i.e., hybridization, cannot create new genetic variability just by utilizing the existing variability for shuffling and reshuffling of genes to create new recombination. On the contrary, induced mutation creates new genetic variability, sometimes with novel traits, and hence, the use of this method in crops like pulses with little genetic variability is more rewarding. The induced mutants with novel traits can be utilized in hybridization programs for development of high-yielding pulse varieties with a broad genetic base.

9.1.2 Spontaneous and Induced Mutations

Spontaneous or natural mutations are a vital force in the evolution process. The mutations that occur naturally without any deliberate use of mutagens are called spontaneous mutations. Spontaneous mutations may be induced by errors in DNA replication, transposable genetic elements, and background radiations. There are some examples of natural mutants that were released as varieties for commercial cultivation, e.g., spontaneous mutant variety Co-9 in lablab (the Co-9 was further used to develop varieties, viz., Co-11, Co-12, Co-13, and Co-GB-14), black gram variety, Vamban-2, is the spontaneous mutant from T9. Several spontaneous mutants were identified for different traits in chickpea, e.g., nonnodulating mutant (Rupela and Sundarshana, 1986), simple and multipennate leaf mutants (Kumar et al., 1999), and open flower mutant OFM-3 (Srinivasan and Gaur, 2012). Male sterile spontaneous mutant in peas (Sing and Sing 1995), obcordate leaf and open flower mutant (Bohra et al., 2016), and early maturing mutant (Verulkar and Singh 1997; Saxena et al., 2019) in pigeon pea were identified. The spontaneous mutations in pulse crops were reported with a very low frequency, i.e., 1×10^{-9} to 1×10^{-6}, while induced mutations enhance the mutation frequency by many folds, i.e., 1×10^{-3} by using various physical and chemical mutagens. Thus, we can enhance the process of mutations artificially, which otherwise happens in nature at a very slow rate. Hence, genetic modification through induced mutation is outside the regularity norms of environmental and health safety.

9.1.3 Traits to Be Targeted for Genetic Improvement of Pulses through Induced Mutation

Induced mutations are sudden heritable changes that occur following mutagenic treatments. These mutations are random and mostly recessive in nature; hence, the genetics of traits (mostly the qualitative trait) under improvement is of the utmost importance in induced mutations. The most common targets for the induced mutations are one or a few specific traits in well-adapted variety or elite line, e.g., disease resistance, early maturity, reduction in plant height, and some novel traits like insect and herbicide resistance (Table 9.1). In pulse crops, the most common trait to be targeted is the disease resistance as these crops are heavily infected by a number of diseases, causing heavy yield losses of up to 100%, e.g., yellow mosaic disease (YMD) in mungbean and black gram (Varma and Malathi 2003 and Kang et al., 2005). While targeting traits like disease resistance, the genetics of resistance gene should be taken into consideration. In chickpea and pigeon pea, *Fusarium* wilt resistance is governed by recessive genes, and therefore, it can be targeted and easily screened in M_2 generation in disease-sick plots (Sharma and Muehlbauer, 2007; Sharma et al., 2019 and Patil et al., 2017). Similarly, YMD in mungbean, black gram, and horse gram (Dhole and Reddy, 2012; Dhole and Reddy, 2013; Sudha et al., 2013; Gupta et al., 2005), cowpea yellow mosaic virus in cowpea (Dinesh et al., 2018), bean common mosaic in common bean (Feng et al., 2018), powdery mildew in pea (Bobkov and Selikhova, 2021), and *Ascochyta* blight and rust in lentil (Chowdhury et al., 2001; Kumar and Gill 2020) are the diseases for which genetic resistance can be developed through induced mutation due to recessive nature of resistance genes. Another most frequently targeted trait in pulse crops is the maturity duration, which is useful to increase the area under pulses through multiple cropping, reduction in inputs like irrigation, and enables escape from terminal drought conditions. In early maturing pulse crops like mungbean, black gram, cowpea, and common bean a target of 5–10 days earliness is desired, while in other long duration crops like chickpea, peas, lentil, and horse gram, earliness of more than 15 days is always desirable without compensating the yield. Maturity duration is of utmost importance in very long duration crops like pigeon pea, where super early (<90 days), extra early (90–120 days), and early (120–160 days) maturing mutants will enable fitting of this crop in multiple cropping systems. The problems of photosensitivity and indeterminate growth habit in pulse crops pose limitations on the cultivation of these crops in all seasons. Determinate growth habit with photo insensitivity could be the target traits in pulses, particularly

TABLE 9.1
Traits to Be Targeted for the Genetic Improvement of Pulse Crops through Induced Mutation

	Crops								
Traits	**Chickpea**	**Pigeon pea**	**Mungbean**	**Black gram**	**Pea**	**Cowpea**	**Lentil**	**Common bean**	**Horse gram**
Disease resistance	*Fusarium* wilt (recessive genes: Sharma &Muehlbauer, 2007; Sharma et al., 2019)	*Fusarium* wilt (Two recessive genes: Patil et al., 2017)	YMD (single/two recessive genes: Dhole & Reddy 2012; Dhole & Reddy 2013)	YMD single recessive genes (Sudha et al., 2013;Gupta et al., 2005)	Powdery mildew (Two recessive genes: Bobkov & Selikhova 2021)	Cowpea yellow mosaic virus (Single recessive gene: Dinesh et al., 2018)	*Ascochyta* blight (single recessive gene: Chowdhury et al., 2001) Rust (Single recessive gene: Kumar & Gill 2020)	Bean common mosaic virus (four recessive genes: Feng et al., 2018)	Yellow Mosaic Virus
Maturity duration (days)	90–100	Super early: <90, Extraearly:90–120	50–55	60–65	80–90	60–65	90–100	60–65	90–100
Plant type	Suitable for MH	Determinate	Suitable for MH	Suitable for MH	Determinate	Determinate	Suitable for MH	Determinate top bearing	Determinate
Quality	High protein and low phytic acid content	Cooking quality and protein content	High protein and low phytic acid content	High protein, less phytic acid and oligo-saccharides content	Lower haemagglutinins, phytic acid and trypsin inhibitor	Lower phytic acid and trypsin inhibitor and oligo-saccharides	Lower haemagglutinins, phytic acid and trypsin inhibitor	Lower α-Amylase inhibitor and phytic acid	Lower oxalic acid and tannins

(*Continued*)

TABLE 9.1 (Continued)
Traits to Be Targeted for the Genetic Improvement of Pulse Crops through Induced Mutation

	Crops								
Traits	**Chickpea**	**Pigeon pea**	**Mungbean**	**Black gram**	**Pea**	**Cowpea**	**Lentil**	**Common bean**	**Horse gram**
Yield	Seed size, seeds pod^{-1}, pod size branching	More number of secondary and tertiary branches	Large seed, seeds pod^{-1} and more number of clusters	More number of cluster and seeds plant^{-1}	More number of secondary and tertiary branches	More number of clusters and pods	Seed size and more number of pods	More number of clusters and pods	More number of pods plants^{-1}
Novel traits	Pod borer tolerance	Tolerance to Pod borer, drought, and water logging	Tolerance to *Maruca* pod borer	Tolerance to *Maruca* pod borer	Supernodulation and tolerance to Pod borer	Tolerance to *Maruca* pod borer	Weedicide (Imazethapyr) resistance	Tolerance to *Maruca* pod borer	Tolerance to *Maruca* pod borer

Notes: *$^{-1}$ denotes "per".
Abbreviations: YMD:Yellow Mosaic Disease, MH: Mechanical Harvesting.

winter season pulses, like chickpea, peas, lentil, and determinate growth habit could be the target trait in black gram, cowpea, common bean, and pigeon pea. High cost of cultivation and nonavailability of labors for harvesting are the major problems faced by farmers in the modern industrial era. Hence, plant types suitable for mechanical harvesting can be restructured through mutation breeding in high-yielding varieties of chickpea, mungbean, black gram, cowpea, and lentil. Improvement of quality in terms of increasing the nutrient contents and cooking quality along with reducing the antinutritional factors like phytic acid are the major objectives in all the pulses (Souframanien et al., 2021). Raffinose-like oligosaccharides in black gram, trypsin inhibitor in cowpea and peas, •-amylase inhibitor in common bean, oxalic acid and tannins in horse gram, and hemagglutinins in peas and lentil are the antinutritional factors that can be reduced through induced mutations (Kumar et al., 2022). Nondestructive analysis of important nutrients and micronutrients of large number of samples from seeds of large M_2 populations is now possible with the availability of instruments like near-infrared spectroscopy (NIR) and energy dispersive X-ray fluorescence (EDXRF) (Guild and Stangoulis 2021). Thus, such high-throughput techniques will boost biofortification programs through mutation breeding. The induced mutation in quantitative traits like yield and other yield components like seed test weight, pod number, number of secondary and tertiary branches, number of clusters per plant, number of seeds per plant, and plant height in pulses can be targeted as the screening of these traits is easy and based on phenotype.

Novel traits like insect tolerance, root traits, tolerance to drought and water logging, and supernodulation for increasing symbiotic nitrogen fixation can be induced through mutations in pulse crops. Breeding for insect tolerance through induced mutations in pulses is generally not a common objective because of complexity of the trait and the difficulty experienced in screening for insect resistance, resulting in very few successful examples in these crops. The pulse crops are very prone to insect pests like all types of pod borers, sucking pests like aphids, jassids, thrips, whiteflies, and mites. Some of the sucking pests are the vectors for some serious viral diseases like yellow mosaic disease and cowpea aphid-borne mosaic virus spread through whiteflies and aphids, respectively. The development of insect resistance through mutation breeding in edible crops like pulses is an alternative to the GMOs, which will reduce the use of pesticides, thereby reducing the cost of cultivation and stabilizing the yield potential of modern varieties. Similarly, the herbicide resistance can be induced in pulse crops to reduce yield losses and also the cost of cultivation by reducing the dependence on manual weeding.

9.1.4 Choice of Mutagens for Pulse Crop Improvement

Mutagens are the physical and chemical agents used for induction of mutations in all living organisms. The X-rays were discovered by Roentgen in 1895, while radioactivity and radioactive elements were discovered by Becquerel in 1896 and by Marie and Pierre Curie in 1898, respectively. The utility of physical mutagens in induced mutagenesis was demonstrated much later in the late 1930s in *Drosophila*, maize and barley, followed by mustard gas in *Drosophila* (Auerbach 1946; Auerbach and Robson 1946). Since then, a large number of physical and chemical mutagens have been identified and used in mutation breeding of various food crops including pulses. These confidence-instilling discoveries gave humans the assurance that they can induce mutations in any organism that can be used for the betterment of mankind.

9.1.5 Physical Mutagens

Physical mutagens included ionizing radiations like X-rays, gamma rays, alpha and beta particles, protons, and neutrons, and also non-ionizing radiations like ultraviolet rays. Among the ionizing radiations, X-rays were used first for inducing mutations. X-rays are electromagnetic radiations emitted as quanta from X-ray machines. For mutation induction, hard X-rays (short wavelength) are generally preferred since their penetration is greater than soft X-rays (which have a longer wavelength) (Spencer-Lopes et al., 2018). Gamma rays are the most potent electromagnetic radiation applied for inducing mutation in plants due to their high energy (1.17 and 1.33 MeV) and high penetration power. The radioisotopes "Cobalt-60" (^{60}Co) and "Caesium-137" (^{137}Cs) are the main sources of gamma rays. Both X-rays and gamma rays induce mutations by directly breaking the DNA strands, while in indirect effects, reactive oxygen species are formed due to the radiolysis of water, which causes the breakages in DNA strands. The alpha and beta particles are the particulate radiations, which have very little penetration power and, therefore, are rarely used for induction of mutations in plants. Neutrons are the neutral particles present in the nucleus of an atom. It is stable inside the nucleus, and once separated from the nucleus the neutron decays with a mean lifetime of about 15 minutes, releasing various kinetic energies. Thermal (0.4–100 eV) and fast neutrons (200 keV to 10 MeV) are useful for plant mutation induction (Spencer-Lopes et al., 2018). In nonionizing radiation, ultraviolet rays have less penetration power and hence can be used only for inducing mutations in plant parts like pollen grains, cells, and/or plant tissue cultures. Electron beams and ion beams are the physical mutagens that have the potential to be used in induced mutations in crop plants due to their high energy deposition. Electron beams predominantly induce single- and double-strand DNA breaks, resulting in nonrepairable genetic changes. Ion beam irradiation and ion beam implantation are the high LET physical mutagens most recently used for induced mutations in various crops (Abe et al., 2007). The ion beam implantation has many advantages, including a lower damage rate and a higher mutation rate with a wider and novel mutational spectrum.

9.1.6 Chemical Mutagens

The most commonly used chemical mutagens are alkylating agents including methyl methanesulfonate (MMS), ethyl methanesulfonate (EMS), diethyl sulfate (DES), methyl nitroso urea (MNH), ethyl nitrosourea (ENH), and ethyleneimine (EI). The alkylating agents are electrophilic, with one or more alkyl groups reacting with DNA by alkylating the phosphate groups in the phosphodiester backbone as well as the various imino- or carbonyl-groups present on the purine (adenine, guanine) or pyrimidine bases (cytosine, thymine) (Ingelbrecht et al., 2018). After alkylating agents, sodium azide (NaN_3; SA) is used frequently for crop improvement including pulses.

It inhibits cellular respiratory processes in living organisms and induces GC to AT transition type of point mutations. Apart from these major chemical mutagens, base analogues, antibiotics, acridines, nitrous acid, hydroxylamine, and colchicine are also used for inducing mutations.

9.1.7 Space Mutagenesis

In space mutagenesis, cosmic rays are used for induction of mutations by sending the seeds in returnable satellites, space shuttles, or high-altitude balloons. Lesser physiological damage was noticed in space mutagenesis over the gamma rays–induced mutagenesis, and it provides a special, superclean, supervacuum environment with strong cosmic radiation, microgravity, and weak geomagnetic fields (Liu et al., 2009). The space mutagenesis program started way back in China in 1987 and has resulted in the development of 66 crop varieties (Liu et al., 2009).

9.1.8 Importance of LD_{50} and Useful Dose Determination

Determination of LD_{50} is the most important step before starting induced mutation breeding in any pulse crop. The radiosensitivity of various plant parts will be different and depends on the moisture content and oxygen. In all the pulses, seed is mainly used for mutagenic treatment, and hence, seed moisture content is very important and should be around 12%–14%. On the contrary, water-soaked seeds for 6–8 hours are used for chemical mutagenic treatments. The previous reports can be used for deciding the appropriate dose of the mutagen, but it is always advised to do the radiosensitivity study to determine the GR_{50} or LD_{50}. In the same experiment, LD_{50} dose for two genotypes may be different due to their inherent ability to respond to mutagen and varying moisture content at the time of treatment. Even a slight difference of 0.2%–0.3% moisture content might greatly alter the radiosensitivity of different genotypes of the same species (Ingelbrecht et al., 2018). The LD_{50}/GR_{50} dose/concentrations for important pulse crops are summarized in Table 9.2. Pigeon pea is the most sensitive crop to gamma rays with LD_{50} dose of 200 Gy (Ariraman et al., 2014), followed by peas (250 Gy) (Verma and Purbiya, 2017), black gram (325 Gy) (Veni et al., 2017), chickpea (400 Gy) (Umavathi and Mullainathan, 2015), lentil (400–500 Gy) (Mejri et al., 2021), common bean (500 Gy) (Bolbhat et al., 2020), cowpea (400–590 Gy) (Roy et al., 2019), and mungbean (620 Gy) (Roslim et al., 2015). The electron beam was used in various pulse crops in recent years with the availability of this facility at Industrial Accelerators Section, Raja Ramanna Centre for Advanced Technology, Indore (MP) in India. The LD_{50} doses for electron beam in black gram and cowpea were estimated to be 150 Gy (Veni et al., 2017; Eswaramoorthy et al., 2021), while the GR_{50} in mungbean was estimated to be much higher (470 Gy) (Dhole et al., 2023). Reports on the use of proton beam in pulses are rare, and the LD_{50} (200–300 Gy) has been estimated only in cowpea (Kang et al., 2020). The chemical mutagens, viz., EMS and SA, were the most commonly used mutagens in pulses, and a lot of reports are available on LD_{50} in various pulse crops with wide range of concentrations. The LD_{50} concentration of EMS is the highest in mungbean (0.5%) (Sofia et al., 2020) and common bean (0.3%–0.5%) (Bolbhat et al., 2020) followed by black gram (0.4%)

(Usharani et al., 2017; Goyal et al., 2019), cowpea (0.4%) (Yaseen et al., 2020; Opoku-Gyamfi et al., 2022), chickpea (0.36%–0.46%) (Bhat et al., 2011; Umavathi and Mullainathan 2015), and pigeon pea (25–30 mM) (Ariraman et al., 2014, Giri 2014). Among the pulses, cowpea is the most sensitive to SA, with LD_{50} of 0.04% (Raina et al., 2022), while field pea can tolerate higher concentrations with LD_{50} of 1.27% (Nepolian et al., 2018). The other chemical mutagens, like HZ and MH, have been less commonly used in pulse crops for the induction of mutations. The mutagenic dose should be decided on the basis of LD_{50}, i.e., the dose below LD_{50} was found to be more efficient and effective for inducing viable mutations due to their higher survival rate in the M_1 generation.

9.1.9 Efficient and Effective Dose

Different mutagenic doses or treatments can be compared on the basis of their efficiency and effectiveness for deciding the optimal dose in mutation breeding experiments. The efficiency is based on the rate of biological damages like the seedling injury and pollen sterility, while effectiveness refers to the number of mutations per unit dose (Konzak et al., 1965). The efficient and effective doses of different mutagens for different pulse crops are presented in Table 9.2. The efficient dose of gamma rays in chickpea was reported to be in the range of 200–400 Gy, while lower doses, i.e., 150–300, were more effective to induce viable mutants (Mabrouk et al., 2018; Brahmi et al., 2014; Joshi-Saha et al., 2015, Pérez-León et al., 2019, Sharma et al., 2018). Same doses of gamma rays were found to be both efficient and effective in pigeon pea (50 Gy) (Etther et al., 2019), mungbean (200 Gy) (Bonde et al., 2020; Dhole et al., 2023), black gram (200–400 Gy) (Souframanien et al., 2016; Usharani et al., 2017; Usharani and Ananda Kumar, 2015), peas (100 Gy) (Dhulgande et al., 2011), cowpea (179–318 Gy) (Kang et al., 2020; Prabhu et al., 2022; Gnankambary et al., 2019; Dorvlo et al., 2022), lentil (200 Gy) (Laskar and Khan, 2017), and common bean (300 Gy) (More & Borkar, 2016). Electron beam is another physical mutagen recently used in some pulse crops in which the efficient and effective doses were found to be similar in mungbean (200 Gy) (Dhole et al., 2023) and cowpea (100–200 Gy) (Priyadharshni et al., 2020). Higher dose of electron beam, i.e., 300 Gy, was efficient, while lower dose of 200 Gy was reported to be effective in chickpea (Joshi-Saha et al., 2015) and black gram (Souframanien et al., 2016). EMS showed higher efficiency at concentrations of 0.1% to 0.3%, while it was effective at concentrations of 0.05%–0.4% in inducing mutations in different pulse crops. Higher concentrations of EMS were found effective in pigeon pea (Ariraman et al., 2014; Giri 2014; Etther et al., 2019) and mungbean (Mahto et al., 2018; Sofia et al., 2020; Vikhe and Nehul 2020). The SA treatments at lower concentrations (0.01%) were efficient and effective in chickpea (Bhat et al., 2011; Khan et al., 2005), pigeon pea (Chaudhary et al., 2021; Sangle and Kothekar 2013), and mungbean (Wani 2020; Mahto et al., 2018; Sofia et al., 2020; Vikhe and Nehul 2020), while 0.02% showed similar results in cowpea (Raina et al., 2022;) and lentil (Solanki 2005; Ali et al., 2014b). Higher concentrations of SA (0.25%) were reported to be most efficient and effective in common beans (Bolbhat et al., 2020) and peas (Nepolian et al., 2018; Bansod et al.,

TABLE 9.2
LD_{50}/GR_{50} Dose/Concentration, Efficient and Effective Dose of Various Mutagens for the Important Pulse Crops

S. No.	Crop	Mutagen	GR_{50}/LD_{50}	Efficient dose	Effective dose	Reference
1	Chickpea	Gamma rays	GR_{50}:150/ LD_{50}: 400 Gy	200–400 Gy	150–300 Gy	Mabrouk et al., 2018; Brahmi et al., 2014; Joshi-Saha et al., 2015; Pérez-León et al., 2019; Umavathi and Mullainathan 2015; Sharma et al., 2018
		Electron beam	300 Gy	300 Gy	200 Gy	Joshi-Saha et al., 2015
		EMS	0.36%–0.46%/30 mM	0.1%	0.05%	Bhat et al., 2011; Umavathi and Mullainathan 2015; Sharma et al., 2018
		SA	0.045–0.58	0.01%	0.01%	Bhat et al., 2011; Khan et al., 2005
		HZ	0.052–0.070	0.03%	0.03%	Bhat et al., 2011; Khan et al., 2005
2	Pigeon pea	Gamma rays	200 Gy	50 Gy	50 Gy	Ariraman et al., 2014; Etther et al., 2019
		EMS	25–30 mM	5 mM	30 mM	Ariraman et al., 2014; Giri 2014; Etther et al., 2019
		SA	1.60–2.69 mM	0.01%	0.01%	Chaudhary et al., 2021; Sangle and Kothekar 2013
3	Mungbean	Gamma rays	GR_{50}:375–440 Gy/ LD_{50}:620	200 Gy	200 Gy	Roslim et al., 2015; Bonde et al., 2020;Dhole et al., 2023
		Electron beam	470 Gy	200 Gy	200 Gy	Dhole et al., 2023
		EMS	0.5%/20 mM	0.2%	0.4%	Mahto et al., 2018; Sofia et al., 2020; Vikhe and Nehul 2020
		SA	1 mM/5 mM	0.01%/ 0.06%	0.01%/ 0.06%	Wani 2020; Mahto et al., 2018; Sofia et al., 2020; Vikhe and Nehul 2020
		HZ	0.01%	0.01%	0.01%	Wani 2020; Das and Baisakh 2020

(Continued)

TABLE 9.2 (Continued)
LD_{50}/GR_{50} Dose/Concentration, Efficient and Effective Dose of Various Mutagens for the Important Pulse Crops

S. No.	Crop	Mutagen	GR_{50}/LD_{50}	Efficient dose	Effective dose	Reference
4	Black gram	Gamma rays	325 Gy	200 Gy/400 Gy	200 Gy/600 Gy	Souframanien et al., 2016; Veni et al., 2017; Usharani et al., 2017; Usharani & Ananda Kumar 2015
		Electron beam	400 Gy	400 Gy	200 Gy	
		EMS	0.4%/60 mM	50 mM	50 mM	Goyal et al., 2019; Devi et al., 2021; Usharani et al., 2017
		MH	0.02%	0.01%	0.01%	Devi et al., 2021
		SA	0.5%	40 mM	40 mM	Ali et al., 2014b; Makeen et al., 2013
5	Pea	Gamma rays	250 Gy	100 Gy	100 Gy	Dhulgande et al., 2011; Verma & Purbiya, 2017
		EMS	0.37%	0.15%	0.05%	Dhulgande et al., 2011; Sing et al., 2017
		SA	1.27%		0.2%–0.25%	Nepolian et al., 2018; Bansod et al., 2022
6	Cowpea	Gamma rays	GR_{50}:200–300, LD_{50}:400–590 Gy	179–318 Gy	179–318 Gy	Kang et al., 2020; Roy et al., 2019; Prabhu et al., 2022; Gnankambary et al., 2019; Dorvlo et al., 2022
		Electron beam	150 Gy	100–200 Gy	100–200 Gy	Priyadharshni et al., 2020; Eswaramoorthy et al., 2021
		Proton beam	200–300	-	-	Kang et al., 2020
		EMS	0.4%/50 mM	0.3%	0.3%	Yaseen et al., 2020; OpokuGyamfi et al., 2022; Dhanavel et al., 2008
		SA	0.04%	0.02%	0.02%	Raina et al., 2022
7	Lentil	Gamma rays	400–500 Gy	200 Gy	200 Gy	Laskar and Khan 2017; Mejri et al., 2021
		EMS	-	0.02%	0.02%	Solanki 2005

TABLE 9.2 (Continued)
LD_{50}/GR_{50} Dose/Concentration, Efficient and Effective Dose of Various Mutagens for the Important Pulse Crops

S. No.	Crop	Mutagen	GR_{50}/LD_{50}	Efficient dose	Effective dose	Reference
		SA	0.3%	0.02%	0.02%	Solanki 2005; Ali et al., 2014a
		HZ	0.4%	0.2%	0.2%	Laskar & Khan 2017; Laskar et al., 2017
8	Common bean	Gamma rays	500 Gy	300 Gy	300 Gy	More & Borkar, 2016; Bolbhat et al., 2020
		EMS	0.3%–0.5%	0.15%	0.15%	Kumar & Dhillon 2020; Bolbhat et al., 2020
		SA	0.25%	0.25%	0.25%	Bolbhat et al., 2020

Abbreviations: MH: Maleic Hydrazide; HZ: Hydrazine hydrate; SA: Sodium azide.

2022). The previous reports on efficient and effective doses/concentrations of various mutagens may be very useful for deciding the appropriate dose to induce the desired genetic change in various pulse crops.

9.1.10 Selection of Parental Material, Population Size, and Mutant Screening

The selection of parental material for induction of mutation is mainly based on objectives or the traits to be improved. The parental material should be a widely adapted genotype, a recently released cultivar, an advanced promising line about to be released, or an introduced variety having very specific lacunae. After selection of parental material, another important step is the selection of mutagen(s) based on previous studies or dose determination based on mutagen sensitivity studies. The parental material should be genetically pure, and as far as possible, nucleus or breeder seeds should be used for the mutagenic treatment. If the parental material is sourced from a third party, the progeny test must be followed to ensure genetic purity for agronomic traits. The other important factors to be considered before mutagenic treatment are the germination percent and moisture content of the seeds. The germination rate in the parental material is very important as it plays a decisive role in maintaining the desirable population size in the M_1 and M_2 generations. The mutational frequency of a specific trait of interest mainly determines the size of the population to be screened. Some mutations like dwarfism and lodging resistance are very frequently induced as compared to disease and pest resistance, and therefore, to get such rare mutants, the population size must be at least 10–20 times larger (Donini and Sonnino 1998). Considering the desirable mutation frequency of 10^{-3} per test unit, i.e., pod in pulses with 90% probability and 80% survival rate in M_1 generation, at least 600 seeds per treatment should be treated to get one desirable mutant (Ghanim et al., 2018). On the basis of such an estimation, 6000 seeds should be treated to get 10 desirable mutants. In most of the previous studies on induced mutation, a very small number of seeds were treated with mutagens. As the mutations are random event, the mutation frequency and optimal mutation treatment to get the desired mutant can never be predicted. Therefore, the size of the population should be decided by the breeder on the basis of the facilities available to screen the M_2 population and how much of the population can be handled efficiently. The M_1 population should be space planted to get as many seeds as possible to grow a sufficiently large M_2 population. The M_1 generations of the photoinsensitive crops like mungbean, black gram, and cowpea can be grown in the offseason to save time and the M_2 populations should be screened in the target environment or in the proper season. The M_1 plants should be harvested individually for testing the plant-to-row progenies in M_2 generation. Plant-to-row progeny screening in M_2 generation is very useful to identify and distinguish the real mutants from the variants resulting from mechanical mixtures and or outcrossing. Other methods like branch-to-row, single seed, or modified bulk can be applied to grow M_2 population, wherein two to three seeds from each branch or each of the M_1 plants are bulked so that each of the M_1 plants is properly represented in the M_2 population. These approaches can be followed in crops like pigeon pea and chickpea, where single plant can produce a large number of pods and seeds. The screening methods are of utmost importance in mutation breeding as the success of identification of desired mutants from large M_2 populations solely depends on the efficient screening methodology. Morphological mutants can be identified easily from the large population through visual observations, while identification of mutants for specific traits like disease resistance, yield, and quality traits is difficult and mostly relies on efficient screening techniques.

9.1.11 Identification of Mutants for Qualitative Traits

In most of the mutation breeding programs, the macromutations are more frequently reported due to conspicuous changes in qualitative traits following mutagenic treatment that are easy to identify in large M_2 populations (Table 9.3). In very early stage of crop development, most commonly found mutations include nonviable mutations like albino and xantha along with a wide range of other chlorophyll mutants like chlorina, viriscens, viridis, flavo-viridis, albo-viridis, chlorina-terminalis, chlorina-viriscens, albo-viriscens, chlorotica, coppery, and variegated (Goyal and Khan, 2010, Bhosale and Hallale, 2011, Dhole et al., 2023). The macromutations can be identified at various developmental stages with frequent visual screening of M_2 population. In pulse crops, the growth habit and plant type play an important role in the overall productivity, and hence, the induction of dwarf, compact, determinate, branching, top bearing, and lodging-resistant mutants is desirable. In chickpea determinate, compact, dwarf, prostrate, and axillary branched mutants were reported frequently (Yildirim et al., 2013; Gaur et al., 2008; Kozgar, 2014). The mutants like tall, dwarf, branching, erect, and basal branching in pigeon pea (Sagle 2016; Etther et al., 2019), mungbean (Dhole et al., 2023), and black gram (Souframanien et al., 2016); dwarf and lodging-resistant mutants in cowpea (Olasupo et al., 2018; Raina et al., 2022); and compact dwarf, prostrate, spreading, and axillary branching mutants in lentil (Laskar and Khan 2017) were commonly noticed. Early flowering mutants have been

TABLE 9.3
Various Mutants Identified for Qualitative Traits through Induced Mutations in Pulse Crops

Trait	Mutant	Crop	Mutagen used	Reference
Growth habit/ plant type	Determinate	Chickpea	Gamma rays	Yildirim et al., 2013
	Compact growth	Chickpea	EMS	Gaur et al., 2008
	Tall, compact, prostrate mutants	Chickpea	Gamma rays + EMS	Kozgar, 2014
	Dwarf, spreading mutants, axillary branched mutants	Chickpea	Gamma rays + EMS	Kozgar, 2014
	Tall	Pigeon pea	Gamma rays	Etther et al., 2019
	Dwarf	Pigeon pea	Gamma rays	Etther et al., 2019
	Branching	Pigeon pea	Gamma rays	Etther et al., 2019
	Erect	Pigeon pea	EMS & SA	Sangle, 2016
	Tall, erect, dwarf, top bearing, basal branching	Mungbean	Gamma rays &EB	Dhole et al., 2023
	Top pod bearing mutant, extreme dwarf mutant	Black gram	EB	Souframanien et al., 2016
	Tendrillar, tall, erect, bushy	Black gram	Gamma rays & SA	Makeen et al., 2013
	Dwarf and lodging resistant	Cowpea	Gamma rays	Olasupo et al., 2018; Raina et al., 2022
	Dwarf, compact/bushy, prostrate, spreading, axillary branches	Lentil	Gamma rays & HZ	Laskar and Khan 2017
	Dwarf, semidwarf, spindly	Common bean	EMS	Porch et al., 2009
Flower	Early flowering	Chickpea	EMS	Kashid and More 2016
	Early flowering	Chickpea	Gamma rays	Shah et al., 2009
	Twin flower	Chickpea	Gamma rays	Shah et al., 2009
	White color flower	Chickpea	Gamma rays + EMS	Kozgar, 2014
	Early flowering	Cowpea	EMS	Girija and Dhanavel, 2009; OpokuGyamfi et al., 2022
	Open flower and sterile	Chickpea	EB	Joshi-Saha et al., 2015
	Flower color mutants	Cowpea	Gamma rays	Horn et al., 2016
	Multiple flower color, open flower	Lentil	Gamma rays & HZ	Laskar and Khan, 2017
	White, red, purple, yellow and blue color flower	Common bean	Gamma rays & EMS	Borkar and More, 2010

(Continued)

TABLE 9.3 (Continued)
Various Mutants Identified for Qualitative Traits through Induced Mutations in Pulse Crops

Trait	Mutant	Crop	Mutagen used	Reference
	Early flowering	Common bean	EMS	Ramandeep et al., 2018; Kumar and Dhillon, 2020
Maturity	Early maturity	Chickpea	Gamma rays	Joshi-Saha et al., 2015
	Early maturing mutants	Chickpea	Gamma rays + EMS	Kozgar, 2014
	Early maturity	Pigeon pea	Gamma rays & EMS	Etther et al., 2019
	Early maturity	Pigeon pea	Gamma rays & EMS	Sangle, 2016
	Early and synchronous maturity	Mungbean	Electron bean	Dhole et al., 2023
	Early maturity	Mungbean	Gamma rays	Yaqoob and Rashid 2001
	Synchronous maturity	Mungbean	Gamma rays	Kumar et al., 2009; Tah, 2009
	Early maturity	Mungbean	Gamma rays, SA & EMS	Khan and Goyal, 2009
	Early maturity	Black gram	Gamma rays	Kumar et al., 2009; Makeen et al., 2013
	Early maturity	Cowpea	EMS	Girija and Dhanavel 2009
	Early maturity	Cowpea	Gamma rays	Olasupo et al., 2018
	Extra early maturity	Cowpea	Gamma rays	Dorvlo et al., 2022
	Early maturity	Lentil	Gamma rays & HZ	Laskar and Khan 2017
	Early maturity	Lentil	SA	Khan et al., 2006
	Late maturity	Common bean	EMS	Porch et al., 2009
Leaf	Missing leaflet, round leaf	Chickpea	Gamma rays	Joshi-Saha et al., 2015
	Smooth leaflet without serration	Chickpea	EB	Joshi-Saha et al., 2015
	Giant leaves, narrow leaves, altered leaf architecture mutants, elongated rachis	Chickpea	Gamma rays + EMS	Kozgar 2014
	Broad leaf, leathery leaves, irregular leaves, dissected leaves, tetra-foliate leaves	Mungbean	Gamma rays, EMS & SA	Auti and Apparao 2009
	Four primary leaves, lanceolated leaves, small leaves	Mungbean	Gamma rays & EB	Dhole et al., 2023

TABLE 9.3 (Continued)
Various Mutants Identified for Qualitative Traits through Induced Mutations in Pulse Crops

Trait	Mutant	Crop	Mutagen used	Reference
	Oval leaf, narrow leaf, simple leaf	Black gram	Gamma rays	Kumar et al., 2009; Jiao et al., 2019
	Penta-foliate	Black gram	Gamma rays & SA	Makeen et al., 2013
	Four primary leaves	Black gram	Gamma rays & EMS	Goyal et al., 2019
	Tetra and penta-foliate leaf	Cowpea	Gamma rays	Girija and Dhanavel, 2009
	Uni-foliate, narrow leaf, lettuce leaf, four primary leaves	Cowpea	Gamma rays	Olasupo et al., 2018
	Broad dark and spinach-like leaves	Cowpea	Gamma rays	Horn et al., 2016
	Broad/narrow/giant leaf, altered leaf architecture, elongated rachis	Lentil	ϖ rays & HZ	Laskar and Khan 2017
	Chlorotic, leaflet number, leaflet shape	Common bean	EMS	Porch et al., 2009
Pod	Elongated pod	Chickpea	EMS	Kashid, 2016
	Twin pod	Chickpea	Gamma rays	Shah et al., 2009
	Small, narrow pods	Chickpea	Gamma rays + EMS	Kozgar, 2014
	Bold pod mutants	Chickpea	Gamma rays + EMS	Kozgar, 2014
	Five seeded	Pigeon pea	Gamma rays & EMS	Sangle, 2016
	Constricted pods, yellow and brown pod, thin and long pod	Mungbean	Gamma rays & EB	Dhole et al., 2023
	Long pod mutant, curved pod, hairy pod, flat pod	Mungbean	Gamma rays, EMS & SA	Auti and Apparao, 2009
	Long Pod Mutant	Mungbean	EMS	Wani et al., 2017
	Bunchy pod	Black gram	EMS	Kumar et al., 2009
	Single seeded pod	Cowpea	Gamma rays	Girija and Dhanavel, 2009
	Short pod and coiled pods	Cowpea	Gamma rays	Horn et al., 2016
	Small, narrow, large pods	Lentil	Gamma rays & HZ	Laskar and Khan, 2017
	Straight pods	Common bean	EMS	Ramandeep et al., 2018
Seed	Antho-seed mutants	Chickpea	EMS & SA	Kashid and More, 2015

(Continued)

TABLE 9.3 (Continued)
Various Mutants Identified for Qualitative Traits through Induced Mutations in Pulse Crops

Trait	Mutant	Crop	Mutagen used	Reference
	Large seed	Chickpea	SA	Barshile, 2018
	Brown and yellow seed coat, large elongated and round seed	Mungbean	Gamma rays & EB	Dhole et al., 2023
	Large seed	Mungbean	EMS	Wani et al., 2017
	Large seed	Cowpea	Gamma rays & EMS	Girija and Dhanavel, 2009; Olasupo et al., 2018
	Seed coat color mutants	Cowpea	Gamma rays	Horn et al., 2016
	Seed coat color, shape of seed, smooth seed coat	Lentil	Gamma rays & HZ	Laskar and Khan, 2017
	Embryo lethal/dormant	Common bean	EMS	Porch et al., 2009
Root	Nodulation-deficient mutant	Chickpea	Gamma rays	Davis et al., 1985
	Long root mutant	Mungbean	Gamma rays	Dhole and Reddy, 2010
	Long root mutant	Mungbean	Gamma rays & EB	Dhole et al., 2023
	Long root mutant	Black gram	EB	Souframanien et al., 2016
	Hypernodulating and supernodulating mutants	Pea	Physical & chemical mutagens	Borisov et al., 1992; Sidorova et al., 2015
	Nodulation-deficient mutant	Common bean	EMS	Porch et al., 2009

identified in chickpea (Kashid and More 2016; Shah et al., 2009), cowpea (Girija and Dhanavel 2009; Opoku Gyamfi et al., 2022), and common bean (Ramandeep et al., 2018; Kumar and Dhillon 2020) involving treatments with various physical and chemical mutagens. Unique mutants with twin flowers and open and sterile flowers were noticed in chickpea and may be useful in the breeding programs (Shah et al., 2009, Joshi-Sah et al., 2015). Flower color mutants are another type of qualitative trait mutants which were induced in crops like chickpea (Kozgar 2014), cowpea (Horn et al., 2016), lentil (Laskar and Khan 2017), and common bean (Borkar and More 2010). The flower color mutants can be utilized as physical markers, especially in the maintenance of genetic purity of cultivars. Early maturity is one of the prime objectives of every plant breeding program. Mutation breeding is helpful in ensuring reduction in maturity duration without affecting other traits. Early maturing mutants have been identified and isolated in chickpea (Joshi-Sah et al., 2015; Kozgar 2014), pigeon pea (Sagle 2016; Etther et al., 2019), mungbean (Kumar et al., 2009; Tah 2009; Khan and Goyal 2009; Yaqoob and Rashid 2001; Dhole et al., 2023), black gram (Makeen et al., 2013), cowpea (Girija and Dhanavel 2009; Olasupo et al., 2018; Dorvlo et al., 2022), and lentil (Laskar and Khan 2017; Khan et al., 2006). The leaf traits–related macromutations, though easy to identify, are mostly sterile and hence of no use in breeding. Still, some of the leaf mutants, like those with missing leaflets, round leaves, smooth leaflets without serrations, giant, narrow leaf, and elongated rachis mutants, were identified in chickpea and are useful in leaf developmental genetics related basic studies (Joshi-Sah et al., 2015; Kozgar 2014). The various types of leaf mutations affecting leaf ontogeny can be ideal experimental materials to generate valuable information on regulation and evolution of leaf morphogenesis. The mutants with tetra-foliate leaves, four primary leaves, narrow, lanceolated leaves, and small leaves are frequently reported in mungbean (Dhole et al., 2023), black gram (Kumar et al., 2009; Makeen et al., 2013; Goyal et al., 2019; Jiao et al., 2019), and cowpea (Girija and Dhanavel 2009; Olasupo et al., 2018; Horn et al., 2016). The pod features, like size and number, directly influence the yield in pulse crops. A lot of genetic variation has been created in various pulse crops for pod characteristics. In chickpea, large, elongated, small, narrow, and twin-pod mutants were induced through physical and chemical mutagens (Shah et al., 2009; Kozgar et al., 2014; Kashid 2016). Similarly, five-seeded pods in pigeon pea (Sangle 2016), yellow pods and long pods in mungbean (Dhole et al., 2023; Auti and Apparao 2009; Wani et al., 2017), bunchy pods in black gram (Kumar et al., 2009), and single-seed pods as well as short and coiled pods in cowpea (Girija and Dhanavel 2009; Horn et al., 2016) were reported. The ultimate commercial product of almost all pulses is the seeds, which are always correlated to yield and nutritional quality of the produce. Various macromutations are reported in pulse crops which are related to seed size, shape, color, number of seeds per pod, and dormancy. The mutants related to seed features are immensely important in pulse breeding programs. The seed coat color in pulses is important for the consumer acceptance and varietal identification. Various mutants for seed coat color were identified in chickpea (Kashid and More 2015; Barshile 2018), mungbean (Dhole et al., 2023), cowpea (Horn et al., 2016), and lentil (Laskar and Khan 2017). Large seed mutants were reported in mungbean (Wani et al., 2017) and cowpea (Girija and Dhanavel 2009; Olasupo et al., 2018). In common bean, a mutant with seed dormancy was reported which can be very useful to reduce the losses due to preharvest sprouting (Porch et al., 2009). The roots are the hidden half of the plants which are always ignored in the crop improvement programs. Strong and long root system helps the plant to absorb water from the deeper zones of the soil in drought condition. For the development of drought resistant varieties, roots traits should be targeted through induced mutation with efficient screening techniques. Long root mutants were identified in mungbean (Dhole and Reddy 2010; Dhole et al., 2023) and black gram (Souframanien et al., 2016). Being legume crops, pulses are bestowed with symbiotic nitrogen fixation ability that can be enhanced through induced mutations like that in supernodulating mutants. The supernodulation/hypernodulating mutants were identified in peas (Borisov et al., 1992; Sidorova et al., 2015) and nodulation-deficient mutants in chickpea (Davis et al., 1985) and common bean (Porch et al., 2009). Identification of nodulation mutants will help in the basic understanding of genes and genetics of nodule development.

9.1.12 Identification of Mutants for Quantitative Traits

The mutations in quantitative traits are termed micromutations, which itself indicates the very small effect of mutations on total phenotype of the plant, and hence are difficult to detect in large M_2 populations. The yield and yield-contributing traits are mainly quantitative in nature and involve large number of genes, and each gene has a small cumulative effect with various gene interactions like additive, nonadditive, and their interactions. Advances in genomics revealed the presence of major and minor quantitative trait loci (QTLs) for these quantitative traits. The mutations at major QTLs can bring notable changes in the phenotype, which can be detected in the later generations like M_5 and M_6 when the mutations become stable. Various micromutations or yield and yield-contributing traits were reported in pulses (Table 9.4). High-yielding mutants are always desirable for plant breeders. Such mutants were reported in most of the pulse crops like chickpea (Kulthe and Kothekar 2011; Kozgar 2014), pigeon pea (Etther et al., 2019, Sangle, 2016), mungbean (Sarwar and Ahmad 2003; Dewanjee and Sarkar, 2018, Auti and Apparao, 2009), cowpea (Olasupo et al., 2018; Dorvlo et al., 2022; Raina et al., 2022), common beans (Kumar and Dhillon 2020), and lentils (Khan et al., 2006). The seed size is highly correlated with seed yield and also with the quality of most of the pulse crops. The large seed mutants can be identified very easily in M_2 generation by visual screening and may breed true as this trait is highly heritable. Large seed mutants were identified and isolated in chickpea (Kozgar 2014), pigeon pea (Etther et al., 2019), and black gram (Souframanien et al., 2016). The pod size is another important quantitative trait, which can be screened visually and positively correlated with yield and number of seeds per pod. Large pod mutants were reported in black gram (Goyal et al., 2021) and common bean (Ramandeep et al., 2018). In chickpea, the mutants were induced for other quantitative traits like number of pods per plant, number of pods bearing branches, and a greater number of seeds per plant (Kozgar 2014, Amri et al., 2018). The mutants with a greater number of pods per plant and seeds per pod were reported in pigeon pea (Almosawi 2019), cowpea (Opoku-Gyamfi et al., 2022), black gram (Goyal et al., 2021), and common bean (Ramandeep et al., 2018). These quantitative traits with low heritability should be closely monitored for four to five generations, and after stabilization, such mutants should be characterized in replicated and multilocation trials.

9.1.13 Identification of Biotic Stress-tolerant Mutants

Diseases and insect pests are the major biotic stresses which are responsible for low productivity of pulse crops. Mutation breeding contributed a lot through the development of disease-resistant mutants which were either directly released as a variety or used in the hybridization program. The disease-resistant mutants were more frequently reported as compared to insect pest-resistant mutants (Table 9.5). The disease screening is quite easy as it can be done in disease-sick plots, in hot spots, and in laboratories in growth chambers, and diseases can be induced by spraying inoculums on large areas. The disease organism can be maintained in the laboratory on culture media except obligate parasites and can be used for screening as and when required. On the other hand, insects are difficult to rear, and it is not possible to confine them to

TABLE 9.4
Various Mutants Identified for Quantitative Traits through Induced Mutations in Pulse Crops

Crop	Mutant	Mutagen used	Reference
Chickpea	High yielding	SA	Kulthe and Kothekar, 2011
	High yielding	Gamma rays + EMS	Kozgar, 2014
	100 seed weight	Gamma rays + EMS	Kozgar, 2014
	Number of pods plant^{-1}	Gamma rays + EMS	Kozgar, 2014
	Number of pods bearing branches	Gamma rays + EMS	Kozgar, 2014
	Plant height	Gamma rays + EMS	Kozgar, 2014
	Total number of pods and seeds plant^{-1}, yield plant^{-1}	Gamma rays	Amri et al., 2018
Pigeon pea	High yielding	Gamma rays	Etther et al., 2019
	100 seed weight	Gamma rays	Etther et al., 2019
	Number of pods plant^{-1} and number of seed pods^{-1}	Gamma rays, EMS & SA	Almosawi, 2019
	High yielding	Gamma rays & EMS	Sangle, 2016
	Number of pods plant^{-1}, 100 seed weight	Gamma rays, EMS & SA	Sangle, 2020
Mungbean	High yielding	Gamma rays	Sarwar and Ahmad, 2003
	High yielding	Gamma rays	Dewanjee and Sarkar, 2018
	High yielding	EMS & SA	Auti and Apparao, 2009
Black gram	More number of clusters plant^{-1}	Gamma rays	Ignacimuthu and Babu, 1989
	Number of seeds pod^{-1}, 100 seed weight, pod length	Gamma rays & EMS	Goyal et al., 2021
	increased pod number	Gamma rays & EB	Souframanien et al., 2016
Cowpea	Increased pod number	Gamma rays + EMS	Girija and Dhanavel, 2009
	High yielding	Gamma rays	Olasupo et al., 2018; Dorvlo et al., 2022; Raina et al., 2022
	For number of pods plant^{-1}, number of seeds pod^{-1}, number of seeds plant^{-1}, number of locules pod^{-1}	EMS	Opoku-Gyamfi et al., 2022
Lentil	High yielding	SA	Khan et al., 2006
Common bean	Number of pods plant^{-1}, green pod yield plant^{-1}, Seed yield plant^{-1}	EMS	Kumar and Dhillon, 2020
	Number of pods plant^{-1}, pod length	EMS	Ramandeep et al., 2018

Note: *$^{-1}$ denotes "per".

specific areas for screening post release. Several disease-resistant mutants were identified in various pulse crops through physical and chemical mutagens. In chickpea, resistant mutants for *Ascochyta* blight, *Fusarium* wilt, and parasite *Orobanche foetida* were developed through gamma rays and EMS (Mabrouk et al., 2018; Shah et al., 2009; Brahmi et al., 2014). Gamma rays–induced yellow mosaic disease-resistant mutants were identified in mungbean (Sing and Chaturvedi 1981; Reddy 2009; Ngampogsai et al., 2009; Vairam et al., 2016; Anuragi et al., 2022) and black gram (Kumar et al., 2009; Makeen et al., 2013; Kundagrami et al., 2009; Vanniarajan et al., 2017). In black gram, *Cercospora*leaf spot disease-resistant mutant was induced through EMS and gamma rays (Kumar et al., 2009). Bean common mosaic virus-resistant mutant was reported to be developed through EMS treatment in common bean (Ramandeep et al., 2018). In cowpea, mutants resistant to cowpea aphid-borne mosaic virus and leaf crinkle diseases have been reported (Dhanasekar and Reddy 2015). Breeding for insect resistance has not been given serious thought in almost all crops including pulses, due to nonavailability of resistance gene sources, complex nature of resistance mechanism, and tedious and difficult screening techniques. Still, few preliminary successes have been reported through gamma rays–induced mutations in cowpea for *Maruca* pod borer (Dhanasekar et al., 2023), aphids, and bruchids (Munyinda and Kanenga 2018) (Table 9.5). Such results are very encouraging to develop insect resistance through induced mutations in pulse crops which are more prone to insect pest damage.

9.1.14 Identification of Abiotic Stress-tolerant Mutants

The pulse crops have been neglected for several decades from genetic improvement as these crops are not regarded as commercial crops. Farmers also grow these crops in marginal soils with minimum inputs. Hence, pulse crops are often encountered with various abiotic stresses like drought and salinity. The induced mutation for abiotic stress tolerance has achieved very limited success in a few pulse crops for drought and freezing tolerance (Table 9.5). Winter season crops like chickpea are very often exposed to freezing temperatures, for which freezing-tolerant mutants induced through gamma rays may reduce the losses due to freezing injury (Akhar et al., 2011). The moisture stress is the most common abiotic stress in rain-fed pulse crops. The mutation breeding was applied successfully which resulted in the development of drought-tolerant mutants in lentil (Lal and Tomar 2009), black gram (Souframanien et al., 2016), mungbean (Dhole and Reddy 2010; Dhole et al., 2023), and cowpea (Olasupo et al., 2018). These mutants possessed traits like long roots, staying green, and thick dark green leaves which help retain greater water potential under moisture stress conditions.

Another abiotic stress is the exposure to herbicides which are used to control the weeds in standing crops. The induced mutation is regarded as an alternative method to transgenic to develop herbicide tolerance. Herbicide-tolerant maize (Newhouse et al., 1991), cotton (Rajasekaran et al., 1996), canola (Ouellet et al., 1992), sunflower (Sala and Bulos, 2012), barley (Lee et al., 2011), etc., have been developed and commercialized by inducing point mutations which lead to altered-enzyme, acetohydroxyacid synthase (AHAS). Enzyme AHAS is a key enzyme for biosynthesis

TABLE 9.5
Various Mutants Identified for Biotic and Abiotic Stress Tolerance through Induced Mutations in Pulse Crops

Type of stress	Trait	Mutant	Crop	Mutagen used	Reference
Biotic stress	Disease resistance	*Ascochyta* blight resistance	Chickpea	Gamma rays	Mabrouk et al., 2018
		Resistant to *Orobanche foetida*	Chickpea	Gamma rays	Brahmi et al., 2014
		Fusarium wilt resistance	Chickpea	Gamma rays & EMS	Shah et al., 2009
		Yellow Mosaic disease resistance	Mungbean	Gamma rays	Reddy et al., 2009
		Powdery mildew resistance	Mungbean	Gamma rays	Ngampongsai et al., 2009
		Yellow Mosaic disease resistance	Black gram	Gamma rays	Kumar et al., 2009; Makeen et al., 2013
		CLS resistance	Black gram	EMS & Gamma rays	Kumar et al., 2009
		Bean common mosaic virus resistance	Common beam	EMS	Ramandeep et al., 2018
		Cowpea aphid-borne mosaic virus and leaf crinkle diseases resistance	Cowpea	Gamma rays	Dhanasekar and Reddy 2015
	Insect pest tolerance	*Maruca* pod borer tolerance	Cowpea	Gamma rays	Dhanasekar et al., 2023
		Aphid and bruchid tolerance	Cowpea	Gamma rays	Munyinda and Kanenga 2018
Abiotic stress	Freezing injury	Freezing tolerance	Chickpea	Gamma rays	Akhar et al., 2011
	Drought	Drought-tolerant mutant	Lentil	Gamma rays, EMS & SA	Lal and Tomar 2009
		Drought-tolerant mutant	Black gram	EB	Souframanien et al., 2016
		Drought-tolerant long root mutant	Mungbean	Gamma rays	Dhole and Reddy 2010
		Drought-tolerant mutant	Mungbean	Gamma rays & EB	Dhole et al., 2023
		Drought-tolerant mutant	Cowpea	Gamma rays	Olasupo et al., 2018
	Herbicide resistance	Metsulfuron methyl, Idosulfuron methyl and Tribenuron methyl herbicides resistance	lentil	Gamma rays, EMS, HH and SA	Rizwan et al., 2017

Abbreviation: HH: Hydrazine Hydrate.

of branched-chain amino acids like leucine, valine, and isoleucine (Bernasconi et al., 1995). The altered AHAS enzyme does not allow its inhibitors like sulfonyl urea, imidazolinones, triazolopyrimidines, pyrimidinyloxybenzoates, and sulfonyl amino carbonyl-triazolinones (all are herbicides) to bind to its active site, and thus, the enzyme retains its activity even in the presence of the herbicides following their spray in the fields (McCourt et al., 2006). There are very few reports on induced mutations for herbicide tolerance in pulse crops because of limited use of herbicides to control weeds in pulse crops. In lentil, resistant mutants to herbicides like metsulfuron methyl, idosulfuron methyl, and tribenuron methyl have been reported to be induced by gamma rays, EMS, hydrazine hydrate, and sodium azide (Rizwan et al., 2017).

9.1.15 High-throughput Screening for Mutants in Genomic Era

The contribution of mutation breeding in crop improvement was limited to varietal development and creation of new genetic variability. The recent advances in molecular breeding techniques and genomics paved the way for various applications of induced mutations in pulse crops for basic understanding of genes, gene expression, and their evolutionary importance. The various DNA markers were applied to trace pedigree of mutants for mutant gene tagging and cloning and identifying tightly linked markers for marker-assisted selection. A novel reverse genetic high-throughput approach, "TILLING" (Targeting Induced Local Lesions in Genomes), has been developed to identify point mutations from EMS or chemical mutagen–treated populations. This approach involved PCR-based mutant screening in the mutagenized population to detect point mutations and recover a range of alleles including knockouts and missense changes in very little time (McCallum et al., 2000). The TILLING has been successfully applied in pulse crops like peas (Triques et al., 2007) and beans (Galeano et al., 2009; Porch et al., 2009). The speed of TILLING will be enhanced with the new high-throughput approaches like HRM (High-Resolution Melting) genotyping, KeyPoint™ Technology, and EMAIL (Endonucleolytic Mutation Analysis by Internal Labelling) (Cross et al., 2008; Rigola et al., 2009; Botticella et al., 2011). The next-generation sequencing (NGS) further boosts up the mutant screening with MutMap (mapping-by-sequencing) and MutChromSeq (helps to assort the desired genes in shortest time span) (Sanchez-Martin et al., 2016). If the prior information on gene sequence is available, the gene functions and mutant gene can be identified through RNAi, TILLING, and EcoTILLING techniques (Till et al., 2007). The new techniques like SHOREmap (SHOrtREad map), NGM (Next-Generation Mapping), MutMap, MutMap+, Mut-Map-Gap, and QTL-Seq can discover the mutant genes for the trait of interest without any prior sequence information (Schneeberger et al., 2009; Austin et al., 2011; Abe et al., 2012; Fekih et al., 2013; Takagi et al., 2013a and 2013b).

9.1.16 Achievements of Mutation Breeding in Pulse Crops

The induced mutation has proven its contribution to the development of improved varieties which were widely cultivated all over the world and, thereby, ensured the nutritional security of underdeveloped and developing countries. The mutant varieties (more than 3200) registered with the mutant variety database (MVD) of IAEA include mutants directly released as varieties, varieties developed from the cross between two mutants, a mutant and a cultivar or a mutant derivative. The varieties of pulse crops developed through induced mutations with improved traits are summarized in Table 9.6.

9.1.17 Early Maturity

Maturity is the most frequently targeted trait through induced mutation, which has resulted in the development of a number of early maturing mutant varieties in pulse crops (Table 9.6). In chickpea, early maturing varieties, viz., Hyprosola (Faridpur-1), Binasola-3, TAEK-SAGEL, and Kiran, were developed following the treatments with gamma rays and fast neutrons. Similarly, with the same mutagens, early maturing varieties like TAT-5, TAT-10 CO (RG-7), and Co-5 were released for commercial cultivation. Early and synchronous maturing mutant varieties of mungbean were developed, of which TMB-37, TJM-3, TM-2000-2, LGG-450, NIAB Mung 19-19, NIAB Mung20-21, BM-4, and Binamung-2 are maturing in just 60 days. The mutant varieties like Binamash-1 and Manikya in black gram, Lukusuzi, and V-38 in cowpea were early maturing. Many early and synchronously maturing mutant varieties (Hans-2, Streletchkii 11, Nemchynovskii 85, Vympel, V-38 [Swarna] and Piast) of peas were developed through various chemical and physical mutagens. The cowpea varieties, viz., Pusa Parvati, Universal, Sanilac, Seafarer, OAC Seaforth, Wesland, Northland, Midland, Suncrest, Stinger, and AC Hensall, were developed through X-ray irradiation.

9.1.18 Disease Resistance

Induced mutations have been applied successfully to develop disease-resistant varieties in the pulses that help to stabilize the yield potential (Table 9.6). Use of gamma rays and EMS resulted in the resistant chickpea varieties to diseases like *Ascochyta* blight (CM 72, NIFA-88 (CM-1918), C-88 and TAEK-SAGEL), *Fusarium* wilt (CM-88, CM-98, Pusa 547, CM-2008 and BM-547), and stunt virus (Pusa 413 and WCG-10). Through fast neutron treatments, multiple disease-resistant pigeon pea varieties, viz., CO (RG)-7, Co-9, TJT-501, PKV-TARA, and TT-401 with resistance to *Fusarium* wilt and sterility mosaic virus were reported. Many YMD tolerant mungbean varieties were developed directly through gamma rays or indirectly (mutants were used as one of the parents), viz., NIAB Mung 19-19, NIAB Mung 20-21, Binamoog-2, Binamoog-3, Pant Moong 2, LGG 450 (Pushkara), LGG-407 (LAM-407), PBM-1, Meha (IPM99-125), TMB-37, and Pant Moong 6. Black gram is another crop heavily infested with various diseases, where gamma rays and MMS treatments resulted in disease-resistant varieties, viz., TU-94-2, PKV-Greengold, Binamash-1, TU-40 and GU-1 (resistant to YMD), GU-1 and Binamash-1 (tolerant to *Cercospora* leaf spot), and Co-4 (tolerant to leaf crinkle virus). The cowpea mosaic-resistant variety TC-901, yellow mosaic virus-resistant Cowpea-88, Co-9, Co (FC-8), and Pusa-578 were developed following treatment with gamma rays. The gamma rays, X-rays, and EMS were successfully used to develop disease-tolerant cowpea varieties like Universal, Sanilac, Unima, Seafarer (resistant to *Colletotrichum lindemuthianum*), Seaway,

TABLE 9.6
Mutant Varieties Released for Commercial Cultivation with Improved Agronomical Traits

Trait	Crop	Mutagen used	Improved trait	Name of variety
Maturity duration	Chickpea	Gamma rays	Early maturity	Hyprosola(Faridpur-1), Binasola-3(L84),TAEK-SAGEL
		Fast neutron	Early maturity	Kiran (RSG-2)
	Pigeon pea	Gamma rays	Early maturity	CO-5
		Fast neutron	Early maturity	TAT-5, TAT-10, CO(RG)-7
	Mungbean	Gamma rays	Early maturity	Co 4, TAP-7, LGG 450 (Pushkara), LGG-407 (LAM-407),TMB-37, TJM-3,NIAB Mung-28, NIAB Mung 19-19, NIAB Mung 121-25, NIAB Mung 13-1, NIAB Mung 20-21, NIAB Mung 92, Binamoog-2, Binamoog-8, Binamoog-9
		EMS	Early maturity	BM 4
		Gamma rays	Synchronous maturity	TM 2000-2 (Pairymung), NIAB Mung 54, NIAB Mung 51, Binamoog-3, Binamoog-4, Binamoog-5, Binamoog-7
	Black gram	Gamma rays	Early and synchronous maturity	Binamash-1, Manikya
	Cowpea	Gamma rays	Early maturity	Lukusuzi
		DMS	Early maturity	V38 (Swarna)
	Pea	EI	Early maturity	Hans-2, Streletchkii 11, Nemchynovskii 85, Vympel
		Gamma rays	Early maturity	Wasata, Paride, Pirro, Start
		Gamma rays + EI	Early maturity	Gorizont
		EMS	Early maturity	Drujba
		NEU and NMU	Early maturity	Olejka
		Gamma rays	Synchronous maturity	Piast
	Common bean	X-ray	Early maturity	PusaParvati, Universal, Sanilac, Seafarer, OAC Seaforth, Wesland, Northland, Midland, Suncrest, Stinger, AC Hensall
		Gamma rays	Early maturity	Harkovskaya 8, CAP-1070

		EMS	Early maturity	Alfa, C-20, Mogano, Norstar, Huron, Plovdiv 11 M, Albion
Disease resistance	Chickpea	Gamma rays	Resistance to chickpea blight (*Ascochyta rabiei*)	CM 72, NIFA-88 (CM-1918), C-88, TAEK-SAGEL
		Gamma rays	Resistance to bacterial blight	NIFA-95,Pusa 408 (Ajay)
		Gamma rays	Resistant to *Fusarium* wilt	CM-88, CM-98, Pusa 547
		EMS	Resistant to *Fusarium* wilt	CM-2008
		Gamma rays	Tolerant to wilt, root rot and stunt diseases	BGM-547
		Gamma rays	Resistance to stunt virus	Pusa 413 (Atul), WCG10 (Pant G-10)
	Pigeon pea	Fast neutrons	Tolerant to Wilt, Sterility Mosaic and *Phytophthora* blight	CO(RG)-7, Co-9
		Fast neutrons	Tolerance to *Fusarium* wilt, SMD	TT401, TJT501, PKV TARA
	Mungbean	Gamma rays	Tolerant to YMD	NIAB Mung 19-19, NIAB Mung 20-21, NIAB Mung 54, NIAB Mung 51, NIAB Mung 92, Binamoog-2, Binamoog-3, Binamoog-4, Binamoog-5, NIAB Mung 98, ML 26-10-3(Pant Moong 2), LGG 450 (Pushkara), LGG-407 (LAM-407), PBM-1, Meha (IPM99-125), TMB-37, Pant Moong 6 (UPM 02-17), VBN(Gg)3

(Continued)

TABLE 9.6 (Continued)
Mutant Varieties Released for Commercial Cultivation with Improved Agronomical Traits

Trait	Crop	Mutagen used	Improved trait	Name of variety
		Gamma rays	Resistant to bacterial leaf spot	AZRI Mung-2018
		Gamma rays	Resistant/tolerant to *Cercospora* leaf spot	NIAB Mung 20-21, NIAB Mung 54, Camar, Binamoog-2, Binamoog-3, Binamoog-4, NIAB Mung 98, Binamoog-5, Binamoog-6, TM-96-2
		EMS	Tolerant to *Cercospora* leaf spot	Binamoog-7
		EMS	Tolerant to YMD	BM 4
	Black gram	Gamma rays	Resistant/Tolerant to YMD	TU-94-2, PDKV Blackgold (AKU 10-1), Binamash-1
		MMS	Resistant/Tolerant to YMD	VBN(bg)4, GU-1, TU-40
		Gamma rays	Tolerant to *Cercospora* leaf spot	GU-1, Binamash-1
		MMS	Tolerant to leaf crinkle virus, tip blight, PM	Co-4
	Cowpea	DMS	Resistance to fungal and bacterial diseases	V16 (Amba), V240
		Gamma rays	Resistant to YMV and anthracnose	Cowpea-88
		Gamma rays	Resistant to YMV	Co-(FC) 8, V578 (Pusa-578), Co-9
		Gamma rays	Tolerant to *Ascochyta* and *Cercospora*	Lukusuzi

	Gamma rays	Resistant to CMV	TC-901
Common bean	X-rays	Resistance to *Colletotrichum lindemuthianum*	Universal, Sanilac, Unima, Seafarer
	X-rays	Resistance to bean common mosaic virus	Seaway, Gratiot, Seafarer, Ouray
	Gamma rays	Resistant to bacterial diseases	Saparke 75
	EMS	Resistance to *Fusarium, Colletotrichum lindemuthianum*	Alfa
	EMS	Resistant to rust	Giza 80
	Gamma rays	Resistance to anthracnose, resistance to leaf spot	FT-Paulistinha
	EMS	Resistance to Golden Mosaic Virus Disease	IAPAR 57, IAPAR 65
Lentil	Gamma rays	Resistance to *Fusarium* and *Botrytis*	Djudje
	Gamma rays	Tolerant to rust and blight	Binamasur-2, NIAB MASOOR-2006
	EMS	Tolerant to rust and blight	Binamasur-3
Horsegram	Gamma rays	Tolerant to PM	Cridaharsha CRGH-19

(*Continued*)

TABLE 9.6 (Continued)
Mutant Varieties Released for Commercial Cultivation with Improved Agronomical Traits

Trait	Crop	Mutagen used	Improved trait	Name of variety
		Gamma rays	Tolerant to YMV	CRIDALATHA (CRHG-4), CRIDA-18R, Paiyur-2
Insect tolerance	Chickpea	Gamma rays	Moderately resistant to pod borer	BGM 547
	Pigeon pea	Gamma rays	Tolerant to pod borer	Co-6
		Fast neutrons	pod fly damage	TT-401
		Gamma rays	Moderately resistant to *Maruca* and pod fly	Co-9
	Cowpea	Gamma rays	Tolerance to aphids and bruchids	Lunkhwakwa
Improvement in yield and yield-contributing traits	Chickpea	Fast neutrons	Increased pod number, high yield	Kiran (RSG-2)
		Gamma rays	Improved plant architecture	Pusa 408 (Ajay), Pusa 413 (Atul)
		Gamma rays	High yield, profuse branching, high pod number	Pusa 417 (Girnar)
		Gamma rays	High yield, large grain size	BGM 547, Pusa 547, CM 72, NIFA-88 (CM-1918), CM-88, Hassan-2K, Binasola-4, Binasola-3 (L-84), TAEK-SAGEL, Binasola-5, Binasola-6, Binasola-7, Binasola-8, Binasola-9, Binasola-10
		Gamma rays	Higher pod number and harvest index and higher yield	Hyprosola (Faridpur-1)

	Gamma rays+ EMS	Higher yield	Line 3
	EMS	Higher yield	CM-2008
Pigeon pea	EMS	High yield, large seeded, higher shelling	Co-3
	Fast neutrons	Large seed size	Trombay Vishakha-1, TAT 5, TAT 10
	Fast neutrons	High yield	TT-401, TJT-501, PKV-TARA
	Gamma rays	Large seed size	Co-9
Mungbean	Gamma rays	High yield	Co-4, ML 26-10-3(Pant Moong 2), TAP-7, LGG 450 (Pushkara), LGG-407(LAM-407), TARM-1, TARM-2, TARM-18, TMB-37, NIAB Mung-28, NIAB Mung 19-19, NIAB Mung 121-25, NIAB Mung 13-1, NIAB Mung 20-21, Binamoog-4, Binamoog-5
	EMS	High yield	BM-4, Binamoog-7
	Gamma rays	High yield and large seed size	NIAB Mung 54, NIAB Mung 51, Binamoog-2, Binamoog-3, Chai Nat 84-1, Binamoog-9
	Gamma rays	Increased number of seed and pod size	Binamoog-6
Black gram	Gamma rays	High yield	TAU 1, DU-1, Binamash-1
	Gamma rays	High yield and high seed weight	TPU-4, TAU-2, TU-94-2, Manikya
Cowpea	DMS	High yield	V16 (Amba), V37 (Shreshtha), V38 (Swarna), V240
	Gamma rays	High yield	Co-5, COCP 702, TRC77-4 (Kalleshwari), ICV-12, Uneca-Gama, CBC-5, Lukusuzi, Lunkhwakwa
	Gamma rays	Large seed size	Cowpea-88
	Gamma rays	High yielding, increased seed size, pod size and pod number	TC-901

(*Continued*)

TABLE 9.6 (Continued)
Mutant Varieties Released for Commercial Cultivation with Improved Agronomical Traits

Trait	Crop	Mutagen used	Improved trait	Name of variety
	Common bean	X-rays	High yield	Pusa Parvati, Universal, Neptune
		Gamma rays	High yield	Saparke 75, Giza 80, Harkovskaya 8, Campeiro
		EMS	High yield	Mogano
	Faba bean	Gamma rays	High yield	Karna, Bronto, Tuwaitha
		NMU	High yield	Severinovskie 1
	Grass pea	Gamma rays	High yield	Binakhesari-1, Bogdan
	Pea	X-rays	High yield	Stral-art
		Gamma rays	High yield and large seed size	Sum, Wasata, Hamil, Paride, Priamo, Agra
		EI	High yield	Nemchynovskii 85, Hans-2
		Gamma rays	Higher number of pods	Kwestor
	Lentil	X-rays	High yield	B-256 (Ranjan), Aurie
		Gamma rays	High yield	Djudje, Elitsa, NIAB MASOOR 2002, Binamasur-2, Binamasur-6, Binamasur-5, Binamasur-8, Binamasur-9, Binamasur-11
		EMS	High yield	Binamasur-3
		Gamma rays	Higher number of pods	NIAB MASOOR-2006
	Moth bean	Gamma rays	High yield	RMO-423, MARU VARDHAN (RMO-225)
Abiotic stress tolerance	Chickpea	Fast neutron	Salt tolerant	Kiran (RSG-2)
Quality traits				
	Pigeon pea	Gamma rays	Drought tolerance	Co-5
	Mungbean	Gamma rays	Tolerance to salinity and acid soil	Camar
		Gamma rays	Drought tolerance	Co-4, K-1

Cowpea	Gamma rays	Tolerant to aluminum toxicity	Lunkhwakwa
Common bean	Gamma rays	Tolerance to salinity and tolerance to P deficit	CIAT 899
Lentil	Gamma rays	Tolerance to low temperatures	Rajendra Masoor 1
	Gamma rays	Drought tolerance	Verzuie
Pea	Gamma rays	Resistance to lodging	Milewska, Mihan, Ramir, Jaran, Miko, Agra
	EI	Resistance to lodging	Streletchkii 11, Nemchynovskii 85
Mothbean	Gamma rays + EMS	Drought tolerance	RMO-40
	Gamma rays + EMS	Resistant to drought and heat	CAZRI MOTH-1 (CZM-79)
Chickpea	•-rays	High seed protein	TAEK-SAGEL
	•-rays	Attractive straw seed coat color	Binasola-6, Binasola-8, Binasola-10
Mungbean	•-rays	High starch	Chai Nat 84-1
Pea	EI	better seed quality	Hans-2
	DES	High seed protein	Moskovsky 73
Common bean	EMS	High protein content and cooking quality	Alfa
	Gamma rays	High protein content and cooking quality	Giza 80
Horse gram	Gamma rays	High protein	Cridaharsha CRGH-19
Faba bean	Gamma rays	High protein content	Tuwaitha, Babylon
Grass pea	Gamma rays	Low BOAA content	Binakhesari-1
Kidney bean	NMU	High protein content	Svetlaya

(*Continued*)

TABLE 9.6 (Continued)
Mutant Varieties Released for Commercial Cultivation with Improved Agronomical Traits

Trait	Crop	Mutagen used	Improved trait	Name of variety
	Lentil	Gamma rays	High protein content (27.9%), good culinary and organoleptic quality	Djudje, Aurie
	White lupin	Gamma rays	Less alkaloid, high protein (44%) and lysine (6.8%) contents	Kievsky Mutant
		Gamma rays+ EI	Low content of alkaloids	Gorizont
		EI	Low content of alkaloids	Dnepr
		NMU, EI and DMS	Low content of alkaloids	Ukrainskii
		NMU and EI	Low content of alkaloids	Pyshevoj
		NEU and NMU	Low content of alkaloids	Olejka
		MNH	High protein content	Siniiparus
	Yellow lupin	Gamma rays	Good fodder quality	Narochanskii

Source: www.mvd.iaea.org

Gratiot, Seafarer, Ouray (resistant to bean common mosaic virus), Alfa (resistant to *Fusarium* wilt), and Giza-80 (resistant to rust).

9.1.19 Insect Tolerance

Insect-tolerant mutant varieties were developed and released for commercial cultivation in a few crops like chickpea, pigeon pea, and cowpea. The pod borer tolerant varieties were induced through gamma rays in chickpea (BGM-547) and pigeon pea (Co-6). The pigeon pea varieties TT-401 and Co-9 developed through fast neutrons and gamma rays, respectively, showed tolerance to pod flies. Aphid and bruchid tolerance were reported in the cowpea variety Lunkhwakwa and were induced through gamma rays.

9.1.20 High Yield

Various physical and chemical mutagenic treatments resulted in the development of high-yielding varieties in pulse crops. Most notable high-yielding mutant varieties of chickpea released all over the world are Kiran, Ajay, Atul, Girnar, BGM-547, Binasola series of varieties, Pusa-547, CM-2008, and TAEK-SAGEL. High-yielding and large seed size varieties were developed through induced mutations in pigeon pea (TT-401, TJT-501, PKV-TARA, Co-9, TV-1, TAT-5, TAT-10 and Co-3). Mungbean mutant varieties with high yield and large seed size, large pod size etc. were reported to be induced by mutagens like gamma rays (NIAB Mung-54, NIAB Mung-51, Binamoong-2, Binamoong-3, TAP-7, TARM-1, TARM-2, TARM-18, TM-2000-2, TMB-37, Co-4, LGG-407) and EMS (BM-4 and Binamoong-7). The gamma rays–induced black gram varieties like TAU-1, DU-1, Binamash-1, TPU-4, TU-94-2, and Manikya were high yielding with large seed sizes, and hence popular among the farmers. The high-yielding cowpea varieties like TRC-77-4, TC-901, Co-4, and Cowpea-88 were developed through gamma rays. Many mutant varieties of pea, common bean, lentil, and moth bean with high yield and yield-contributing traits were induced through various physical and chemical mutagens (Table 9.6).

9.1.21 Abiotic Stress Tolerance

Mutation breeding also contributed to the development of abiotic stress-tolerant varieties in various pulse crops. Treatment with fast neutrons resulted in the salt-tolerant chickpea variety Kiran (RSG-2), while pigeon pea variety, Co-5, with drought tolerance was developed using gamma rays. Mungbean variety (Camar) developed by gamma rays can tolerate both saline and acid soils. The cowpea variety Lunkhwakwa has been reported to be tolerant to aluminum toxicity, while common bean variety CIAT-899 tolerates salinity and phosphorus deficiency and is released following gamma ray irradiation. Lodging-resistant mutant varieties of pea, viz., Milewska, Mihan, Ramir, Jaran, Miko, Agra, Streletchkii-11, and Nemchynovskii-85, were developed through gamma rays and EI.

9.1.22 Quality Traits

Mutant varieties with better quality in terms of increase in nutrient contents and reduction in antinutritional compounds were registered in Mutant Variety Database (Table 9.6). The most important nutrient in pulse crops is the protein, for which mutant varieties with high protein content were reported in chickpea (TAEK-SAGEL), pea (Moskovsky 73), faba bean (Tuwaitha and Babylon), horse gram (Cridaharsha CRGH-19), kidney bean (Svetlaya), lentil (Djudje and Aurie), and white lupin (Sniiparus). The common bean varieties Alfa and Giza 80, with high protein content and better cooking quality, were developed through EMS and gamma ray treatments, respectively. Mutant varieties with reduced alkaloid contents in white lupin (Kievsky mutant, Gorizont, Dnepr, Ukrainskii, Pyshevoj and Olejka) were released for cultivation. The grass pea variety Binakhesari-1 having less BOAA content was induced through gamma rays.

9.1.23 Impact of Mutant Varieties in Boosting Pulses Production

The mutant pulse varieties released for commercial cultivation proved to be very important in increasing the production and productivity of various pulse crops. Most notable among these is the black gram variety, TAU-1, once reported to occupy more than 90% of the area under black gram cultivation in Maharashtra state of India. Even after 37 years of its release, this variety is still under cultivation in India, indicating its popularity owing to its stability, adaptability, and high yield. The mungbean variety TM-96-2 was the first mutant variety released for rice fallow conditions and is popular in Andhra Pradesh state of India. The YMD tolerant mungbean variety TMB-37 is popular in Northeast Plane zone of India and was rereleased in Punjab state as it can fit into wheat-mungbean-rice cropping systems due to early and synchronous maturing traits. Pigeon pea is the second most important pulse crop of India and mutant varieties in this crop covered considerable areas of central India. The varieties TJT-501 and PKV-TARA occupy most of the pigeon pea cultivating areas of Madhya Pradesh and Vidarbha region of Maharashtra, respectively. The gamma rays–induced cowpea variety TC-901 is the first variety released for summer cultivation in central zone of India which has the potential to change the scenario of cowpea cultivation in summer season due to its determinate growth habit and tolerance to various diseases. Thus, the mutant varieties of pulse crops have contributed substantially to increasing productivity and ensuring nutritional security of the country.

9.1.24 Conclusion and Future Prospects

The pulse crop production and productivity can be enhanced only through the development of high-yielding varieties with broad genetic base and high buffering capacity with potential to combat biotic and abiotic stresses which are more likely to prevail in the existing scenario of climate change. Induced mutation is a potential plant breeding method which can create new genetic variability with macro- and micromutations and novel mutations like pests and abiotic tolerance. The induced mutation is proving very useful especially for traits for which (various biotic and abiotic stress tolerance) genetic source are absent in the existing germplasm. This method is also equally applicable to enhancing the quality by increasing the nutrient content and reducing the antinutritional factors. Induced mutants are very useful for basic studies like gene expression, gene regulation, and developmental genetics. The advent of molecular techniques and genomics gave rise to high-throughput mutant screening and mutant identification at DNA level, which will enable precision mutation breeding in pulse crops.

REFERENCES

Abe A., Kosugi S., Yoshida K., et al., 2012. Genome sequencing reveals agronomically important loci in rice using MutMap. *Nat. Biotechnol.* 30:174–178. Doi:10.1038/nbt.2095

Abe T., Y. Kazama, H. Ichida, Y. Hayashi, H. Ryuto and N. Fukunishi. 2007. Plant breeding using the ion beam irradiation in RIKEN. In *Proceedings of the 18th International Conference on Cyclotrons and Their Applications,* 222–224.

Akhar F. K., A. Bagheri, N. Moshtaghi and A. Nezami. 2011. The effect of gamma radiation on freezing tolerance of chickpea (*Cicer aretinum* L.) at in vitro culture. *J. Biol. Environ. Sci.* 5(14): 63–70.

Ali A., S. Hayat and U. Kr. Deka. 2014a. Mutagenic sensitivity in early generation in black gram (*Vigna mungo* L. *Hepper). IOSR J. Agric. Vet. Sci.* 7(8) Ver. II: 35–40.

Ali A., K. Yubey, U. Kr. Deka and S. M. S. Tomar. 2014b. Effect of sodium azide on seed germination and related agro-metrical traits in M_1 Lentil (*Lens culinaris* Medik.) *generation. World J. Agric. Sci.* 10 (3): 95–102.

Almosawi M. B. H. 2019. Effect of ethyl methane sulfonate, sodium azide and gamma rays on yield performance of pigeon pea (*Cajanus cajan* L.) in M_3 generation. *Plant Arch.* 19 (2): 1572–1574.

Amri W., M. Laouar, A. Abdelguerfi, J. Jankowicz, L. Jankuloski and B. J. Till. 2018. Genetic variability induced by gamma rays and preliminary results of low-cost tilling on M_2 generation of chickpea (*Cicer arietinum* L.). *Front Plant Sci.* 9: 1568–1582.

Anonyms. 2000. Economic survey of India 1999–2000, DES, Government of India, Ministry of Agriculture, cooperation and Farmer Welfare.

Anonyms. 2022. Economic survey of India-2021–22, DES, Government of India, Ministry of Agriculture, cooperation and Farmer Welfare.

Anuragi H., R. Yadav and R. Sheoran. 2022. Gamma-rays and EMS induced resistance to mungbean yellow mosaic India virus in mungbean (*Vigna radiata* L. Wilczek) and its validation using linked molecular markers. *Int. J. Rad. Biol.* 98(1): 69–81.

Ariraman M., S. Gnanamurthy, D. Dhanavelb, T. Bharathi, S. Murugan. 2014. Mutagenic effect on seed germination, seedling growth and seedling survival of Pigeon pea (*Cajanus cajan* L. Millsp). *Int. Lett. Nat. Sci.* 16: 41–49.

Auerbach C. 1946. Chemically Induced Mosaicism in Drosophila melanogaster. *Proceedings of the Royal Society of Edinburgh, Section B: Biological Sciences* 62(2): 211–222.

AuerbachC. And J. M. Robson. 1946. Chemical production of mutations. *Nature* 157: 302.

Austin R.S., D. Vidaurre, G. Stamatiou et al., 2011. Next-generation mapping of *Arabidopsis* genes. *Plant J.* 67(4): 715–725. Doi:10.1111/j.1365-313X.2011.04619.x

Auti S. G. and B. J. Apparao. 2009. Induced Mutagenesis in Mungbean (*Vigna radiata* L. Wilczek). In: Q.Y. Shu (ed.), *Induced Plant Mutations in the Genomics Era.* Food and Agriculture Organization of the United Nations, Rome, 2009, 97–100.

Bansod S. V., J. P. Gupta, B. K. Pandey and S. R. Awachar.2022. Effect of chemically induced mutation on yield contributing and seed quality attribute in pea (*Pisum sativum* L.). *Biological Forum—An International Journal* 14(3): 1497–1500.

Barshile J. D. 2018. Characterization of bold seed mutant in chickpea cultivar Vijay. *Int. J. Res. Biosci. Agric. Technol.* 6(2): 17–20.

Barshile J. D., S. G.Auti, S. C.Dalve and B. J.Apparao. 2006. Mutagenic sensitivity studies in chickpea employing SA, EMS and gamma rays. *Indian J. Pulses Res.* 19(1): 43–46.

Bernasconi P., A. R. Woodworth, B. A. Rosen, M. V. Subramanian and D. L. Siehl. 1995. A naturally occurring point mutation confers broad range tolerance herbicides that target acetolactate synthase. *J. Biol. Chem.* 270: 17381–17385.

Bhat Mehraj-U-Din, SamiullahKhan and M. ImranKozgar. 2011. Studies on induced mutations in chickpea (*Cicer arietinum* L.) I. Responses of the mutagenic treatments in M_1 biological parameters. *Elec. J. Plant Breed.* 2(3): 422–424.

Bhosale U. P. and B. V. Hallale.2011. Gamma radiation induced mutations in black gram (*Vigna mungo* (L.) Hepper). *Asian J. Plant Sci. Res.* 1(2): 96–100.

Bobkov S. V. and T. N. Selikhova. 2021. Introgession of powdery mildew resistance into cultural pea from wild accession of *P. fulvum. IOP Conf. Ser. Earth Environ. Sci.* 650: 012091.

Bohra A., I. P. Singh, S. Mehandi, D. Singh and N. Nadarajan. 2016. Investigating inheritance patterns of a spontaneous mutant in *pigeon pea* and its future implications for CGMS based hybrid breeding. *Legume Res.* 39 (1): 35–38.

Bolbhat S.N. and K. N. Dhumal. 2023. Induced macromutations in horse gram (*Macrotyloma uniflorum* Lam. Verdc). *Legume Research* 32(4): 278–281.

Bolbhat S.N., A. B.Lande, P. D.Naik, S. S.Gaikwad and S. Shaikh. 2020. Induced genetic variability in black turtle bean (*Phaseolus vulgaris* L.). *Int. J. Creat. Res. Thoughts* 8(4): 258–262.

Bonde P., B. S. Thorat and V. J. Gimbhavnekar. 2020. Effect of gamma radiation on germination and seedling parameters of mungbean (*Vigna radiata*). *Int. J. Curr. Microbiol.* 11: 1582–1587.

Borisov A., E. V. Morzhina, O. A. Kulikova, S. A. Tchetkova, V. K. Lebsky and I. A. Tikhonovich. 1992. New symbiotic mutants of pea (*Pisum sativum* L.) affecting either nodule initiation or symbiosome development. *Symbiosis* 14: 297–313.

Borkar A. T. and A. D. More. 2010. Induced flower colour mutations in *Phaseolus vulgaris* Linn. Through physical and chemical mutagen. *Advances in Bioresearch* 1(1): 22–28.

Botticella E., F. Sestili, A. Hernandez-Lopez, A. Phillips and D. Lafiandra. 2011. High resolution melting analysis for the detection of EMS induced mutations in wheat SbeIIa genes. *BMC Plant Biol.* 11: 156.

Brahmil Y. Mabrouk, K. Charaabi, P. Delavault, P. Simier and O. Belhadj. 2014. Induced mutagenesis through gamma radiation in chickpea (*Cicer arietinum* L.): developmental changes and improved resistance to the parasitic weed *Orobanchefoetida* Poir. *Int. J. Adv. Res.* 2(11): 670–684.

Chaudhary L., Sharma, R. and M. Kumar. 2021. Estimation of LD_{50} and effect of sodium azide on germination and seedling parameters of different cultivars of *Cajanus cajan* L. *Millspaugh. Toxicol. Environ. Health Sci.* 13: 279–285.

Chowdhury M.A., C. P. Andrahennadi, A. E. Slinkard and A. Vandenberg. 2001. RAPD and SCAR markers for resistance to *Ascochyta* blight in lentil. *Euphytica* 118: 331–337.

Cooper H. D., C. Spillane and T. Hodgkin. 2001. Broadening the genetic base of crops: an overview. In: Cooper H. D., C. Spillane and T. Hodakin (Eds.). *Broadening the Genetic Base of Crop Production*, CAB International, Oxon, UK, pp. 1–23.

Cross M. J., D. L. Waters, L. S. Lee and R. J. Henry. 2008. Endonucleolytic mutation analysis by internal labeling (EMAIL). *Electrophor.* 29(6): 1291–1301.

Dahiya P. K., A. R. Linnemann, M. J. R. Nout, M. A. J. S. Boekel and R. B. Grewal. 2013. Nutrient composition of selected newly bred and established mungbean varieties. *LWT—Food Sci Tech* 54: 249–256.

Das T. R. and B. Baisakh. 2020. Effectiveness and efficiency of physical and chemical mutagens in greengram (*Vigna radiata* L. *Wilczek). E-planet* 18 (1): 6–12.

Davis T.M., K. W. Foster and D. A. Phillips. 1985. Nodulation mutants in chickpea. *Crop Sci.* 25: 345–348.

Devi M., Sreevalli, D. Mohan Reddy, V. L. N. Reddy, D. Lokanadha Reddy, P. Sudhakar and S. Sofia. 2021. Effect of mutagenesis on growth and yield related traits in black gram (*Vigna mungo* (L.) Hepper). *J. Res. ANGRAU* 49(4): 14–29.

Dewanjee S. and K. K. Sarkar. 2018. Evaluation of performance of induced mutants in mungbean (*Vigna radiata* L. Wilczek). *Legume Res.* 41(2): 213–217.

Dhanasekar P. and K. S. Reddy. 2015. Serological screening of cowpea genotypes for resistance against cowpea aphid borne mosaic virus using DAS-ELISA. *Asian J. Plant Pathol.* 9(2): 83–90.

Dhanasekar P., J. Souframanien, V. J. Dhole, A. J. Hingane and S. Sivasankar. 2023. Physical mutagenesis for induction of resistance against legume pod borer (*Marucavitrata* F.) in cowpea (*Vigna unguiculata* L. Walp). In: *International Conference on, "Pulses: Smart Crops for Agricultural Sustainability and Nutritional Security*", Feb. 10–12, 2023 at NASC, New Delhi-110012.

Dhanavel D., P. Pavadai, L. Mullainathan, D. Mohana, G. Raju, M. Girija and C. Thilagavathi. 2008. Effectiveness and efficiency of chemical mutagens in cowpea (*Vigna unguiculata* (L.) Walp). *African J. Biotechnol.* 7(22): 4116–4117.

Dhole V. J. and K. S. Reddy. 2010. Gamma rays induced moisture stress tolerant long root mutant in mungbean (*Vigna radiata* L. *Wilczek). Elec. J. Plant Breed.* 1: 1299–1305.

Dhole V. J. and K. S. Reddy. 2012. Genetic analysis of resistance to mungbean yellow mosaic virus in mungbean (*Vigna radiata*). *Plant Breed.* 131: 414–417.

Dhole V. J., and K. S. Reddy. 2013. Development of a SCAR marker linked with a MYMV resistance gene in mungbean (*Vigna radiata* L. Wilcek). *Plant Breed.* 132: 127–132. Doi: 10.1111/pbr.12006

Dhole V. J., J. Souframanien, K. S. Reddy and V. C. Petwal. 2023. Comparison of effectiveness and efficiency of electron beam over gamma rays to induce novel mutations in mungbean (*Vigna radiata* L. *Wilczek). Appl. Radiat. Isot.* 194: 110719.

Dhulgande G. S., D. A. Dhale, G. L. Pachkore and R. A. Satpute. 2011. Mutagenic effectiveness and efficiency of gamma rays and ethyl methanesulphonate in pea (*Pisum sativum* L.). *J. Exp. Sci.* 2(3): 07–08.

Dinesh H. B., H. C. Lohithaswa, K. P. Viswanatha, Poonam Singh, L. Manjunatha, D. S. Ambika and M. K. Prasanna Kumar. 2018. Genetic analysis and marker assisted backcrossing for transfer of mosaic virus resistance in cowpea (*Vigna unguiculata* L. *Walp.). Legume Res.* 41: 663–668.

Donini P. and A. Sonnino. 1998. Induced mutation in plant breeding: current status and future outlook. *Somaclonal variation and induced mutations in crop improvement*, pp. 255–291. Springer.

Dorvlo I. K., G. Amenorpe, H. M. Amoatey, S. Amiteye, J. T. Kutufam, E. Afutu, E. Asare-Bediako and A. A. Darkwa. 2022. Improvement in cowpea variety Videza for traits of extra earliness and higher seed yield. *Heliyon* 8: e12059.

Eswaramoorthy V., T. Kandasamy, K. Thiyagarajan, C. Vanniarajan and J. Souframanien. 2021. Effectiveness and efficiency of electron beam in comparison with gamma rays and ethyl methane sulfonate mutagens in cowpea. *Appl. Radiat.* Isot. 171: 109640. Doi: 10.1016/j.apradiso.2021.109640

Etther Y., S. J. Gahukar, A. Akhare, A. N. Patil, S. J. Jambhulkar and M. Gawande. 2019. Mutagenic effectiveness and efficiency of gamma rays, ethyl methyl sulfonate and their synergistic effects in *pigeon pea* (*Cajanus cajan* L.). *J. Pharmacogn. Phytochem.* 8(3): 489–493.

Fekih R., H. Takagi, M. Tamiru, et al., 2013. MutMap+: genetic mapping and mutant identification without crossing in rice. *PloS One* 8: 1–10. Doi:10.1371/journal. Pone.0068529

Feng X., G. E. Orellana; J. R. Myers, and A. V. Karasev. 2018. Recessive resistance to Bean common mosaic virus conferred by the bc-1 and bc-2 genes in common bean (*Phaseolus vulgaris* L.) affects long distance movement of the virus. *Phytopath.* 108(8): 1011–1018.

Galeano C. H., M. Gomez, L. M. Rodriguez, M. W. Blair and I. CEL. 2009. Nuclease digestion for SNP discovery and marker development in common bean (*Phaseolus vulgaris* L.). *Crop Sci.* 49: 381–394.

Gaur P. M., V. K. Gaur and S. Srinivasan. 2008. An induced brachytic mutant of chickpea and its possible use in ideotype breeding. *Euphytica* 59: 35–41.

Ghanim M. A. A., Spencer-Lopes, M. M., and W. Thomas. 2018. Mutation breeding in seed propagated crops: parental selection, mutant generation development, mutation detection, mutant evaluation and factors influencing success. In: *Manual on Mutation Breeding*. Edited by Spencer-Lopes, M. M., B. P. Forster and L. Jankuloski. Food and Agriculture Organization of the United Nations International Atomic Energy Agency, Vienna, 2018, pp. 119–156.

Giri S. P. 2014. Studies of mutagenic sensitivity in *Pigeon pea* (*Cajanus Cajan* L. Mill sp.). *Bio. Disc.* 5(2): 227–229.

Girija M. and D. Dhanavel. 2009. Mutagenic effectiveness and efficiency of gamma rays ethyl methane sulphonate and their combined treatments in cowpea (*Vigna unguiculata* L. Walp). *Global Journal of Molecular Sciences* 4(2): 68–75.

Gnankambary K., T. B. J. Batieno, N. Sawadogo, M. Sawadogo, D. Yonli and T. J. Ouedraogo. 2019. Assessment of radio-sensitivity for three cowpea genotypes to gamma irradiation. *Int. J. Genet. Mol. Biol.* 11(2): 29–33.

Goyal S. and S. Khan. 2010. Induced mutagenesis in urdbean (*Vigna mungo* L. Hepper): A review. *Int. J. Bot.* 6: 194–206.

Goyal S., M. R. Wani and S. Khan. 2019. Comparative mutagenic analysis of gamma rays, EMS and their combination treatments in black gram (*Vigna mungo* L. Hepper). *Thai J. Agric. Sci.* 52(1): 20–33.

Goyal S., M. R. Wani, A. Raina, R. A. Laskar and S. Khan. 2021. Phenotypic diversity in mutagenized population of urd bean (*Vigna mungo* L. Hepper). *Heliyon.* 7(5): e06356.

Guild G. E. and J. C. R. Stangoulis. 2021. EDXRF for screening micronutrients in lentil and sorghum biofortification breeding programs. *Plant Soil.* 463: 461–469.

Gumber, R. K. and S. Sarvjeet. 1996. Genetics of flowering time in chickpea—A preliminary report. *Crop Improv.* 23: 295–296.

Gupta S. K., R. A. Singh and S. Chandra. 2005. Identification of a single dominant gene for resistance to mungbean yellow mosaic virus in black gram (*Vigna mungo* L. Hepper). *SABRAO J. Breed. Genet.* 37(2): 85–89.

Horn L. N, H. M. Ghebrehiwot and H. A. Shimelis. 2016. Selection of novel cowpea genotypes derived through gamma irradiation. *Front. Plant Sci.* 7: 262. Doi: 10.3389/fpls.2016.00262

Ignacimuthu S. and C. R. Babu. 1989. Induced variation in protein quantity and quality in the wild and cultivated urd mungbeans. *Indian. J. Genet.* 49: 173–181.

Ingelbrecht I., J. Jankowicz-Cieslak, M. Szurman, B. J. Till and I. Szarejko. 2018. Chemical mutagenesis. In: Manual on Mutation Breeding. Edited by Spencer-Lopes, M. M., B.

P. Forster, and L. Jankuloski. Food and Agriculture Organization of the United Nations *International Atomic Energy Agency*, Vienna, 2018, pp. 51–82.

Jiao K., X. Li, S. Su, W. Guo, Y. Guo, Y. Guan, Z. Hu, Z. Shen and D. Luo. 2019. Genetic control of compound leaf development in the mungbean (*Vigna radiata* L.). *Hortic. Res.* 6: 23.

Joshi-Saha A., K. S. Reddy, V. C. Petwal and J. Dwivedi. 2015. Identification of novel mutants through electron beam and gamma irradiation in chickpea (*Cicer arietinum* L.). *J. Food Legum.* 28(2): 1–6.

Kang B. C., I. Yeam, and M. M. Jahn. 2005. Genetics of plant virus resistance. *Ann. Rev. Phytopath.* 43: 581–621.

Kang R., E. Seo, G. Kim, et al., 2020. Radio sensitivity of cowpea plants after gamma-ray and proton-beam irradiation. *Plant Breed. Biotechnol.* 8: 281–292.

Kashid N. G. 2016. Mutagenic effect of ethyl methane sulphonate and sodium azide on pod characters in chickpea (*Cicer arietinum* L.). *Int. J. Sci. Res.* 5(2): 1535–1537.

Kashid N. G. and S. B. More. 2015. Genetic variability induced by ethyl methane sulphonate and sodium azide on seed characters in chickpea (*Cicer arietinum* L.). *Int. J. Recent Sci. Res.* 6(10): 6676–6679.

Kashid N. G. And S. B. More. 2016. A study of effect of induced mutation on flowering of plant in M_2 & M_3 generations in chickpea (*Cicer arietinum* L.). *Trop. Plant Res.* 3(1): 182–185.

Khan S. and S. Goyal. 2009. Improvement of mungbean varieties through induced mutations. *African J. Plant Sci.* 3: 174–180.

Khan S., M. R. Wani and K. Parveen. 2006. Sodium azide induced high yielding early mutant in lentil. *Agric. Sci. Digest* 26(1): 65–66.

Khan S., Mohd. Rafiq Wani, Mehraj-Ud-Din Bhat and K. Parveen. 2005. Induced chlorophyll mutations in chickpea (*Cicer arietinum* L.). *Int. J. Agri. Biol.*, 7(5): 764–767.

Konzak C. F., R. A. Nilan, J. Wagner, et al., 1965. Efficient chemical mutagenesis. *Rad. Bot.* 5(Suppl.): 49–70.

Kozgar M. I. 2014. *Mutation Breeding in Chickpea: Perspectives and Prospects for Food Security*. De Gruyter Open Ltd, Warsaw/Berlin.

Kulthe M. P. and V. S. Kothekar. 2011. Effect of sodium azide on yield parameters of chickpea (*Cicer arietinum* L.). *Journal of phytology and Phytophysiology* 3(11): 39–42.

Kumar A. and R. K. Gill. 2020. Inheritance of rust resistance, yield performance in irrigated condition and identification of high yielding rust resistant recombinant inbred lines in lentil. *Int. J. Curr. Microbiol. App. Sci.* 9(2): 2393–2399.

Kumar A. and T. S. Dhillon. 2020. Responses of EMS on growth and yield of French bean (*Phaseolus vulgaris*). *Indian J. Agric. Sci.* 90(2): 436–438.

Kumar S., S. Gupta, S. Chandra, et al., 2004. How wide is the genetic base of pulse crops? In: Ali M. et al. (Eds.) *Pulses in New Perspective. Proceedings of the National Symposium on Crop Diversification and Natural Resources Management*, 20–22 December 2003. Indian Society of Pulses Research and Development, Indian Institute of Pulses Research, Kanpur, pp. 211–221.

Kumar V., A. K. Sharma, V. P. Singh and M. Kumar. 2009. Characterization of pre-breeding genetic stocks of urdbean (*Vigna mungo* L. Hepper) induced through mutagenesis. In: Q.Y. Shu (ed.), *Induced Plant Mutations in the Genomics Era*. Food and Agriculture Organization of the United Nations, Rome, 2009, 391–394.

Kumar Y., S. Basu, D. Goswami, M. Devi, U. S. Shivhare and R. K. Vishwakarma. 2022. Anti-nutritional compounds in pulses: Implications and alleviation methods. *Legum. Sci.* 4: e111.

Kumar, J., D. S. Mathur, S. S. Yadav, and N. S. Hooda. 1999. Two new leaf mutants in chickpea. *Indian J. Genet. Plant Breed.* 59(03): 381–382.

Kumar, J., K. C. Arbind, K. S. Ramesh, and P. Aditya. 2011. Towards marker assisted selection in pulses: a review. *Plant Breed.* 130: 297–313.

Kundagrami J., S. Basak, B. Maiti, T. K. Das, A. Gose and A. Pal. 2009. Agronomic, genetic and molecular characterization of MYMV tolerant mutant lines of *Vigna mungo*. *Int. J. Plant Breed. Genet.* 3(1): 1–10.

Ladizinsky O. 1985. Founder effect in crop-plant evolution. *Econ. Bot.* 39: 191–199.

Lal J. P. and A. K. Tomer. 2009. Genetic Enhancement of Lentil (*Lens culinaris* Medikus) for drought tolerance through induced mutations. In: Q.Y. Shu (ed.), *Induced plant mutations in the genomics era*. Food and Agriculture Organization of the United Nations, Rome, 2009 , 151–154.

Laskar R. A. and S. Khan. 2017. Mutagenic effectiveness and efficiency of gamma rays and phenotyping of induced mutations in lentil cultivars. *Int. Lett. Nat. Sci.* 64: 17–31.

Laskar R., Ruhul Amin and S. Khan. 2017. Evaluation of optimal doses for gamma rays and hydrazine hydrates in lentil genotypes. *Trends in Biosciences* 10(27): 5822–5825.

Lee H., S. Rustgi, N. Kumar, I. Burke, J. P. Yenish, K. S. Gill and D. V. Wettstein. 2011. Single nucleotide mutation in the barley acetohydroxy acid synthase gene confers resistance to imidazolinone herbicides. *PNAS* 108: 8909–8913.

Liu L. X., H. J Guo, L. S. Zhao et al., 2009. Achievements and perspectives of crop space breeding in China. In: Shu, Q. Y. (ed) *Induced plant mutations in genomics era*. IAEA, Vienna, Austria, pp. 213–215.

MabroukY., K. Charaabi, D. Mahiout, M. Rickauer and O. Belhadj. 2018. Evaluation of chickpea (*Cicer arietinum* L.) irradiation-induced mutants for resistance to *Ascochyta* blight in controlled environment. *Brazilian Journal of Botany*. https://doi.org/10.1007/s40415-018-0458-8

Mahto C. S., S. Suman, N. Kumar and S. Sen. 2018. Mutagenic effectiveness and efficiency of EMS and sodium azide in mungbean (*Vigna radiata* L. Wilczek). *Int. J. Bioresour. Stress Manag.* 9(4): 537–540.

Makeen K., B. G. Suresh, G. R. Lavanya and A. Kumari. 2013. Study of chlorophyll and macromutations induced by gamma rays and sodium azide in urd bean (*Vigna mungo* L. Hepper). *Afr. J. Agric. Res.* 8(47): 5958–5961.

McCallum C. M., L. Comai, E. A. Greene, et al., 2000. Targeted screening for induced mutations. *Nat. Biotechnol.* 18: 455–457.

McCourt J. A., S. S. Pang, J. King-Scott, L. W. Guddat and R. G. Duggleby. 2006. Herbicide binding sites revealed in the structure of plant acetohydroxy acid synthase. *PNAS* 103: 569–573.

Mejri S., I. Hemissi, C. Brinsi, A. Asmi, Mouldi, Saidi and Y. Mabrouk. 2021. Dose dependent effects of gamma radiation on growth parameters of *Lens culinaris Medikus subsp. Culinaris*. *Acta Sci. Agric.* 5(7): 83–89.

More A. D. and A. T. Borkar. 2016. Mutagenic effectiveness and efficiency of gamma rays and EMS in *Phaseolus vulgaris* L. *Int. J. Curr. Microbiol. App. Sci.* 5(10): 544–554.

Mubarak A. E. 2005. Nutritional composition and anti-nutritional factors of mungbean seeds 40 (*Phaseolus aureus*) as affected by some home traditional processes.*Chem.* 89: 489–495.

Munyinda K. and K. Kanenga. 2018. *Creating variability in cowpea for adaptation and value addition through induced mutation*. In: FAO/IAEA International Symposium on Plant Mutation Breeding and Biotechnology; 27–31 Aug 2018 at Vienna, Austria. p. 17.

MVD, IAEA. 2023. www.mvd.iaea.org

Nene, Y. L. 2006. Indian pulses through the millennia. *Asian Agri-History* 10(3): 179–202.

Nepolian Th., M. S. Jeberson, N. B. Singh, Ph. R. Sharma and K. S. Shashidhar. 2018. Study of biological effects of sodium azide in M_1 generation and estimation of LD_{50} for growth and reproduction parameters. *Pharma. Innov. J.* 7(10): 291–293.

Newhouse K. E., B. K. Singh, D. Shanner and M. Stidham. 1991. Mutations in corn (*Zea mays* L.) conferring resistance to imidazolinone herbicides. *Theor. Appl. Genet.* 83: 65–70.

Ngampongsai S., A. Watanasit, S. Srisombun, P. Srinives and A. Masari. 2009. Current status of mungbean and the use of mutation breeding in Thailand. In: Q. Y. Shu (ed.), *Induced Plant Mutations in the Genomics Era*. Food and Agriculture Organization of the United Nations, Rome, 355–357.

Olasupo F., C. Ilori, E. Stanley, T. Owoeye and D. Igwe. 2018. Genetic analysis of selected mutants of cowpea (*Vigna unguiculata* L. Walp) using simple sequence repeat and *rcb*L markers. *American J. Plant Sci.* 9: 2728–2756.

Opoku Gyamfi M., J. S. Y. Eleblu, L. G. Sarfoa, I. K. Asante, F. Opoku-Agyemang and E. Y. Danquah. 2022. Induced variations of ethyl methane sulfonate mutagenized cowpea (*Vigna unguiculata* L. Walp) *plants. Front. Plant Sci.* 13: 952247.

Ouellet T., R. G. Rutledge and B. L. Miki. 1992. Members of acetohydroxy acid synthase multigene family of *Brassica napus* have divergent pattern of expression. *Plant J.* 2: 321–330.

Patil P. G., A. Bohra, J. Dubey, P. R. Saabale, R. K. Mishra, G. Pange, A. Das, M. Rathore, F. Singh and N. P. Singh. 2017. Genetic analysis and molecular resistance to race 2 of *Fusarium* wilt in *pigeon pea* (*Cajanus cajan* L. Millsp.). *Crop Prot.* 100: 117–123.

Perez-Leon N. J., R. I. Castro-Menduina and A. Echevarria-Hernandez. 2019. Genetic improvement of Chickpea cultivars (*Cicer arietinum* L.) by gamma irradiation. *Biotecnol. Apl.* 36(2): 2231–2234.

Porch T. G., M. W. Blair, P. Lariguet, C. Lariguet, C. E. Pankhurst and W. J. Broughton. 2009. Generation of a mutant population for TILLING common bean genotype BAT 93. *J. Amer. Soc. Hort. Sci.* 134(3): 348–355.

Prabhu S.C., A. Mothilal, P. Anantharaju, V. Rajan Babu, P. Jeyaprakash and C. Vanniarajan. 2022. Effect and sensitivity of gamma irradiation to various biometrical traits of cowpea (*Vigna unguiculata* (L.) Walp.). *Biological Forum-An International Journal* 14(3): 711–723.

Priyadharshni S., S. Saravanan, K. Elanchezyan, A. Kavitha Pushpam and M. Arumugam Pillai. 2020. Efficiency and effectiveness of physical and chemical mutagens in cowpea (*Vigna ungiculata* L. Walp). *Elec. J. Plant Breed.* 11(3): 803–808.

Raina A., Laskar R. A., M. R. Wani, B. L. Jan, S. Ali and S. Khan. 2022. Gamma rays and sodium azide induced genetic variability in high-yielding and biofortified mutant lines in cowpea (*Vigna unguiculata* L. Walp.). *Front. Plant Sci.* 13: 911049. Doi: https://doi.org/10.3389/fpls.2022.911049.

Rajasekharan K., J. W. Grula and D. M. Anderson. 1996. Selection and characterization of mutant cotton (*Gossypium hirsutum* L) cell lines resistance to sulfonylurea and imidazolinone herbicides. *Plant Sci.* 199: 115–124.

Ramandeep T. S. Dhillon, R. K. Dhall and B. S. Gill. 2018. Effect of mutagen-ethyl methane sulphonate on yield increasing parameters of French bean (*Phaseolus vulgaris* L.). *Genetika* 50(1): 199–207.

Reddy K. S. 2009. A new mutant for yellow mosaic virus resistance in mungbean (*Vigna radiata* L. Wilczek) variety SML-668 by recurrent gamma-ray irradiation. In: Q. Y. Shu (ed.), *Induced Plant Mutations in the Genomics Era*. Food and Agriculture Organization of the United Nations, Rome, 361–362.

Rigola D., J. van Oeveren, A. Janssen, A. Bonne, H. Schneiders, H. J. A. van der Poel, N. J. van Orsouw, R. C. J. Hogers, M. T. J. de Both and M. T. J. van Eijk. 2009. High-throughput detection of induced mutations and natural variation using KeyPoint™ technology. *PloS One* 4(3): e4761.

Rizwan M., M. Aslam, M. J. Asghar, G. Abbas, T. M. Shah, H. Shimelis. 2017. Pre-breeding of lentil (*Lens culinaris* Medik.) for herbicide resistance through seed mutagenesis. *PloS One* 12(2): e0171846. Doi:10.1371/journal.pone.0171846

Roslim D. I., Herman and I. Fiatin. 2015. Lethal dose 50 (LD_{50}) of mungbean (*Vigna radiata* L. *Wilczek) cultivar Kampar. SABRAO J. Breed. Genet.* 47(4): 510–516.

Roy U., D. Basak and S. Nath. 2019. Mutagenic sensitivity analysis of gamma irradiations in Cowpea (*Vigna unguiculata* L. Walp). *Emer Life Sci. Res.* 5(2): 12–16.

Rupela O. P. and M. R. Sundarshana. 1986. Identification of a nonnodulating spontaneous mutant in chickpea. *International Chickpea Newsletter,* 15: 13–14.

Sala C. A. and M. Bulos. 2012. Inheritance and molecular characterization of broad range tolerance to herbicides targeting acetohydroxyacid synthase in sunflower. *Theor. Appl. Genet.* 124: 355–364.

Sanchez-Martin J., B. Steuernagel, S. Ghosh, et al., 2016. Rapid gene isolation in barley and wheat by mutant chromosome sequencing. *Genome Biol.* 17(1): 221.

Sangle S. M. 2016. Induced mutagenesis in pigeon pea (*Cajanus cajan*L. *Millsp). Int. J. Curr. Res.* 8(07): 33953–33957.

Sangle S. M. 2020. Induced genetic variability for quantitative traits in *pigeon pea. Indian J. Agric. Res.* 54(2): 139–146.

Sangle S. M. and V. S. Kothekar. 2013. Mutagenic effectiveness and efficiency in *pigeon pea. Int. J. Adv. Sci. Techn. Res.* 3(4): 40–44.

Sarwar G. and M. Ahmad. 2003. Development of new high yielding mungbean variety "AEM-96" through induced mutation. *SAARC J. Agric.* 1: 173–180.

Saxena K., A. K. Choudhary, R. K. Srivastava, A. Bohra, R. K. Saxena, R. K. Varshney. 2019. Origin of early maturing *pigeon pea* germplasm and its impact on adaptation and cropping systems. *Plant Breed.* 138: 243–251.

Schneeberger K., S. Ossowski, C. Lanz, et al., 2009. SHOREmap: Simultaneous mapping and mutation identification by deep sequencing. *Nat. Methods* 6 : 550–551. Doi:10. 1038/nmeth0809-550.

Shah T. M., B. M . Atta, J. I. Mirza and M. A. Haq. 2009. Screening of chickpea (*Cicer arietinum*) induced mutants against *Fusarium* wilt. *Pakistan J. Bot.* 41: 1945–1955.

Sharma K. D. and F. J. Muehlbauer. 2007. Fusarium wilt of chickpea: physiological specialization, genetics of resistance and resistance gene tagging. *Euphytica* 157:1–14.

Sharma K. D., G. Katna, N. Sharma, R. Nag, B. K. Sharma and A. Joshi-Saha. 2018. Mutagenic effectiveness and efficiency of gamma rays, ethyl methane sulphonate and their combination treatments in chickpea (*Cicer arietinum* L.). *Int. J. Curr. Microbiol. App. Sci.* 7(11): 509–515.

Sharma M., R. Ghosh, A. Tarafdar, et al., 2019. Exploring the genetic cipher of chickpea (*Cicer arietinum* L.) through identification and multi-environment validation of resistant sources against *Fusarium* wilt (*Fusarium oxysporum* f. *sp. Ciceris). Front. Sustain. Food Syst.* 3: 78.

Sidorova K. K., M. N. Glyanenko, T. M. Mishchenko, E. Y. Vlasova, V. K. Shumny. 2015. Symbiotic nitrogen fixation in legumes as a genetic and selection trait. *Vavilov J. Genet. Breed.* 19: 50–57.

Singh B. B. and D. P. Singh. 1995. Inheritance of spontaneous male sterility in peas. *Theoret. Appl. Genet.* 90: 63–64.

Singh R. R., M. S. Jeberson, N. B. Singh and Ph. Ranjit Sharma. 2017. Effect of EMS on M_1 parameters of field pea (*Pisum sativum* L. var. *arvense*). *Journal of Global Biosciences* 6(5): 4966–4970.

Singh V. P. and S. N. Chaturvedi. 1981. Gamma-ray induced male sterile mutants in mungbean. *Pulse Crops Newsletter* 1: 41–42.

Sofia S., D. M. Reddy, K. H. P. Reddy, P. Latha and B. R. Reddy. 2020. Effect of gamma rays, ethyl methane sulphonate and sodium azide on seedling traits, fertility and varietal sensitivity in mungbean (*Vigna radiata* (L.) Wilczek). *Int. J. Chem. Stud.* 8(4): 1109–1114.

Solanki I. S. 2005. Isolation of macromutations and mutagenic effectiveness and efficiency in lentil (*Lens culinaris* Medik.). *Indian J. Genet.* 65(4): 264–268.

Souframanien J., K. S. Reddy, V. C. Petwal and J. Dwivedi. 2016. Comparative effectiveness and efficiency of electron beam and ^{60}Co γ-rays in induction of mutations in black gram (*Vigna mungo* L. Hepper). *J. Food Leg.* 29(1): 1–6.

Souframanien J., V. J. Dhole and K. S. Reddy. 2021. Breeding for low phytates and oligosaccharides in mungbean and *black gram*. In *Breeding for Enhanced Nutrition and Bio-Active Compounds in Food Legumes.* Edited by Debjyoti Sen Gupta, Sanjeev Gupta, Jitendra Kumar. Springer International Publishing pp. 99–130.

Spencer-Lopes M. M., L. Jankuloski, A. M. A. Ghanim, M. Matijevic and KODYM. 2018. Physical mutagenesis. In: *Manual on mutation breeding*. Edited by Spencer-Lopes, M. M., B. P. Forster and L. Jankuloski. Food and Agriculture Organization of the United Nations International Atomic Energy Agency, Vienna, pp. 5–49.

Srinivasan S. and P. M. Gaur. 2012. Genetics and characterization of an open flower mutant in chickpea. *J. Hered.* 103(2): 297–302.

Srivastava A. K., S. K. Chaturvedi and N. P. Singh. 2017. Genetic base of Indian chickpea (*Cicer arietinum* L.) varieties revealed by pedigree analysis. *Legume Res.* 40(1): 22–26.

Sudha M., A. Karthikeyan, V. G. Shobhana, et al., 2015. Search for *Vigna species* conferring resistance to mungbean yellow mosaic virus in mungbean. *Plant Genet. Resour.* 13(2): 162–167.

Tah P. 2009. Induced synchrony in pod maturity in mungbean. *ARPN J. Agri. Boil. Sci.* 4(1): 42–44.

Takagi H., Abe A., Yoshida K., et al., 2013a. QTL-Seq: Rapid mapping of quantitative trait loci in rice by whole-genome resequencing of DNA from two bulked populations. *Plant J.* 74: 174–183. Doi:10.1111/tpj.12105

Takagi H., A. Uemura, H. Yaegashi, et al., 2013b. MutMap-Gap: Whole-genome re-sequencing of mutant F_2 progeny bulk combined with de novo assembly of gap regions identifies the rice blast resistance gene Pii. *New Phytol.* 200: 276–283.

Tamilzharasi M., K. Dharmalingam, T. Venkatesan, J. Souframanien and J. Palaniappan. 2022. Mutagenic effectiveness and efficiency of gamma rays and combinations with EMS in the induction of macro mutations in black gram (*Vigna mungo* L. Hepper). *Appl. Radiat. Isot.* 188: 110382. Doi: 10.1016/j.apradiso.2022.110382

Taunk J., N. R . Yadav, R. C. Yadav and R. Kumar. 2012. Genetic diversity among green gram (*Vigna radiata* L. Wilczek) genotypes varying in micronutrients (Fe and Zn) content using RAPD markers. *Indian J. Biotech* 11: 48–53.

Till B. J., J. Cooper, T. H. Tai et al., 2007. Discovery of chemically induced mutation in rice by TILLING. *BMC Plant Biol.* 7: 19. Doi:10. 1186/1471-2229-7-19

Triques K., B. Sturbois, S. Gallais, M. Dalmais, S. Chauvin, C. Clepet, S. Aubourg, C. Rameau, M. Caboche and A. Bendahmane. 2007. Characterization of *Arabidopsis thaliana* mismatch specific endonucleases: Application to mutation discovery by TILLING in pea. *Plant J.* 51: 1116–1125.

Umavathi S. and L. Mullainathan. 2015. Physical and chemical induced mutagenesis study for identifying lethality dose in chick pea (*Cicer arietinum* L.) *Var. Co-4. Int. Lett. Nat. Sci.* 8: 1–5.

Usharani K. S. and C. R. Ananda Kumar. 2015. Mutagenic efficiency and effectiveness of gamma rays and EMS and their combination in inducing chlorophyll mutations in M_2 generation of urd bean (*Vigna mungo* (L.) Hepper). *Elec. J. Plant Breed.* 6(1): 210–217.

Usharani K. S., C. R. Ananda Kumar and C. Vanniarajan. 2017. Fixation of Lethal Dose 50 and effect of mutagens in m_1 generation under laboratory condition. *Int. J. Curr. Microbiol. App. Sci.* 6(7): 1356–1365.

Vairam N., S. A. Lavanya, M. Muthamilan and C. Vanniarajan. 2016. Screening of M_3 mutants for yellow vein mosaic virus resistance in greengram (*Vigna radiata* L. Wilczek). *Int. J. Plant Sci.* 11(2): 265–269.

Vanmathi S., D. Arulbalachandran, V. Soundarya. 2021. Effects of gamma radiation on quantitative traits and genetic variation of three successive generations of cowpea (*Vigna unguiculata* L. Walp.). *Plant Sci.* 8(3): 578–589.

Vanniarajan C., S. Ganeshram, J. Souframanien, K. Veni, S. Anandhi Lavanya and V. Kuralarasan. 2017. Gamma rays induced urd bean (*Vigna mungo* L. Hepper) mutants with YMV resistance, good batter quality and bold seeded type. *Legume Res.* doi: 10.18805/LR-3824

Varma A. and V. G. Malathi. 2003. Emerging Gemini virus problems: a serious threat to crop production. *Ann. Appl. Biol.* 142: 145–164.

Varshney R. K., T. J. Close, N. K. Singh, D. A. Hoisington and D. R. Cook. 2009. Orphan legume crops enter the genomics era. *Curr. Opin. Plant Biol.* 12: 202–210.

Veni K., C. Vanniarajan and J. Souframanien. 2017. Probit analysis and effect of electron beam and gamma rays in *Black gram* (*Vigna mungo* L. Hepper). *Elec. J. Plant Breed.* 8(3): 950–955.

Verma R. C. and R. Purbiya. 2017. Effects of gamma radiations on seed germination and morphological characteristics of pea (*Pisum sativum* L.). *Indian J. Plant Sci.* 6(3): 21–25.

Verulkar S. and D. Singh. 1997. Inheritance of spontaneous male sterility in pigeon pea. *Theor. Appl. Genet.* 94: 1102–1103.

Vikhe A. C. and J. N. Nehul. 2020. Determination of LD_{50} Value of *Vigna radiata* L. *Wilczek Cultivar- Naval. Adv. Zool. Bot.* 8(1): 6–11.

Wani M. R. 2020. Characterization of chlorophyll deficient mutants in mungbean (*Vigna radiata L.* Wilczek). *Bangladesh J. Bot.* 49(4): 1013–1019.

Wani M. R., A. R. Dar, A. Tak, I. Amin, N. H. Shah, R. Rehman, M. Y. Baba, A. Raina, R. Laskar, M. I. Kozgar and S. Khan. 2017. Chemo-induced pod and seed mutants in mungbean (*Vigna Radiata* L. Wilczek). *SAARC J. Agri.* 15(2): 57–67.

Yaqoob M. and A. Rashid. 2001. Induced mutation studies in some mungbean (*Vigna radiata* L. Wilczek). *J. Biol. Sci.* 1: 805–808.

Yaseen M. S. K., K. Sravan Naga Parimala Kumar, S. Prabakaran and S. Suresh. 2020. Ethyl methane sulphonate (EMS) induced mutations in M_1 generation of cowpea (*Vigna unguiculata* L. Walp). *J. Pharmacogn. Phytochem.* 9(2): 1545–1547.

Yildrim T., H. Canci, I. N. Ertoy, B. F. O. Ceylan, C. Ikten and C. Toker. 2013. Inheritance of female sterility in induced *Cicer species. Turkish J. Field Crops.* 18: 78–81.

10 Variants of CRISPR/Cas9 Technology and Their Role in Rice Genetic Improvement

Rajveer Singh,[1] *Ramanjot Kaur,*[1] *Satwinderpal Singh,*[2] *Krishna Sai Karnatam,*[1] *Umesh Preethi Praba,*[1] *Gurwinder Kaur,*[1] *Gautam Chhabra,*[1] *Shabneek Kaur,*[1] *Jessica Kaur,*[1] *Aanchal Bhatia,*[1] *Muskan Tiwari,*[1] *and Yogesh Vikal*[1*]
[1]School of Agricultural Biotechnology, Punjab Agricultural University, Ludhiana, India
[2]College of Agriculture, Punjab Agricultural University, Ludhiana, India
[*]Corresponding author email; yvikal-soab@pau.edu

10.1 INTRODUCTION

Rice (*Oryza sativa* L.) no doubt is an important staple food crop which feeds more than 3 billion people and provides 20% of world's dietary energy (Mishra et al., 2018). Its global consumption has been estimated to increase from 490 million tons to 650 million tons in the span of years 2020 to 2050 (Rejesus et al., 2012). On the other hand, the world population is continuously increasing and is expected to reach 9.7 billion by 2070 (Bin Rahman and Zhang, 2022). Moreover, the current global rice production is facing enormous challenges due to biotic and abiotic stresses, which lead to almost 10%–15% lower yield than its potential yield (Kumar et al., 2021). Thus rice genetic improvement is a key mandate for global food security, and it involves various strategies for developing new high-yielding, nutritionally improved, and stress-tolerant rice varieties.

To date improvement of rice, especially in yield, nutritional quality, and resistance to biotic/abiotic stresses, has been achieved widely through conventional breeding strategies while utilizing natural genetic variation (Siddiq and Vemireddy, 2021). Rice yield enhancement through various breeding methods (conventional hybridization, ideotype breeding, hybrid breeding, and wide hybridization) has been achieved; however, all such methods have their own limits and are thus difficult to exploit to a major extent (Khan et al., 2015). Similarly breeding rice for tolerant to various biotic and abiotic factors is a routine practice where resistance genes from wild germplasm

DOI: 10.1201/9781003392897-10

are introgressed into elite cultivars; however, lack of genetic variation (Singh et al., 2016), repressed recombination (Fayos et al., 2019), and interspecific hybrid sterility through wide hybridization crosses (Yu et al., 2016) limit the process.

Along with the rice genome sequencing and annotation project, greater leaps in molecular biotechnology, bioinformatics, and integrated Omics platforms (genomics, transcriptomics, proteomics and metabolomics) have allowed a deep understanding of the role of genetic elements and their role in various metabolic pathways, physiological traits, and stress tolerance (Jia et al., 2021; Iqbal et al., 2021). Tailoring crop architecture through generating target-specific change directly into the genome is the key idea of the present century crop improvement. Creating the target-specific DNA mutation started with the advent of DNA binding proteins such as zinc fingers (ZF) and transcription activator-like effector (TALE) motifs fused with bacterial nuclease *FokI* (having deleted DNA recognition sequence but possess cleavage domain) enabled to produce double-strand break (DSB) in DNA at target locus followed by DNA repair mechanism (Puchta et al., 2005). There are two major DNA repair pathways, nonhomologous end joining (NHEJ) and homology repair (HR) mechanisms, that are endogenous to both prokaryotes and eukaryotes (Ran et al., 2013). Among both, NHEJ repair mechanism is common and occurs independent of homology template or sequence; rather, it is a simple mechanism which rejoins the two broken ends and often results in deletion or error that, if they occur in the coding region of a gene, frequently result in frame-shift mutations, commonly inserting premature stop codons. With such capabilities of creating desired changes in target regions a new term called genome editing (GE) came into being. Both custom-engineered ZFNs and TALENs have been used for gene editing in many crops including rice for various trait improvements (Zhang et al., 2018b).

The history of GE has totally changed with the introduction of a third tool named CRISPR/Cas9 in year 2013. CRISPR/Cas9, unlike ZFNs or TALENs, does not require complex protein array engineering specific to target DNA; rather it utilizes custom designing of 20 bp CRISPR RNA or guide RNA strand which is transcribed along with tracer fragment (tracer-RNA) as single transcript or single guide RNA (sgRNA). The sgRNA do Watson-crick pairing with 20 bp target DNA site and allow Cas9 nuclease to produce DSB 3–4 bp upstream of protospacer adjacent motif (PAM), which is specific for different Cas9 orthologs or variants (in case of *Streptococcus pyogenes* Cas9, i.e., SpCas9 it is 5′NGG3′) (Ran et al., 2013). All it requires is customizing the 20 bp CRISPR RNA (crRNA) in the sgRNA for different multiple experiments or multiple target edits (multiplexing) (Ran et al., 2013). CRISPR/Cas9 has been widely adopted in many crop species including rice for various qualitative and quantitative trait improvements (Jaganathan et al., 2018; Wada et al., 2020). In rice itself, CRISPR/Cas9 has been used to improve yield, nutritional quality, storability, eating and cooking quality, and biotic and abiotic stress tolerance (Zafar et al., 2020a; Tabassum et al., 2021; Premnath et al., 2022). After the discovery of this new genome editing tool, it was soon realized that window of SpCas9 applicability to cover maximum genomic regions of crop species is limited due to NGG PAM. Thus new Cas9 orthologs were discovered, and new Cas9 variants with diverse PAMs were studied. The application of this technology was not restricted to targeted insertion/deletions but extended up to large fragment deletions, targeted insertions, single nucleotide change (base editing), and also changing the epigenetic landscape (epigenome editing). With this brief introduction to this technology, the current chapter will discuss briefly CRISPR/Cas9 as a plant genome editing tool, workflow of genome editing in plants, variants of Cas9 and their role in rice genetic improvement, target traits improvement in rice, some technological advances, limitations, and future prospects.

10.2 CRISPR/CAS9 AS PLANT GENOME EDITING TOOL: A BRIEF ACCOUNT

CRISPR/Cas9 has an interesting journey spanning from its discovery to its use as genome editing tool. It was accidentally discovered in *Escherichia coli* as series of repeat sequences (Ishino et al., 1987), and later on, its role as bacterial adaptive immune system against infecting bacteriophages was established (Mojica et al., 2005). With the progressive gain of knowledge about underlying mechanism, CRISPR was cleverly brought into use for GE in 2012 by two individual groups: First is Feng Zhang and George Church and the second is Jennifer Doudna, Emmanuelle Charpentier for which Noble Prize for Chemistry, 2020 was awarded (https://bitesizebio.com/47927/history-crispr/). Recently, CRISPR/Cas systems have been categorized into two classes (class I and class II); class I consist of three subtypes (type 1, e.g., Cas3; type III, e.g., Cas10; type IV, e.g., Csf1), and similarly class II consist of three subtypes (type II, e.g., Cas9; type V, e.g., Cas12; type VI, e.g., Cas13) (Khan et al., 2021). Among these classes, type II (type II Cas9 and type V Cas12) system has been identified for use in genome engineering. Cas9 is a multifunctional protein that has the ability to process pre-crRNA into crRNA and induce cleavage at the target site. The Cas9 protein contains two domains, i.e., HNH domain and RucV-like domain. The HNH domain cuts the complementary strand of crRNA, while the RucV-like domain cleaves the opposite strand of the double-stranded DNA. Naturally in bacteria, CRISPR interference is acquired by the integration of short fragments of the invading DNA known as spacers between two adjacent repeats at the proximal end of a CRISPR locus. The CRISPR arrays, including the spacers, are transcribed during subsequent encounters with invasive DNA and are processed into small interfering CRISPR RNAs (crRNAs) approximately 40 nt in length, which combine with the transactivating CRISPR RNA (tracrRNA, encoded upstream of Cas operon) to activate and guide the Cas9 nuclease (Barrangou et al., 2007). This cleaves homologous double-stranded invading DNA sequences homologous to crRNA (Barrangou et al., 2007). A prerequisite for the activity of Cas9 is the presence of a conserved protospacer-adjacent motif (PAM) 3-4 nucleotide downstream of the target cutting site, which usually has the sequence 5′-NGG-3′ (N, any nucleotide; G, guanine) (Jinek et al., 2012) but less frequently NAG (Hsu et al., 2013). The Cas9 enzyme recognizes the PAM sequence and triggers the melting of dsDNA, allowing crRNA strand invasion and base pairing with target site. Specificity is provided by the so-called 'seed sequence' approximately 12 bases upstream of the PAM, which must match between the crRNA and target DNA. The PAM sequences are specific to Cas nucleases isolated from different bacterial species. For instance, the SpCas9 nuclease cuts upstream of the PAM sequence 5′-NGG-3′ while the PAM sequence 5′-NNGRR(N)-3′ is required for SaCas9 (from *Staphylococcus aureus*) to target a DNA region for editing.

Practically 3′ end of CRISPR RNA (crRNA) fused with 5′ end of *trans*-encoded CRISPR RNA (tracrRNA) to form a chimeric single guide RNA (sgRNA) (Mali et al., 2013). The sgRNA is a synthetic RNA with a length of about 100 nt, with dual-tracrRNA and crRNA secondary structure and is used favorably in GE. Its 5′-end has a 20-nt sequence that acts as a guide sequence to identify the target sequence accompanied by a PAM sequence. PAM is not contained within the sgRNA DNA template or transcribed RNA but is required to be immediately adjacent to the target sequence in the genome for the optimized Cas9 nuclease activity. The participation of tracrRNA in DNA cleavage and presence of PAMs are unique features of the type II system. Together with the Cas9 protein, sgRNA is able to form an RNA-guided endonuclease that mediates sequence-specific cleavage in the genome (Jinek et al., 2012). Right at the beginning, CRISPR system used for GE includes use of Cas9 isolated from *S. pyogenes* (spCas9). In context to the first application of CRISPR/Cas9-based GE in plants, five independent reports were published in 2013 (Feng et al., 2013; Li et al., 2013; Nekrasov et al., 2013; Shan et al., 2013; Xie and Yang, 2013). This first group of studies were demonstrated in model plants such as *Arabidopsis thaliana* and *Nicotiana benthamiana* and rice, targeting both endogenous genes and transgenes using a range of transformation platforms.

10.3 WORKFLOW OF CRISPR/CAS9-BASED GENOME EDITING IN PLANTS: GENERAL OVERVIEW

The major prerequisites for designing genome or gene editing experiment are the availability of genomic or gene sequence (could be partial or full) to be targeted, well-established tissue regeneration protocol of the crop species, and proof of concept study which includes testing the efficiency of system while targeting reporter genes like *GFP*, *YFP*, or *RFP* (Ma et al., 2016), endogenous phenotypic marker genes (phytoene desaturase, i.e., *PDS*) (Shan et al., 2013), and testing the editing efficiency while targeting endogenous genes in protoplasts (Lin et al., 2017). The other important consideration for designing a GE experiment for crop genetic improvement (nutritional, biotic and abiotic stresses and yield) usually requires a fundamental knowledge of gene sequence, gene structure and function, novel genes, and quantitative trait loci (QTL) responsible for traits of interest (Jiang et al., 2012). The knowledge of genes encoding enzymes or proteins involved in metabolic pathways where they act as negative regulator or inhibitors to desired metabolic product could help us to knock down their effect. Moreover, the knowledge of intermediary metabolic products could help us to block the metabolic cascade for desired product (metabolic engineering) and thus enable us to target endogenous pathways and produce novel phenotype that are otherwise absent in natural genetic pools.

The CRISPR-based genome editing is being used not only for genetic improvement but also for validating the functions of novel genes and regulatory elements (reverse genetics) (Malzahn et al., 2017). The basic workflow of genome editing in all monocots and dicots is almost similar, and it requires certain considerations such as selection of target site, selection of gRNA, selection of CRISPR vector, target tissue, and transformation method, and screening of edited events.

10.3.1 Selection of Target Site(s)

The target site selection is the first crucial step in GE experiment. Usually for gene knockouts, targeting a region close to the start codon or 5′ end (initial exons that code for N terminal of protein) is preferred in order to make protein nonfunctional at the very beginning (Van Campenhout et al., 2019). For instance, In initial rice knockout studies, second exon of the *CHLOROPHYLL A OXYGENASE 1* (*CAO1*) gene and the third exon of the *LAZY1* gene was targeted (Miao et al., 2013), and in case of *OsSWEET13* gene, 4 bp downstream of the translation start codon was used as target site (Zhou et al., 2015). If the target gene is well characterized for its active domains, then targeting such functional domains region is the best target site (Shi et al., 2015). Moreover, this strategy can also be applied to whole gene families using a sgRNA against exons that are conserved between all family members (Van Campenhout et al., 2019).

10.3.2 Selection of gRNA(s) and Their Validation

The second most important part of GE experiment is selection of suitable 20 bp crRNA or gRNA(s). Mohr et al. (2016) defined that the choice of gRNA for genome editing experiment needs to consider many factors which include whether target region or gene is single copy or has many isoforms, type of Cas9 and PAM required, type of promoter to be used for driving the expression of gRNA so that terminator sequence for that promoter could be excluded from the gRNA sequence, and cloning strategy which might put additional constraints on gRNA sequence. Moreover, the choice of gRNA depends upon intended application of GE; for instance, in case of gene knockout(s), the off-targets on same chromosome that of target gene will be more difficult to replace through backcrossing than those on different chromosome(s). In case of CRISPR activation (CRISPRa) designing of gRNA is generally 500–50 bp upstream of transcription start site (TSS); on the other hand, in case of CRISPR interference (CRISPRi) gRNA is designed near the TSS. The use of multiple or pair of gRNAs for single target region in case of both gene knockouts (cause large fragment deletion) and CRISPRa or CRISPRi has also been suggested (Mohr et al., 2016).

The gRNA sequence can be designed as complementary to the + or − strand of the target DNA region. Up until now, dozens of online tools have been developed to devise efficient and specific sgRNA including E-CRISPR, Cas-OFFinder, Cas-Designer, CasOT, SSFinder, CCTop, CHOPCHOP, COD, CRISPR MultiTargeter, CRISPR Primer Designer, CRISPRdirect, CRISPR-ERA, CRISRP-P, CRISPR-PLANT, E-CRISP, GT-Scan, MIT Optimized CRISPR Design, Off-Spotter, sgRNA Designer, sgRNAcas9, and ZiFiT, which make the design of sgRNA become easier (Miglani, 2017). The most important factor governing gRNA efficiency is its G/C content which in case of plants ranges from 30% to 80%; however, the guides with G/C of more than 50% have higher editing efficiency than the ones having G/C below

50% (Ma et al., 2015). The stretch of poly-T (i.e., TTTT) should be avoided as it is a signal of termination for polymerase III (Cui et al., 2018). The other most important aspect is the secondary structure formed by sgRNAs which could affect target efficiency is well discussed (Liang et al., 2016).

10.3.3 Construction of sgRNA/Cas9 Expression Cassette

There have been many special expression vectors designed for CRISPR/Cas9-based genome editing in plants, which are well described by Alok et al. (2021) and can also be checked out from www.addgene.org/crispr/plant/. The major key features of these vectors are the promoters that drive the expression of sgRNA and Cas9, gRNA(s) cloning strategies, and selectable/scorable markers. Generally gRNA expression is driven by U3 or U6 Pol III promoters; U6 has a preference for a guanine and U3 for an adenine (Shan et al., 2013). In case of monocots, rice *U6* promoters such as *OsU6a*, *OsU6b*, and *OsU6c* are most commonly used, whereas for dicots *Arabidopsis AtU6*-1 and *AtU6*-29 promoters are commonly used. However, it has been recommended to use endogenous or species-specific U6 promoters for higher gRNA expression (Montecillo et al., 2020). For multiple gRNAs expression (multiplexing) various strategies like Cys4 endoribonucleoprotein, ribozyme, and t-RNAs loops processing RNAses are used, which needs processing of spacer sequences in polycistronic gRNAs transcript (expressed by Pol III promoter) and releases individual sgRNAs (Gao et al., 2014).

For the Cas9 expression initially CaMV 35S promoter with optimized codon sequence was used, but later on, *Ubiquitin* gene promoters were found to be better than the *CaMV 35S* promoter (Feng et al., 2013). Apart from these, exceptionally high genome editing efficiencies were reported when RPS5a or DD45 promoters were used for Cas9 expression (Ordon et al., 2018). Maize dmc1 promoter has been recognized for its high activity and leads to high Cas9 expression. There has also been report of combined effect of U3 and dmc1 promoters, and it is better than U6-Cas9/sgRNA expression (Montecillo et al., 2020).

10.3.4 Delivery Methods

The success of GE experiment depends entirely on delivery of gRNA: Cas9 machinery into target tissue using suitable approach. There are a plethora of delivery methods available for genome editing in plants, including *Agrobacterium*-mediated construct delivery, particle gun–mediated construct as well as Ribonucleoprotein (RNP) delivery, PEG-mediated RNP delivery, floral dip method, delivery using nanoparticles, and pollen magnetofection (Sandhya et al., 2020). *Agrobacterium*-mediated gRNA:Cas9 construct delivery is the natural and widely used method as it is more efficient and showed high editing efficiency than particle gun. Moreover, *Agrobacterium*-mediated transformation is broadly applicable, cost-effective, and known for stable transgene integration. Another cost-effective and straightforward technique is floral dip method, in which no tissue culture facility is required and showed 0.01% to 1% transformation efficiency in different plant species (Zlobin et al., 2020). However, there are certain limitations of floral dip or in-planta methods; chimeric plants in T0, false positive and unreproducible plants, screening of large number of T1 plants, low transformation efficiency, and low applicability to broad range of plant species (Zlobin et al., 2020). The use of different nanoparticles, including mesoporous silica nanoparticles (MSNs), carbon nanotubes (CNTs), lipid-like nanoparticles (LNPs), polymer-based nanoparticles, and pollen magnetofection, has also advanced the delivery of CRISPR cargo in plant cells (Vats et al., 2022).

10.3.5 Screening and Molecular Characterization of GE Events

The screening of putative GE events is almost the last but critical and important aspect of GE experiment in plants. The batteries of methods have been developed and utilized for screening of mono- and biallelic mutants, which includes T7 endonuclease-based heteroduplex digestion, Sangar sequencing, next-generation sequencing, restriction enzyme digestion, and RNP assay (Gupta and Li, 2021). Apart from these routine methods, some alternative methods have been developed, which include high-resolution melting analysis (HRMA) (Denbow et al., 2018), single-strand conformation polymorphism (SSCP) (Zeng et al., 2016), nested PCR (Gupta and Li, 2021), Mutation Sites–Based Specific Primers Polymerase Chain Reaction (MSBSP-PCR) (Guo et al., 2018), and Bindel-PCR (Sakurai et al., 2019). Once the mutation is detected in T0 plants, the positive edited mutants are further advanced to next generation and characterized for desired traits along with segregation analysis of transgenes (generally generations are tested for desired mutation and PCR negative gene cassette).

10.4 VARIANTS OF CAS9 EXPLORED FOR GENOME EDITING

The CRISPR/Cas9 system has relentlessly evolved much for the reason that it could be utilized limitlessly to create every possible genetic variation ranging from small to large insertions/deletions, targeting multiple genetic loci (multiplexing), single nucleotide change (base editing), targeted gene replacement (prime editing), and rewriting the epigenetic landscape (epigenome editing), all with precession and minimal off-target effects. All this has become possible with continuous efforts to design or reprogram the initially discovered CRISPR system and also discover new variants and advancement of the technology.

Type II-A *Streptococcus pyogenes* Cas9 (SpCas9) was the first reported Cas9 and is responsible for the majority of the current CRISPR applications (Jinek et al., 2012). However, there are certain limitations of SpCas9 system, which include stringent NGG PAM requirement that limits the genome editing window, significant off-target cleavages in the genome, and larger size of the SpCas9 that limits its packaging into virus-based vectors and delivery into plant cells (Hassan et al., 2021). In order to combat packaging problem in viral vectors, split Cas9 approach was demonstrated where SpCas9 nuclease protein coding domains were functionally split into two inactive fragments delivered across dual vectors, and when coexpressed to dimerize in cells, its activity reconstituted (Wright et al., 2015; Zetsche et al., 2015; Yuan et al., 2022). However, split SpCas9 is less active than the intact SpCas9. In order to enhance the editing choices with flexible PAM requirement and to reduce off-target effects, variants of Cas9 with different PAMs were searched and engineered. All the variants of Cas9 which have been discovered and engineered are discussed below with ascending timeline.

A first SpCas9 ortholog isolated from *Staphylococcus aureus* (SaCas9) recognizes a 6-mer PAM sequence (5′-NNGRRT) (Kleinstiver et al., 2015). In 2015, Zhang and his group at MIT and Broad Institute (USA) discovered a new subtype of Cas9 named cpf1 (CRISPR from *Prevotella and Francisellal*) or Cas12a. Cpf1 recognizes a 5′-TTTV-3′ PAM in a DNA target (Safari et al., 2019). The other class of cpf1 is *Francisella novicida* ATCC 15482 (FnCas12a), *Acidaminococcus* sp. BV3L6 (AsCas12a), *Lachnospiraceae bacterium* ND2006 (LbCas12a), and *Moraxella bovoculi* AAX11_00205 (MbCas12a) (Meliawati et al., 2021). Apart from SpCas9, LbCas12a and AsCas12a have also been used to enhance genome editing in living cells (Jacobsen et al., 2019). The major differences between SpCas9 and Cpf1 are the following: Cpf1 is smaller in size and contains only RuvC endonuclease domain which produces staggered cut instead of blunt cut and thus causes larger deletions than SpCas9; moreover, Cpf1, unlike SpCas9, uses only mature CrRNA without tracer-RNA (Xu et al., 2017). Two different variants of Cas9s obtained from *Streptococcus thermophilus* (st1cas9 and st3cas9) (Muller, 2016), and one from *Neisseria meningitidis* (NmeCas9), recognize an 8-mer PAM sequence (5′-NNNNGATT) (Lee et al., 2016) are the orthologs of SpCas9 which showed reduced off-targets compared to SpCas9.

Kleinstiver et al. (2016) designed SpCas9-HF1 by introducing four mutations (N497A/R661A/Q695A/Q926A) in SpCas9 which showed more than 70% of wild-type SpCas9 activity with reduced or no off-target effects at genome-wide level. Slaymaker et al. (2016) rationally engineered enhanced specificity SpCas9 variant (eSpCas9) while neutralization of positively charged grove of SpCa9 with alanine residues resulting in its weak interaction with nontarget strand and enhanced interaction with target strand and thus enhanced its specificity. The combinatorial alanine substitutions were used to develop eSpCas9 (K855A), eSpCas9 1.0 (K810A/K1003A/R1060A), and eSpCas9 1.1 (K848A/K1003A/R1060A) without compromising specificity along with minimal off-targets. Hirano et al. (2016) studied the largest ortholog of SpCas9 from *Francisella novicida* called fnCas9 having 1629 amino acids that recognizes the 5′-NGG-3′ PAM; however, rationally designed RHA variant (E1369R/E1449H/R1556A) recognize 5′-YG-3′ PAM.

A smallest Cas9 ortholog from *Campylobacter jejuni* (CjCas9) having 984 amino-acid residues and 2.95 kbp was characterized for PAM recognition and genome editing (Kim et al., 2017). The study found that CjCas9 recognizes 5′-NNNNACAC-3′ or 5′-NNNNRYAC-3′ (where R and Y stand for purines and pyrimidines, respectively), and also it induces indels at high frequency. Chen et al. (2017) studied novel domains of Cas9 named REC3 and REC2 which allosterically regulate the threshold of active confirmation of HNH domain to finally dock at target site and induce cleavage, otherwise remaining trapped in inactive state upon binding with mismatch target. In order to further enhance the specificity of SpCas9 over SpCas9-HF1 and eSpCas9 1.1, they designed hyperaccurate Cas9 (HypaCas9) by inducing N692A/M694A/Q695A/H698A mutation in REC3 domain. Casini et al. (2018) further discovered SpCas9 variant named VNEQ (M495V/Y515N/K526E/R661Q) with more specificity even more than HypaCas9 by screening mutant library of REC3 domain.

Hu et al. (2018) from David Liu's lab used directed evolution to generate new SpCas9 variant with seven mutations (A262T/R324L/S409I/E480K/E543D/M694I/E1219V) called xCas9 3.7 also referred to as xCas9 that recognize more relaxed PAMs (NG, GAA and GAT). Nishimasu et al. (2018) rationally designed a SpCas9-NG variant while introducing R1335V/L1111R/D1135V/G1218R/E1219F/A1322R/T1337R mutations in SpCas9 that can also recognize relaxed NG PAMs (TGA, TGT, and TGC). A comparative cleavage kinetics study of SpCas9-NG and xCas9 variants revealed that SpCas9-NG outperforms xCas9 in recognizing NGH PAMs in vitro. Another modified version of SpCas9 (Cas9-VQR) which required 5′NGA as PAM for target gene editing has also shown better efficiency with minimal off-targets (Hu et al., 2018).

Chatterjee et al. (2018) bioinformatically searched SpCas9 ortholog called ScCas9 from *Streptococcus canis* genome. Global pairwise alignment of all curated Cas9 protein sequences revealed that ScCas9 not only showed 89.2% sequence homology to SpCas9 but also had insertion of 10 positively charged amino acids within the highly conserved REC3 domain forming a loop that makes close contact with target DNA. Apart from that another insertion of two amino acids (KQ) upstream of PAM binding site was also found. These insertions played significant role in expanding ScCas9 PAM, which is 5′-NNG-3′ with one more nucleotide choice than SpCas9 PAM, i.e., 5′-NGG-3′. Recently Chatterjee et al. (2020) used evolutionary information from orthologs and engineered three ScCas9 variants (Cas9-Sc^{+}, Cas9-Sc^{++}, and a higher-fidelity mutant Cas9Sc^{++++} or HiFi-Sc^{++}) with broad PAM range and specificity. The Thr1227Lys mutation into the ORF of ScCas9 generates Cas9-Sc^{+}, the substitution of the loop sequence (containing extra lysine residue instead of histidine and special SG motif) from *S. anginosus* into the Cas9-Sc^{+} ORF generated ScCas9^{++}, and induction of Arg701Ala mutation in ScCas9^{++} leads to generation of Cas9Sc^{++++} or HiFi-Sc^{++} variant.

Lee et al. (2018) developed a method for screening of Cas9 variant named Sniper-Cas9 with high specificity, i.e., low off-targets without compromising on-target activities. This method involves selection of *E. coli* cells in which Cas9 with distinct sequence (Cas9 mutant library) integrates into ccdB lethal gene and interrupts its function (leading to cell survival or called positive selection pressure) and also Cas9 off-target sequence (a sequence that differs from on-target sequence by few mismatches) introduced into *E. coli* genome, whose DSB leads to cell death (negative selection pressure). This combined positive and negative selection leads to selection of Cas9 variant with increased specificity without killing its on-target activity (Sniper screen).

Walton et al. (2020) developed SpCas9 variant SpG (SpCas9 carrying mutation: D1135L/S1136W/G1218K/E1219Q/R1335Q/T1337R) with expanded NGN PAMs and near-PAMless SpCas9 variant called SpRY (A61R/L1111R/N1317R/A1322R/R1333P) having NRN>NYN PAMs. Chatterjee et al. (2020) discovered ortholog of Cas9 from *Streptococcus macacae* (SmacCas9) which can target A-rich regions with 5′-NAA-3′ PAM recognition.

Recently three new SpCas9 mutations have led to engineering of SpCas9-NRRH, SpCas9-NRCH, and SpCas9-NRTH variants (Miller et al., 2020). Pausch et al. (2020) discovered a hypercompact Cas12j2 (CasΦ) from huge Biggiephage virus. It is smaller in size (70–80 Kda), almost half the size of SpCas9 and Cas12a. CasΦ prefers T-rich PAM (VTTV/VTTTV), and upon target recognition, it produces staggered 5′-overhangs of 8–12 nt, similar to Cas12a (Liu et al., 2022). Contrary to other Cas proteins, CasΦ produces mature crRNA without assistance of tracer-RNA and uses single RuvC active site for both pre-crRNA processing and DNA cleavage (Pausch et al., 2020). In addition, a growing number of CRISPR genome-editing tools have been developed and discussed in recent review (Hassan et al., 2021).

10.5 RICE GENE/GENOME EDITING THROUGH VARIANTS OF CAS9

The rice was among the few crops in which CRISPR/Cas9 tool was applied as proof of concept soon after its discovery as genome editing tool (Shan et al., 2013). The small genome size, diploid nature, annotated genome sequence, applicability of cell and genome engineering techniques, standardized tissue culture, and selection protocols make rice one of the most preferred crops for genome editing (Endo and Toki 2020; Zafar et al., 2020). Further in the year 2013 three genes, namely, *OsBADH2*, *Os02g23823,* and *OsMPK2,* were targeted in rice using codon-optimized SpCas9 (Shan et al., 2013). Similarly genes (*OsSWEET14* and *OsSWEET11*) were targeted for bacterial leaf blight (BLB) resistance (Jiang et al., 2013). In rice most widely used Cas9 is SpCas9 for target trait improvement (discussed in next section); however, the exploitation of new Cas9 orthologs and engineered variants are under testing stage and more or less restricted to random genetic loci and certain genes (discussed below).

The search for new Cas9 variants and their utility in rice have allowed more precise editing with minimal off-targets and wider window of target editing. In rice, Hu et al. (2016) codon-optimized spCas9 to design VQR (NGA PAM sequence) and VRER (NGCG PAM sequence) variants which were validated by targeting narrow leaf gene (*NAL1*) and other agronomic genes. Quin et al. (2019) reoptimized SaCas9 (named SaKKH) by introducing three mutations (E782K/N968K/R105H), which recognize NNNRRT PAM and were used to target *PDS*, *DL,* and *NAL1* genes in *japonica* rice.

The use of another variant called xCas9 in rice was first demonstrated by Wang et al. (2018) in rice calli; however, no stable edited mutant line was produced. Later on, two independent studies were conducted to use xCas9 in rice; Li et al. (2019) used xCas9 3.7 while selecting target sites proximal to NGG, NGC, NGT, NGA, GAA, GAT, CAA, and NNG PAMs within *OsSBEIIb* (encoding a starch branching enzyme IIb) and *OsPDS* (encoding a phytoene desaturase) genes. The editing was reported only with NGG, NGC, NGT, NGA, and NNG, but not with GAA, GAT and CAA. In case of *OsSBEIIb*, xCas9 showed slightly lower editing activity (70.8%) than SpCas9 (88.9%) at NGG PAM. The xCas9 3.7 enables editing of sequences with expanded PAM sites with varied editing frequency for different PAMs: NGT (38.1%), NGA (6.7%), and ACG (1.2%). However, in case of *OsPDS*, target mutations were observed with NGG and NGT PAM. Similarly, Hua et al. (2019) also demonstrated the use of xCas9 3.6 and xCas9 3.7 in rice and found that xCas9 variants generate targeted mutations at all tested PAM sites, viz. CGG, AGC, TGA, CGT, GAT except GAA PAM site. Both xCas9 3.6 and xCas9 3.7 exhibited mutation rates up to 85.3% and 89.3% at canonical CGG PAM, respectively, which were comparable with that achieved by SpCas9 (87.3%). However, at the AGC and GAT PAM sites, both xCas9 variants showed significantly lower editing activity, with mutation rates ranging from 3.2% to 23.8%.

Hua et al. (2019) also tested SpCas9-NG in rice and found that it performed less efficiently (65.7%) compared with SpCas9 (87.9%) at the CGG PAM site. Moreover, xCas9 variants performed more efficiently (1.3-fold) than SpCas9-NG at the CGG PAM site. However, SpCas9-NG (80%) showed an editing efficiency comparable with xCas9 variants at the TGA PAM site, 74.2% at CGT and 81.8% at AGC. The major advantage of both xCas9 variants and SpCas9-NG is that both showed less or no off-target effects as compared to SpCas9.

Following SpCas9, Cas12a or Cpf1 is the second most widely exploited Cas9 in rice for targeted genome editing. Endo et al. (2016) first reported the use of FnCpf1 in rice while targeting Osdrooping leaf (*OsDL*) and OsAcetolactone synthase (*OsALS*) genes with mutation frequencies of 8.3%–25% and 15%, respectively. Xu et al. (2017) used codon-optimized LbCpf1 to target *OsPDS* and *OsBEL* genes and observed mutation frequencies of 21.4% and 41.2%, respectively. Banakar et al. (2020) compared the efficiency of WT Cas9, HiFi Cas9, AsCas12a, and LbCas12a in rice calli for ribonucleoprotein complex (RNP) delivery system for targeting *PDS* gene. The study found that LbCas12a is more efficient than rest of the nucleases, and it mostly produces large deletions (7 bp to 20 bp) proximal to PAM sequence (except few deletions in which PAM is also deleted) instead of insertion deletions as produced by SpCas9.

Xu et al. (2021) used SpG (NGD PAM), SpCas9-NG (NG PAM), and SpRY (NGD and NAN PAM) for genome editing in rice. The comparative study of nuclease activity and target efficiency revealed that SpCas9-NG outperforms SpG-Cas9 when five genetic loci (*OsPAL5*, *OsGSK4*, *OsCERK1*, *OsETR2*, and *OsRLCK185*) were targeted consisting of four types of NGN PAMs. The use of SpRY over 32 genomic sites bearing 16 possible alternative PAM sequences has been demonstrated. Over NGN PAMs, SpRY showed highest editing efficiency at NGG (83%), followed by NGA (> 74%) and NGT (47%), whereas NGC (2.3%) is least recognized. Thus, SpRY nuclease favors the recognition of NGD over NGC PAM in rice. In case of NAN PAMs, SpRY showed high preference for NAT (51.2%), NAA (33%), NAC (29.79%), and NAG (20%). This suggested that SpRY moderately recognizes NAN PAMs but recognizes NGN PAMs much better.

Xu et al. (2022) also tested type II-C Cas9 orthologs, Nm1Cas9 (N_4GATT PAM) and Nm2Cas9 (N_4CC PAM) in rice. A codon-optimized Nm1Cas9 along with 24-nt gRNAs was posed to eight target sites, where it showed mutation at only two loci out of eight. Thus, to improve the efficiency, a 121-nt extended sgRNA (esgRNA) scaffold was fused with sgRNA and tested; it showed mutation frequency (60.4%) at all the selected targets without any off-targets for two target sites (*IPA1-1* and *WX-1*) predicted through Cas-OFFinder. It was also observed that most of the mutations induced by Nm1Cas9 are single base insertions (3 bp upstream of PAM) and their frequency is 4.5-fold higher than insertions induced by SpCas9. Similarly codon-optimized Nm2Cas9 along with 24-nt gRNAs fused with 121-nt esgRNA scaffold were directed to target six genomic sites. The mutation rate of variable frequency (6.25%–77.1%) was observed at four loci without any mutation at other two sites (*PDS-4* and *ALS-3* sgRNAs). It is apparent that in comparison to Nm1Cas9, Nm2Cas9 efficiency is restricted due to the length of protospacer and variation at PAM interaction domain. The use of ScCas9, ScCas9^{+}, and ScCas9^{++} has also been demonstrated in rice to test their efficiency (Liu et al., 2021a). The codon-optimized ScCas9 variants were tested against 12 target sites in the rice genome. ScCas9^{++} showed higher GE activity than ScCas9 and ScCas9^{+}, with target efficiency of 46% at NGG PAM and with low editing efficiency (5.6%) at NAG PAM.

Recently, Cas12j2 (CasΦ) which recognizes VTTV PAM has been tested in rice protoplasts for 35 target sites and found that the majority of target sites containing NTTA, NTTC, and NTTG PAMs showed detectable genome editing (up to 25%), whereas most NTTT PAM sites showed no editing (Liu et al., 2022). Apart from that, CasΦ has also been tested in rice for multiplex GE of four target sites and activation of *OsER1*, *OsNRT1.1A*, *OsCHS,* and *OsGBSS1* genes. Two CasΦ variants, nCasΦ and vCasΦ, were also tested on 15 target sites in rice protoplasts and found that nCasΦ had significant editing efficiency than vCasΦ; however, the overall editing efficiency of CasΦ has been reported to be lower than other Cas proteins.

10.6 TARGET TRAITS IMPROVEMENT IN RICE USING CRISPR/CAS9

Rice is the most widely targeted crop for traits improvement using CRISPR/Cas9 (Abdullah et al., 2021). The various traits targeted in rice for overall genetic improvement are discussed below.

10.6.1 Grain Yield and Other Related Traits

Grain yield is an important and quantitative trait governed by many genes falling under quantitative trait loci (QTLs). The rice grain yield trait has been targeted in three distinct ways: improving plant architecture, panicle architecture, and improving the Abscisic acid (ABA) signaling pathway (Tabassum et al., 2021). For improving rice grain yield through plant architecture, CRISPR/Cas9 has been used to target *semidwarf 1* (*SD1*), resulting in reduced plant height and increased grain yield (Hu et al., 2019); *STRONG CULM3/TEOSINTE BRANCH1/FINE CULM1 (SCM3/OsTB1/FC1)* showed enhanced tiller number (Cui et al., 2020); *Gibberellin-20 oxidase-2* (*OsGA20ox2*) and *SD1* lead to reduced flag leaf length, reduced gibberellins level, reduced plant height (22% reduction), and increased grain yield by 6% (Han et al., 2019); *Ideal Plant Architecture 1* (*IPA1*) QTL resulted in improved plant architecture (Li et al., 2016a) and rice *fruit weight 4*; and *OsFWL4* resulted in large flag leaf, grain length, tiller number, and grain yield (Gao et al., 2020). In case of tailoring panicle architecture in rice, panicle architecture regions *DENSE AND ERECT PANICLE (DEP1)*, *Grain Size 3* (*GS3*), *Grain number 1a* (Gn1a), and *IPA1* lead to erect panicle, improved grain size and their number, more/less number of tillers, and reduced plant height, leading to improved grain yield (Li et al., 2016a). Similarly editing of PIN *auxin efflux carrier-like* gene 5b (*OsPIN5b*) resulted in increased panicle length and thus increased yield (Zeng et al., 2020b). Additionally genes related to grain width/weight (*GW2*, *GW5*, *GW6*, and *GW8*) along with *GS3* resulted in enhanced grain traits (Xu et al., 2016, Zhou et al., 2019, Usman et al., 2021, Usman et al., 2020a), and for grain shape, *GS9* gene was targeted (Zhao et al., 2018). Recently *miRNA396*, a negative regulator of *GS2/GRF4* gene expression, was targeted using CRISPR/Cas9 and resulted in increased grain size and yield and bigger grain length/width ratio (Wang et al., 2022). Further, the exploitation of ABA pathway for enhancing grain yield while targeting ABA receptor gene *Pyrabactin Resistance 9* (*OsPYL9*) (Usman et al., 2020b) and *PYL1*, *PYL4*, and *PYL6* genes (Miao et al., 2018) is another novel strategy.

Apart from enhancing the yield, rice grain has also been targeted for milling, eating, and cooking quality traits using CRISPR/Cas9 (Zafar et al., 2020, Tabassum et al., 2021). Eating and cooking quality of rice seed depends upon the physiochemical properties of its endosperm which includes amylose content (AC), gel consistency, and gelatinization temperature (GT). The ease of cooking, firmness, and stickiness is determined by AC which itself is regulated by *Wx* genes (Wang et al., 1995), whereas *ALK/SSIIa* and *RSR1* control GT (Zhang et al., 2011). CRISPR/Cas9 was used to target *Wx* gene in rice *japonica* cultivars Xiushui134 and Wuyunjing7, which showed reduced AC and showed better cooking trait (Zhang et al., 2018a). Similarly, many novel alleles of *Wx* genes were created through promoter editing using CRISPR/Cas9 (Huang et al., 2020). The increasing the content of resistant starch has also been achieved by targeting starch branching enzymes *SBE1* and *SBEII* (Sun et al., 2017). The other interesting desired cooking trait which could catch consumers' interest is inducing aroma in nonbasmati cultivars which has been achieved by targeting *Betaine aldehyde dehydrogenase 2* (*Badh2*) gene (Ashokkumar et al., 2020). As per milling of rice grain, chalkiness is an undesired trait which negatively affects the milling process. The gene *GS9* mutant generated by CRISPR/Cas9 not only improved grain shape and size but also significantly reduced chalkiness (Zhao et al., 2018).

10.6.2 Herbicide, Biotic, and Abiotic Stress Resistance

To control the growth of a wide variety of grass and weeds in rice fields, herbicides such as bispyribac sodium and chlorsulfuron are used. However, these herbicides have a negative effect on rice growth due to their blocking activity of OsALS enzyme, resulting in blocking of the biosynthesis of branched-chain amino acids. Sun et al. (2016) induced point mutations in *OsALS* gene using CRISPR/Cas9 and phenotypically screened by spraying bispyribac sodium and found that wild-type plants died, whereas edited plants grew normally after 36 days of spraying. Herbicides such as glyphosate tolerance in rice have also been achieved by single sgRNA:Cas9 construct while replacement of second exon of 5-enolpyruvylshikimate-3-phosphate synthase (*EPSPS*) by introducing donor exon template bearing multiple substitutions (Li et al., 2016b).

As per engineering the biotic stress resistance in rice, CRISPR/Cas9 has been utilized against bacterial leaf blight (BLB), rice blast, and tungro virus to generate resistance against them. The BLB is a well-known disease of rice caused by *Xanthomonas oryzae* pv. *oryzae* (*Xoo*) which causes serious threat to global food security. Xanthomonas consist of transcription activator-like effectors (TALEs) and type III effector proteins which target *SWEET* transporter gene family, helping in transport of sugar into apoplast and promoting the growth of pathogen (Cohn et al., 2014). Many independent studies have been conducted to target different *SWEET* genes (especially *OsSWEET14*) along with some others using CRISPR/Cas9: *OsSWEET14* and *OsSWEET11* (Jiang et al., 2013), *OsSWEET11* or *Os8N3* (Kim et al., 2019), *OsXa13/SWEET11* (Li et al., 2019, 2020), *OsSWEET11,13,14* (Oliva et al., 2019), *OsSWEET14* (Zeng et al., 2020a, Zafar et al., 2020b), and *OsSWEET11,14* (Xu et al., 2019a). Similarly for resistance to bacterial leaf streak disease, *OsSULTR3; 6* genes were targeted (Xu et al., 2021). The most destructive disease of rice is rice blast caused by fungus *Magnaporthe oryzae,* which has a serious impact on rice production globally. In order to combat blast disease, genes for generating resistance have been targeted such as *OsERF922* (Wang et al., 2016), *OsALB1*, *OsRSY1* (Foster et al., 2018), *OsSEC3A* (Ma et al., 2018), *OsPi21* (Li et al., 2019), and *OsPi21* (Nawaz et al., 2020). Apart from BLB and blast, CRISPR/Cas9 has also posed to generate resistance in rice against *Rice tungro spherical virus* (RTSV) by targeting *eIF4G,* a candidate gene for RTSV (Macovei et al., 2018). Surprisingly CRISPR/Cas9 has not been much exploited for insect-pest resistance in rice; however, a single report where knockdown of *CYP71A1* gene suppressed the conversion of tryptamine to serotonin by tryptamine 5-hydroxylase resulted in resistance against brown plant hopper (BPH) (Lu et al., 2018).

The second major toll on rice production is taken by abiotic stress factors such as drought, cold, salinity, and osmotic stress. Many genes have been targeted for improving drought tolerance: *OsSAPK2* (Lou et al., 2017), *OsPYL9* (Usman et al., 2020), *OsERA1* (Ogata et al., 2020), *OsSRL1* and *OsSRL2* (Liao et al., 2019), *OsDST* (Kumar et al., 2020), and miRNA gene *OsmiR535* (Yue et al., 2020). For salinity and osmotic stress tolerance, *OsSAPK2* (Lou et al., 2017), *OsRR22* (Zhang et al., 2019), *OsDST* (Kumar et al., 2020), and *OsmiR535* (Yue et al., 2020) have been targeted. For cold tolerance, *OsAnn3* (Shen et al., 2017) and *OsMYB30* (Zeng et al., 2020b) have been targeted.

10.6.3 Nutritional Quality

The biofortification of rice for iron, zinc, vitamins, and essential amino acids is considered as most important aspect to tackle malnutrition. Conventional breeding and transgenic strategies have been extensively exploited in rice biofortification; however, former strategy is constrained due to limited genetic variation, being time and labor intensive, and regulatory issues concerned with GMO (Garg et al.,

2018). CRISPR/Cas9 has emerged as a new and straightforward tool to improve the nutritional quality of rice. In rice many nutritional traits have been targeted which include: improved anthocyanin content (red rice) through targeting *Rc* gene (Zhu et al., 2019), targeting rice orange (*OsOr*) gene for carotenoid enhancement (Endo et al., 2019), marker-free insertion of *Calreticulin 1* (*SSU-crtI*) and *Phytoene synthase* (*ZmPsy*) genes using CRISPR/Cas9-HDR (Dong et al., 2020), targeting *FAD2* genes for high oleic/low linoleic in rice bran oil (Abe et al., 2018), improved phytic acid content in grain by targeting *OsPLDα1* gene (Khan et al., 2019), increased γ-aminobutyric acid while targeting *GAD3* gene (Akama et al., 2020), reducing the accumulation of toxic metals like cesium by targeting *OsHAK1* transporter gene (Nieves-Cordones et al., 2017), and cadmium by targeting *OsNramp5* transporter gene (Tang et al., 2017).

10.6.4 Genome Editing for Breeding, Genetic, and Agronomic Traits

The GE through CRISPR/Cas9 has also been exploited for breeding traits (haploid induction, synthetic apomixes), genetic traits (photosynthesis and flowering time), and agronomic traits (seed shattering and sprouting). For haploid induction in rice, *OsMATL* gene was targeted, resulting in 2%–5% haploid induction rate (Liu et al., 2021b). Apomixis has huge scope in plant breeding, for fixing of F1 hybrid vigor by avoiding fertilization, leading to meiosis but allowing asexual production of clonal seeds (parthenogenesis), which is a holy grail in agriculture. In rice mutating *OsSPO11-1*, *OsREC8*, *OsOSD1*, and *OsMATL* through a CRISPR/Cas9 system has been reported to produce clonal seeds (Xie et al., 2019). In order to increase the crossovers (COs), genes for meiotic recombination such as *PAIR1*, *OsREC8,* and *OsOSD1* have also been targeted in rice (Khanday et al., 2019). Recently, Caddell et al. (2023) enhanced photosynthesis in rice by CRISPR/Cas9-mediated editing of two truncated light-harvesting antennae (TLA) via knockout mutations of *CpSRP43*, *CpSRP54a*, and its paralog, *CpSRP54b* genes. For delayed flowering time, *Hd2* uORFs were targeted which resulted in decreased expression of flowering-promoting genes (*Ehd1*, *Hd3a*, and *RFT1*) (Liu et al., 2021c). For agronomic traits such as seed shattering, *qSH1* gene (Sheng et al., 2020), to reduce postharvest sprouting, *MFT2* gene (Song et al., 2020), to increase postharvest sprouting, *VP1and Sdr4* genes (Chen et al., 2021), and to increase the preharvest sprouting, *PHS9* gene (Xu et al., 2019b) were targeted.

10.7 TECHNOLOGICAL ADVANCEMENTS ADOPTED IN RICE GENOME EDITING

There have been lots of technological advancements in CRISPR/Cas9 tools pertaining to gRNAs and Cas9 (epigenome editing, base editing, prime editing and CRISPR-Combo), specific promoters, delivery methods, optimal editing conditions, tissues used, easy mutant screening, and development of transgene-free methods.

10.7.1 Epigenome Editing in Rice

Owing to adverse global climate challenges, epigenetics plays an important role in crop development and tolerance. Heritable epigenetic changes refer to changes independent of DNA sequence which is associated with altered gene expressions and transmitted phenotypic changes (Kakoulidou et al., 2021). Harnessing epigenetic diversity has now become an important aspect for breeders, as most important development traits like dormancy, flowering time, fruit ripening, disease resistance, and tolerance to various abiotic stresses are controlled epigenetically (www.mdpi.com/journal/agronomy/special_issues/epigenetic_mechanisms_in_crop_plants).

In the past few years a global perspective has emerged to target modulating gene expression in plants through epigenome editing (Miglani and Singh, 2021). Epigenome editing tool consists of DNA binding domain, i.e., sgRNA:dCas9 (where Cas9s, both HNH H840A and RuvC D10A nuclease domains, are inactive and are thus abbreviated as dead-Cas9 or dCas9) linked with effector (could be activator or repressor) which could recruit the epigenetic changes at target site. In rice, dCas9-VP64, as well as more effective CRISPR-Act2.0 system, was used to transcriptionally activate three genes (*Os03g01240*, *Os04g39780*, and *Os11g35410*) (Lowder et al., 2018).

10.7.2 Base Editing in Rice

The other CRISPR/Cas9-based editing system that allows single point mutation or single nucleotide substitution in the target DNA site without the need for double-strand DNA break, donor DNA template, and cellular HDR is called base editing (Rees and Liu, 2018). CRISPR as base editing tool includes Cas9 nickase (Cas9 with one of its two nuclease domain inactive, i.e., RuvC D10A or HNH H840A) to produce single-strand nick linked with a base-modifying enzyme, i.e., deaminase that operates on single-strand DNA. There are two classes of base editors (BEs): cytosine base editors (CBEs) convert a C•G base pair into a T•A base pair, and adenine base editors (ABEs) convert an A•T base pair to a G•C base pair. There are many excellent articles and reviews covering the exploitation of base editing in rice for enhancing yield, nitrogen use efficiency, high amylose, grain traits, herbicide resistance, and disease resistance (Quin et al., 2019; Mishra et al., 2018; Yarra and Sahoo 2021; Tabassum et al., 2021).

10.7.3 Prime Editing in Rice

The so-called base editing can only help to induce four transition mutations (C→T, G→A, A→G, and T→C) but cannot perform eight transitions (C→A, C→G, G→C, G→T, A→C, A→T, T→A, and T→G). Thus a new technological intervention called prime editing was reprogrammed from the basic CRISPR/Cas9 tool to allow the targeted DNA replacement with desired edits without double-strand DNA break or donor DNA (Anzalone et al., 2019). The prime editing system consists of Cas9 nickase (H840A) coupled with reverse transcriptase (RT) and prime editing guide RNA (pegRNA). Prime editing has been successfully adopted in plants such as rice and wheat by Lin and coworkers (Lin et al., 2020). In rice protoplasts, reporter gene was targeted to convert blue fluorescent protein (BFP) to green fluorescent protein (GFP) by changing codon 66 from CAC (histidine) to TAC (tyrosine). Moreover, six endogenous genes of rice were targeted to test the efficiency. The potential applications of prime editing in rice include engineering herbicide tolerance by targeting *ACETOLACTATE SYNTHASE* (*OsALS*), targeting *IDEAL PLANT ARCHITECTURE 1* (*OsIPA*) and *TEOSINTE BRANCHED 1* (*OsTB1*) (Butt et al., 2018).

10.7.4 CRISPR-Combo System in Rice

Another important repurposing of CRISPR/Cas9 system which enables simultaneous orthogonal activation and knockout of genes and boosts efficiency was first developed by Dahlman and coworkers in mammalian cells (Dahlman et al., 2015). However, Pan and coworkers from University of Maryland, USA, first practically used in plant genome editing and was named the CRISPR-Combo system (Pan et al., 2022). In this system, active Cas9 is used along with two different gRNAs: gRNA 1.0 (active) and gRNA 2.0 (dead). The gRNA 1.0 is of 20 nt, which along with active Cas9 is used to create target indels, whereas gRNA 2.0 is of 14-15 nt (truncated), linked with transcriptional activators, and allows gene activation without DSB. In rice this system has been applied to edit *OsGW2* and *OsGN1a* genes while activating *OsBBM1* (morphogenic gene which enhances regeneration).

10.7.5 Special Delivery Methods of CRISPR Reagent

One of the major constraints of Cas9:gRNA vector construct accompanied by selectable marker gene (SMG)–based GE is its integration and persistence in the genome just like transgenes in transgenic crops and hence has become a subject of GMO regulation. Still, the major advantage of GE over transgenic crops, especially sexually reproduced crops, is that once the desired target edit is attained, the vector construct along with SMG can be segregated out through backcrossing (Hare et al., 2002). However, segregating process is time as well as labor intensive and also not a feasible option where species have lengthy juvenile growth period (e.g., trees). In order to combat this issue, delivery of preassembled gRNA:Cas9 ribonucleoprotein (RNP) complex has been adopted in different crops including rice (Zhang et al., 2021).

The delivery of RNP was first adopted in *Arabidopsis* and lettuce and rice protoplasts using polyethylene glycol (PEG) by Woo et al. (2015) and has been mostly transfected using PEG in plant protoplasts because of hindrance posed by rigid plant cell wall (Zhang et al., 2021). Since the regeneration from protoplast is difficult, PEG-induced transfection of rice zygote with RNP was established (Toda et al., 2019). Apart from protoplasts, there have also been reports of cationic lipid-assisted codelivery of gold nanoparticles–coated RNP and plasmid with SMG in immature embryos of maize and wheat (Savitashev et al., 2016; Liang et al., 2017) and rice calli (Banakar et al., 2020) using particle gun bombardment.

10.7.6 Inducible and Tissue-specific Genome Editing in Rice

The constitutive sustained expression of Cas9 generally provides a wider time window for off-targets to happen in the genome. Thus inducible and tissue-specific or spatiotemporal genome editing has now been adopted in crop plants using tissue-specific and inducible promoters for driving Cas9 expression (Rahman et al., 2022). Various tissue specific promoters used in genome editing of crop plants include: dividing cell-active (*YAO, ICU, EFα1, H4, RPS5A, CDC45, Pcubi4, Slp16, SlEF1α*), meiosis-specific (*MGE1, 2, 3* and *DMC1*), egg cell-specific (*EC1.1, EC1.2, AtP5, AtEC1.2e1.1, GmEC1.1, GmEC1.2, SPL, DD45*, and *LAT52*), fiber-specific (*NSD3/SND1*), fruit-specific (*PPC2*), and root cap-specific (*SMB*). Similarly various inducible promoters have also been used in genome editing which include chemically induced promoter such as estrogen-inducible (*G10-90, WOL, WOX5, SCR* and *WER*), heat-inducible (Soybean *HSP17.5E*), and pathogen-inducible (*V86* and *C86*) promoters (Rahman et al., 2022). In rice, Nandy et al. (2019) described an inducible expression system for controlling CRISPR-Cas9 mutagenesis in rice. The system used a heat-shock-inducible promoter to express *cas9* in tandem with the rice *U3* promoter for sgRNA.

10.7.7 Selectable Marker–free Engineered Rice

The major hurdle block in the commercialization of biotech or transgenic rice is the regulatory concerns related to selectable marker genes (SMGs) present in vector construct apart from transgene (Fraiture et al., 2016). Contrary to transgenics, genome editing does not use overexpression product of transgene for trait improvement; rather it relies on creating the permanent site-specific DNA change (indel). Once the site-specific DNA change is created and inherited in further generations, the vector construct and off-target mutations are segregated out through backcross breeding (Hare et al., 2002). The use of SMGs is common in both transgenic and GE constructs; however, removal of SMGs and/or transformation without SMGs has also been reported (Singh et al., 2022). Recently many traditional and advanced methods for developing SMGs-free engineered rice have been reviewed: CRISPR/Cas9-based marker excision, marker-free gene insertion, programmed self-elimination of gene cassette by lethal barnase gene, and delivery of DNA-free RNP complex into rice protoplasts or embryos are the recent methodologies for making transgene and SMGs-free rice (Singh et al., 2022). Recently Endo et al. (2020) used a novel marker elimination system using an I-*Sce*I break and subsequent single-strand annealing (SSA)–mediated DNA repair system.

REFERENCES

Abe, K., Araki E., Suzuki, Y., Toki, S., Saika, H., 2018. Production of high oleic/low linoleic rice by genome editing. *Plant Physiol Biochem.* 131:58–62.

Akama, K., Akter, N., Endo, H., Kanesaki, M., Endo, M., Toki, S., 2020. An In Vivo Targeted Deletion of the Calmodulin-Binding Domain from Rice Glutamate Decarboxylase 3 (OsGAD3) Increases γ-Aminobutyric Acid Content in Grains. *Rice.* 13(1):20.

Alok, A., Chauhan, H., Upadhyay, S.K., Pandey, A., Kumar, J. and Singh, K., 2021. Compendium of plant-specific CRISPR vectors and their technical advantages. *Life*, 11(10), 1021.

Anzalone, A.V., Randolph, P.B., Davis, J.R., Sousa, A.A., Koblan, L.W., Levy, J.M., Chen, P.J., Wilson, C., Newby, G.A., Raguram, A., Liu, D.R., 2019. Search-and-replace genome editing without double-strand breaks or donor DNA. *Nature*. 576(7785):149–157.

Ashokkumar, S., Jaganathan, D., Ramanathan, V., Rahman, H., Palaniswamy, R., Kambale, R., Muthurajan, R., 2020. Creation of novel alleles of fragrance gene OsBADH2 in rice through CRISPR/Cas9 mediated gene editing. *PLoS One*. 15, e0237018.

Banakar, R., Schubert, M., Collingwood, M., Vakulskas, C., Eggenberger, A. L., Wang, K., 2020. Comparison of CRISPR-Cas9/Cas12a Ribonucleoprotein Complexes for Genome Editing Efficiency in the Rice Phytoene Desaturase (OsPDS) Gene. *Rice*. 13(4), 1–7.

Barrangou, R., Fremaux, C., Deveau, H., Richards, M., Boyaval, P., Moineau, S., et al., 2007. *CRISPR provides acquired resistance against viruses in prokaryotes Science*. 315, 1709–1712.

Bin Rahman, A. N. M., & Zhang, J., 2022. Trends in rice research: 2030 and beyond. *Food Energy Secur.* 12, e390. doi:10.1002/fes3.390.

Butt, H., Rao, G.S., Sedeek, K., Aman, R., Kamel, R., Mahfouz, M., 2018. Engineering herbicide resistance via prime editing in rice. *Plant Biotechnol J*. 18(12):2370–2372.

Caddell, D., Langenfeld, N.J., Eckels, M.J., et al., 2023. Photosynthesis in rice is increased by CRISPR/Cas9-mediated transformation of two truncated light-harvesting antenna. *Frontiers in Plant Sci.* 14, 1050483.

Casini, A., Olivieri, M., Petris, G., Montagna, C., Reginato, G., Maule, G., Lorenzin, F., Prandi, D., Romanel, A., Demichelis, F., Inga, A., Cereseto, A.A., 2018. highly specific SpCas9 variant is identified by in vivo screening in yeast. *Nat Biotechnol*. 36(3):265–271.

Chatterjee, P., Jakimo, N., Jacobson, J.M., 2018. Minimal PAM specificity of a highly similar SpCas9 ortholog. *Sci Adv*. 4(10):eaau0766.

Chatterjee, P., Jakimo, N., Lee, J., Amrani, N., Rodríguez, T., Koseki, S.R.T., Tysinger, E., Qing, R., Hao, S., Sontheimer, E.J., Jacobson, J., 2020. An engineered ScCas9 with broad PAM range and high specificity and activity. *Nat Biotechnol*. 38(10):1154–1158.

Chen, J. S., Dagdas, Y.S., Kleinstiver, B.P., Welch, M.M., Sousa, A.A., Harrington, L. B., … Doudna, J.A., 2017. Enhanced proofreading governs CRISPR-Cas9 targeting accuracy. *Nature,* 550(7676), 407–410. doi:10.1038/nature24268

Chen, W., Wang, W., Lyu, Y., 2021. OsVP1 activates Sdr4 expression to control rice seed dormancy via the ABA signaling pathway. *Crop J*. 9(1): 68–78.

Cui, Y., Xu, J., Cheng, M., Liao, X., Peng, S., 2018. Review of CRISPR/Cas9 sgRNA Design Tools. *Interdiscip Sci*. 10(2):455–465.

Cui, Y., Hu, X., Liang, G., Feng, A., Wang, F., Ruan, S., Dong, G., Shen, L., Zhang, B., Chen, D., et al., 2020. Production of novel beneficial alleles of a rice yield-related QTL by CRISPR/Cas9. *Plant Biotechnol. J.* 18, 1987–1989.

Dahlman, J.E., Abudayyeh, O.O., Joung, J., Gootenberg, J.S., Zhang, F., Konermann, S., 2015 Nov. Orthogonal gene knockout and activation with a catalytically active Cas9 nuclease. *Nat Biotechnol*. 33(11):1159–1161.

Denbow C., Ehivet, S.C., Okumoto, S., 2018. High Resolution Melting Temperature Analysis to Identify CRISPR/Cas9 Mutants from *Arabidopsis*. *Bio Protoc*. 8(14):e2944.

Dong, O.X., Yu, S., Jain, R., Zhang, N., Duong, P.Q., Butler, C., Li, Y., Lipzen, A., Martin, J.A., Barry, K.W., et al., 2020. Marker-free carotenoid-enriched rice generated through targeted gene insertion using CRISPR-Cas9. *Nat. Commun.* 11:1178.

Endo, A., Saika, H., Takemura, M., Misawa, N., Toki, S., 2019. A novel approach to carotenoid accumulation in rice callus by mimicking the cauliflower Orange mutation via genome editing. *Rice*. 12(1):81.

Endo, M., Toki, S., 2020. Genome editing in rice. *Rice*. 13:27.

Fayos, I., Mieulet, D., Petit, J., Meunier, A. C., Périn, C., Nicolas, A., Guiderdoni, E., 2019. Engineering meiotic recombination pathways in rice. *Pl. biotechnol. J.*, 17(11):2062–2077.

Feng, Z., Zhang, B., Ding, W., Liu, X., Yang, D.L., Wei, P., Cao, F., Zhu, S., Zhang, F., Mao, Y. and Zhu, J.K., 2013. Efficient genome editing in plants using a CRISPR/Cas system. *Cell Res*. 23(10):1229–1232.

Foster, A.J., Martin-Urdiroz, M., Yan, X., Wright, H.S., Soanes, D.M., Talbot, N.J., 2018. CRISPR-Cas9 ribonucleoprotein-mediated co-editing and counterselection in the rice blast fungus. *Sci. Rep.* 8, 14355.

Fraiture, M.A., Roosens, N.H.C., Taverniers, I., De Loose, M., Deforce, D., and Herman, P., 2016. Biotech Rice: Current Developments and Future Detection Challenges in Food and Feed Chain. *Trends Food Sci. Technol.* 52, 66–79.

Gao, J., Wang, G., Ma, S., Xie, X., Wu, X., Zhang, X., et al., 2014. CRISPR/Cas9-mediated targeted mutagenesis in *Nicotiana tabacum. Plant Mol. Biol.* 10 (2014), 1007.

Gao, Q., Li, G., Sun, H., Xu, M., Wang, H., Ji, J., Wang, D., Yuan, C., Zhao, X., 2020. Targeted Mutagenesis of the Rice FW 2.2-Like Gene Family Using the CRISPR/Cas9 System Reveals OsFWL4 as a Regulator of Tiller Number and Plant Yield in Rice. *Int. J. Mol. Sci.* 21, 809.

Gao, Y., and Zhao, Y., 2014. Self-processing of ribozyme-flanked RNAs into guide RNAs in vitro and in vivo 540 for CRISPR-mediated genome editing. *J. Integr Plant Biol.* 56, 343–349.

Garg, M., Sharma, N., Sharma, S., Kapoor, P., Kumar, A., and Ch, V., 2018. Biofortified Crops Generated by Breeding, Agronomy, and Transgenic Approaches Are Improving Lives of Millions of People Around the World. *Front. Nutr.* 5, 12.

Guo, J., Li, K., Jin, L., Xu, R., Miao, K., Yang, F., Qi, C., Zhang, L., Botella, J.R., Wang, R., Miao, Y.A., 2018. Simple and cost-effective method for screening of CRISPR/Cas9-induced homozygous/biallelic mutants. *Pl. Methods*. 29(14), 40.

Gupta, A., Li, W., 2021. Genotyping strategies for detecting CRISPR mutations in polyploid species: a case study-based approach in hexaploid wheat. *bioRxiv,* 2021–11. doi:10.1101/2021.11.18.469120

Han, Y., Teng, K., Nawaz, G., Feng, X., Usman, B., Wang, X., Luo, L., Zhao, N., Liu, Y., Li, R., 2019. Generation of semi-dwarf rice (*Oryza sativa* L.) lines by CRISPR/Cas9-directed mutagenesis of OsGA20ox2 and proteomic analysis of unveiled changes caused by mutations. *3 Biotech.* 9, 387.

Hare, P.D., Chua, N.H., 2002. Excision of selectable marker genes from transgenic plants. *Nat. Biotechnol.* 20, 575–580.

Hassan, M.M., Zhang, Y., Yuan, G., De, K., Chen, J.G., Muchero, W., Tuskan, G.A., Qi Y., Yang, X., 2021. Construct design for CRISPR/Cas-based genome editing in plants. *Trends Plant Sci.* 26, 1133–1152.

Hirano, H., Gootenberg, J.S., Horii, T., Abudayyeh, O.O., Kimura, M., Hsu, P.D., Nakane, T., Ishitani, R., Hatada, I., Zhang, F., Nishimasu, H., Nureki, O., 2016. Structure and Engineering of *Francisella novicida* Cas9. *Cell*. 164(5):950–61.

Hsu, P.D., Scott, D.A., Weinstein, J.A., Ran, F.A., Konermann, S., Agarwala, V., Li, Y., Fine, E.J., Wu, X., Shalem, O., et al., 2013. DNA targeting specificity of RNA-guided Cas9 nucleases. *Nat. Biotechnol.* 31, 827–832.

Hu, J.H., Miller, S.M., Geurts, M.H., Tang, W., Chen, L., Sun, N., Zeina, C.M., Gao, X., Rees, H.A., Lin, Z., Liu, D.R., 2018. Evolved Cas9 variants with broad PAM compatibility and high DNA specificity. *Nature*. 556(7699):57–63.

Hu, X., Cui, Y., Dong, G., Feng, A., Wang, D., Zhao, C., Zhang, Y., Hu, J., Zeng, D., Guo, L., et al., 2019. Using CRISPR-Cas9 to generate semi-dwarf rice lines in elite landraces. *Sci. Rep.* 9, 19096.

Hu, X., Wang, C., Fu, Y., Liu, Q., Jiao, X., Wang, K., 2016. Expanding the range of CRISPR/Cas9 genome editing in rice. *Mol Plant.* 9, 943–945.

Hua, K., Tao, X., Han, P., Wang, R., Zhu, J.K., 2019. Genome Engineering in Rice Using Cas9 Variants that Recognize NG PAM Sequences. *Mol. Plant.* 12, 1003–1014.

Huang, L., Li, Q., Zhang, C., Chu, R., Gu, Z., Tan, H., Zhao, D., Fan, X., Liu, Q., 2020. Creating novel Wx alleles with fine-tuned amylose levels and improved grain quality in rice by promoter editing using CRISPR/Cas9 system. *Plant Biotechnol. J.* 18, 2164–2166.

Iqbal, Z., Iqbal, M.S., Khan, M.I.R., Ansari, M.I., 2021. Toward Integrated Multi-Omics Intervention: Rice Trait Improvement and Stress Management. *Front. Plant Sci.* 12, 741419.

Ishino, Y., Shinagawa, H., Makino, K., Amemura, M., Nakata, A., 1987. Nucleotide sequence of the *iap* gene, responsible for alkaline phosphatase isozyme conversion in *Escherichia coli*, and identification of the gene product. *J Bacteriol.* 169, 5429–5433.

J.F., Li, J.E., Norville, J., Aach, M., McCormack, D., Zhang, J., Bush, et al., 2013. Multiplex and homologous recombination-mediated genome editing in Arabidopsis and *Nicotiana benthamiana* using guide RNA and Cas9. *Nat. Biotechnol.* 31 (2013), 688–691.

Jacobsen, T., Liao, C., Beisel, C.L., 2019. The Acidamino coccus sp. Cas12a nuclease recognizes GTTV and GCTV as non-canonical PAMs. *FEMS Microbiol Lett.* 366:fnz085.

Jaganathan, D., Ramasamy, K., Sellamuthu, G., Jayabalan, S., Venkataraman, G., 2018. CRISPR for Crop Improvement: *An Update Review. Front. Plant Sci.* 9:985.

Jia, L., Xie, L., Lao, S., Zhu, Q.H., Fan, L., 2021. Rice bioinformatics in the genomic era: Status and perspectives. *The Crop J.* 9(3): 609–621.

Jiang, W., Zhou, H., Bi, H., Fromm, M., Yang, B., Weeks, D.P., 2013. Demonstration of CRISPR/Cas9/sgRNA-mediated targeted gene modification in Arabidopsis, tobacco, sorghum and rice. *Nucleic Acids Res.* 41, e188.

Jinek, M., Chylinski, k., Fonfara, I., Hauer, M., Doudna, J.A., Charpentier, E., 2012. **A** Programmable Dual-RNA–Guided DNA Endonuclease in Adaptive Bacterial Immunity. *Science*. 337(6096), 816–821.

Kakoulidou, I., Avramidou, E.V., Baránek, M., Brunel-Muguet, S., Farrona, S., Johannes, F., Kaiserli, E., Lieberman-Lazarovich, M., Martinelli, F., Mladenov, V., Testillano, P.S., Vassileva, V., Maury, S., 2021. Epigenetics for Crop Improvement in Times of Global Change. *Biology.* 10, 766.

Khan, M. H., Zahoor, A. D., Sher, A. D., 2015. *Breeding Strategies for Improving Rice Yield—A Review. Agricultural Sciences.* doi:10.4236/as.2015.65046.

Khan, M.S.S., Basnet, R., Ahmed, S., Bao, J., Shu, Q., 2020. Mutations of OsPLDa1 Increase Lysophospholipid Content and Enhance Cooking and Eating Quality in Rice. *Plants.* 9(3):390.

Khan, Z. et al., 2022. History and Classification of CRISPR/Cas System. In: Ahmad, A., Khan, S.H., Khan, Z. (eds) *The CRISPR/Cas Tool Kit for Genome Editing*. Springer, Singapore. https://doi.org/10.1007/978-981-16-6305-5_2

Khanday, I., Skinner, D., Yang, B., Mercier, R. and Sundaresan, V. (2019) A male-expressed rice embryogenic trigger redirected for asexual propagation through seeds. *Nature.* 565, 91–95.

Kim, E., Koo, T., Park, S.W., Kim, D., Kim, K., Cho, H.Y., Song, D.W., Lee, K.J., Jung, M.H., Kim, S., Kim, J.H., Kim, J.H., Kim, J.S., 2017. In vivo genome editing with a small Cas9 orthologue derived from Campylobacter jejuni. *Nat Commun.* 21(8), 14500.

Kim, Y.A., Moon, H., Park, C.J., 2019. CRISPR/Cas9-targeted mutagenesis of Os8N3 in rice to confer resistance to *Xanthomonas oryzae pv. oryzae. Rice.* 12, 67.

Kleinstiver, B.P., Prew, M.S., Tsai, S.Q., Nguyen, N.T., Topkar, V.V., Zheng, Z., Joung, J.K., 2015. Broadening the targeting range of Staphylococcus aureus CRISPR-Cas9 by modifying PAM recognition. *Nat Biotechnol.* 33: 1293–1298.

Kleinstiver, B.P., Pattanayak, V., Prew, M.S., Tsai, S.Q., Nguyen, N.T., Zheng, Z., Joung, J.K., 2016. High-fidelity CRISPR-Cas9 nucleases with no detectable genome-wide off-target effects. *Nature.* 529:490–495.

Kumar, M., Sandhya, S., Kumar, P., Singh, A. G., Jukanti, A. K., 2021. Abiotic Stress Tolerance in Rice: Insight in Climate Change Scenario. In (Ed. Min Huang), *Integrative Advances in Rice Research. IntechOpen.* https://doi.org/10.5772/intechopen.98909

Kumar, S., Verma, V.V., Yadav, R.K., Yadav, S.K., Watts, P., Rao, A., Chinnusamy, M.V., 2020. CRISPR-Cas9 mediated genome editing of drought and salt tolerance (OsDST) gene in indica mega rice cultivar MTU1010. *Physiol. Mol. Biol. Plants.* 26, 1099–1110.

Lee, C.M., Cradick, T.J., Bao, G., 2016. The Neisseria meningitidis CRISPR-Cas9 System Enables Specific Genome Editing in Mammalian Cells. *Mol Ther.* 24(3):645–54.

Lee, J.K., et al., 2018. Directed evolution of CRISPR-Cas9 to increase its specificity. *Nat Commun.* 9, 3048.

Li, M., Li, X., Zhou, Z., Wu, P., Fang, M., Pan, X., Lin, Q., Luo, W., Wu, G., Li, H., 2016a. Reassessment of the Four Yield-related Genes Gn1a, DEP1, GS3, and IPA1 in Rice Using a CRISPR/Cas9 System. *Front. Plant Sci.* 7, 377.

Li, J., Meng, X.B., Zong, Y., Chen, K.L., Zhang, H.W., Liu, J.X., Li, J.Y., Gao, C.X., 2016b. Gene replacements and insertions in rice by intron targeting using CRISPR-Cas9. *Nat Plants.* 2, 16139.

Li, J., Luo, J., Xu, M., Li, S., Zhang, J., Li, H., et al., 2019a. Plant genome editing using xCas9 with expanded PAM compatibility. *J. Genet. Genomics.* 46, 277–280.

Li, S., Shen, L., Hu, P., Liu, Q., Zhu, X., Qian, Q., Wang, K., Wang, Y., 2019b. Developing disease-resistant thermosensitive male sterile rice by multiplex gene editing. *J. Integr. Plant Biol.* 61, 1201–1205.

Li, J., Xu, R., Qin, R., Liu, X., Kong, F., Wei, P., 2020. Genome editing mediated by SpCas9 variants with broad non-canonical PAM compatibility in plants. *Mol Plant.* 14(2):352–360.

Liang, G., Zhang, H., Lou, D., Yu, D., 2016. Selection of highly efficient sgRNAs for CRISPR/Cas9-based plant genome editing. *Sci Rep.* 19(6), 21451.

Liang, Z., Chen, K., Li, T., 2017. Efficient DNA-free genome editing of bread wheat using CRISPR/Cas9 ribonucleoprotein complexes. *Nat Commun* 8, 14261.

Liao, S., Qin, X., Luo, L., Han, Y., Wang, X., Usman, B., Nawaz, G., Zhao, N., Liu, Y., Li, R., 2019. CRISPR/Cas9-Induced Mutagenesis of Semi-Rolled Leaf1,2 Confers Curled Leaf Phenotype and Drought Tolerance by Influencing Protein Expression Patterns and ROS Scavenging in Rice (*Oryza sativa* L.). *Agronomy.* 9, 728.

Lin, Q., Zong, Y., Xue, C., Wang, S., Jin, S., Zhu, Z., Wang, Y., Anzalone, A.V., Raguram, A., Doman, J.L., Liu, D.R., Gao, C., 2020. Prime genome editing in rice and wheat. *Nat Biotechnol.* 38(5):582–585.

Liu, J., Liang, D., Yao, L., Zhang, Y., Liu, C., Liu, Y., Wang, Y., Zhou, H., Kelliher, T., Zhang X, Bandyopadhyay, A., 2021a. Rice Haploid Inducer Development by Genome Editing. *Methods Mol Biol.* 2238:221–230.

Liu, X., Liu, H., Zhang, Y., et al., 2021b. Fine-tuning Flowering Time via Genome Editing of Upstream Open Reading Frames of Heading Date 2 in Rice. *Rice* 14, 59.

Liu, T., Zeng, D., Zheng, Z., Lin, Z., Xue, Y., Li, T., Xie, X., Ma, G., Liu, Y.G., Zhu Q., 2021c. The ScCas9^{++} variant expands the CRISPR toolbox for genome editing in plants. *Journal of integrative plant biology*. doi:10.1111/jipb.13164

Liu, S., Sretenovic, S., Fan, T., Cheng, Y., Li, G., Qi, A., Tang, X., Xu, Y., Guo, W., Zhong, Z., He, Y., Liang, Y., Han, Q., Zheng, X., Gu, X., Qi, Y., Zhang, Y., 2022. Hypercompact CRISPR-Cas12j2 (CasΦ) enables genome editing, gene activation, and epigenome editing in plants. *Plant Commun.* 3(6), 100453.

Lou, D., Wang, H., Liang, G., Yu, D., 2017. OsSAPK2 Confers Abscisic Acid Sensitivity and Tolerance to Drought Stress in Rice. *Front. Plant Sci.* 8, 1–15.

Lowder, L.G., Zhou, J., Zhang, Y., Malzahn, A., Zhong, Z., Hsieh, T.F., Voytas, D. F., Zhang, Y., Qi, Y., 2018. Robust Transcriptional Activation in Plants Using Multiplexed CRISPR-Act2.0 and mTALE-Act Systems. Mol. *Pl.* 11(2), 245–256.

Lu, H.P., Luo, T., Fu, H.W., et al., 2018. Resistance of rice to insect pests mediated by suppression of serotonin biosynthesis. *Nat. Plants.* 4, 338–344.

Ma, J., Chen, J., Wang, M., Ren, Y. L., Wang, S., Lei, C. L., Cheng, Z. J., Sodmergen, 2018. Disruption of *OsSEC3A* increases the content of salicylic acid and induces plant defense responses in rice. *J Exp Bot.* 69(5), 1051–1064.

Ma, X., Zhu, Q., Chen, Y., Liu, Y.G., 2016. CRISPR/Cas9 Platforms for Genome Editing in Plants: Developments and Applications. *Mol. Plant.* 9, 961–974.

Ma., X., Zhang, Q., Zhu, Q., Liu, W., Chen, Y., Qiu, R., Wang, B., Yang, Z., Li, H., Lin, Y., Xie, Y., Shen, R., Chen, S., Wang, Z., Chen, Y., Guo, J., Chen, L., Zhao, X., Dong, Z., Liu, Y.G., 2015. A Robust CRISPR/Cas9 System for Convenient, High-Efficiency Multiplex Genome Editing in Monocot and Dicot Plants. *Mol. Plant.* 8(8), 1274–1284.

Macovei, A., Sevilla, N.R., Cantos, C., Jonson, G.B., Slamet-Loedin, I., Čermák, T., Voytas, D.F., Choi, I.-R., Chadha-Mohanty, P., 2018. Novel alleles of rice eIF4G generated by CRISPR/Cas9-targeted mutagenesis confer resistance to Rice tungro spherical virus. *Plant Biotechnol. J.* 16, 1918–1927.

Mali, P., Yang, L., Esvelt, K.M., Aach, J., Guell, M., DiCarlo, J.E., et al., 2013. RNA-guided human genome engineering via Cas9. *Science*. 339, 823–826.

Malzahn, A., Lowder, L. & Qi, Y. Plant genome editing with TALEN and CRISPR. *Cell Biosci.* 7, 21.

Martin Jinek, Krzysztof Chylinski, Ines Fonfara, Michael Hauer, Jennifer A. Doudna, and Emmanuelle Charpentier. 2012. A Programmable Dual-RNA–Guided DNA Endonuclease in Adaptive Bacterial Immunity. *Science.* 337(6096), 816–821.

Meliawati, M., Schilling, C., Schmid, J., 2021. Recent advances of Cas12a applications in bacteria. *Appl Microbiol Biotechnol.* 105(8), 2981–2990.

Miao, C., Xiao, L., Hua, K., Zou, C., Zhao, Y., Bressan, R.A., Zhu, J.K., 2018. Mutations in a subfamily of abscisic acid receptor genes promote rice growth and productivity. *Proc. Natl. Acad. Sci. USA.* 115, 6058–6063.

Miao, J., Guo, D., Zhang, J., et al., 2013. Targeted mutagenesis in rice using CRISPR-Cas system. *Cell Res.* 23, 1233–1236.

Miglani, G., & Singh, R., 2020. *Epigenome editing in crop improvement.* 10.1079/9781789240214.0044

Miglani, G.S., 2017. Genome editing in crop improvement: Present scenario and future prospects, *J Crop Improv*. 31(4), 453–559.

Miller, S.M., Wang, T., Randolph, P.B., Arbab, M., Shen, M.W., Huang, T.P., Matuszek, Z., Newby, G.A., Rees, H.A., Liu, D.R., Continuous evolution of SpCas9 variants compatible with non-G PAMs. *Nat Biotechnol.* 38(4), 471–481.

Mishra, R., Joshi, R.K., Zhao, K., 2018. Genome Editing in Rice: Recent Advances, Challenges, and Future Implications. *Front. Plant Sci.* 9:1361. doi: 10.3389/fpls.2018.01361

Mohr, S.E., Hu, Y., Ewen-Campen, B., Housden, B.E., Viswanatha, R., Perrimon, N., 2016. CRISPR guide RNA design for research applications. *The FEBS journal.* 283(17), 3232–3238.

Mojica, F.J.M., Díez-Villaseñor, C., García-Martínez, J., Soria, E., 2005. Intervening sequences of regularly spaced prokaryotic repeats derive from foreign genetic elements. *J Mol Evol.* 60:174–182.

Montecillo, J.A.V., Chu, L.L., Bae, H., 2020. CRISPR-Cas9 System for Plant Genome Editing: Current Approaches and Emerging Developments. *Agronomy.* 10(7), 1033.

Müller, M., Lee, C.M., Gasiunas, G., Davis, T.H., Cradick, T.J., Siksnys, V., Bao, G., Cathomen, T., Mussolino, C., 2016. Streptococcus thermophilus CRISPR-Cas9 Systems Enable Specific Editing of the Human Genome. *Mol Ther.* 24(3), 636–644.

Nandy, S., Pathak, B., Zhao, S., Srivastava, V., 2019. Heat-shock-inducible CRISPR/Cas9 system generates heritable mutations in rice. *Plant Direct.* 3(5), e00145.

Nawaz, G., Usman, B., Peng, H., Zhao, N., Yuan, R., Liu, Y., Li, R., 2020. Knockout of Pi21 by CRISPR/Cas9 and iTRAQ-Based Proteomic Analysis of Mutants Revealed New Insights into M. oryzae Resistance in Elite Rice Line. *Genes (Basel).* 11, 735.

Nekrasov, V., Staskawicz, B., Weigel, D., Jones, J.D., Kamoun, S., 2013. Targeted mutagenesis in the model plant *Nicotiana benthamiana* using Cas9 RNA-guided endonuclease. *Nat. Biotechnol.* 31 (2013), 691–693.

Nieves-Cordones, M., Mohamed, S., Tanoi, K., Kobayashi, N.I., Takagi, K., Vernet, A., Guiderdoni, E., Périn, C., Sentenac, H., Véry, A.A., 2017. Production of low-Cs^+ rice plants by inactivation of the K^+ transporter OsHAK1 with the CRISPR-Cas system. *Plant J.* 92(1), 43–56.

Nishimasu, H., Shi, X., Ishiguro, S., Gao, L., Hirano, S., Okazaki, S., Noda, T., Abudayyeh, O.O., Gootenberg, J.S., Mori, H. and Oura, S., 2018. Engineered CRISPR-Cas9 nuclease with expanded targeting space. Science, 361(6408), 1259–1262.

Ogata, T., Ishizaki, T., Fujita, M., Fujita, Y., 2020. CRISPR/Cas9-targeted mutagenesis of OsERA1 confers enhanced responses to abscisic acid and drought stress and increased primary root growth under nonstressed conditions in rice. *PLoS One.* 15, e0243376.

Oliva, R., Ji, C., Atienza-Grande, G., Huguet-Tapia, J.C., Perez-Quintero, A., Li, T., Eom, J.S., Li, C., Nguyen, H., Liu, B., et al., 2019. Broad-spectrum resistance to bacterial blight in rice using genome editing. *Nat. Biotechnol.* 37, 1344–1350.

Ordon, J., Bressan, M., Kretschmer, C., Dall'Osto, Marillonnet, L. S., Bassi, R., Stuttmann, J., 2018. Optimized Cas9 expression systems for highly efficient Arabidopsis genome editing facilitate isolation of complex alleles in a single generation. Funct Integr Genomics. 2020 Jan;20(1):151–162. doi: 10.1007/s10142-019-00665-4. Epub 2019 Feb 23. PMID: 30796544. *bioRxiv preprint.* doi:10.1101/393439

Ordon, J., Bressan, M., Kretschmer, C., Dall'Osto, L., Marillonnet, S., Bassi, R., Stuttmann, J., 2020. Optimized Cas9 expression systems for highly efficient Arabidopsis genome editing facilitate isolation of complex alleles in a single generation. *Functional & integrative genomics,* 20, 151–162.

Pan C, Li G, Malzahn AA, Cheng Y, Leyson B, Sretenovic S, Gurel F, Coleman GD, Qi Y. *Boosting plant genome editing with a versatile CRISPR-Combo system.* Nat Plants. 2022. 513–525.

Pausch, P., Al-Shayeb, B., Bisom-Rapp, E., Tsuchida, C.A., Li, Z., Cress, B.F., Knott, G.J., Jacobsen, S.E., Banfield, J.F., Doudna, J.A., 2020. CRISPR-CasΦ from huge phages is a hypercompact genome editor. *Science*. 369(6501), 333–337.

Premnath, A., Ramalingam, A.P., Ramasamy, S.P., Karnatam, K.S. and Ramadoss, B.R., 2022. Genome editing and miRNA-based approaches in cereals under abiotic stress. In Sustainable remedies for abiotic stress in cereals (pp. 647–673). Singapore: Springer Nature Singapore.

Puchta, H., 2005. The repair of double-strand breaks in plants: mechanisms and consequences for genome evolution. *J. Exp. Bot.* 56, 1–14.

Qin, R., Li, J., Li, H., Zhang, Y., Liu, X., Miao, Y., Zhang, X., Wei, P., 2019. Developing a highly efficient and wildly adaptive CRISPR-SaCas9 toolset for plant genome editing. *Plant Biotechnol J.* 17(4):706–708.

Rahman, F., Mishra, A., Gupta, A., Sharma, R., 2022. Spatiotemporal Regulation of CRISPR/Cas9 Enables Efficient, Precise, and Heritable Edits in Plant Genomes. *Front Genome Ed.* 26(4), 870108.

Ran, F., Hsu, P., Wright, J. et al., 2013. Genome engineering using the CRISPR-Cas9 system. *Nat Protoc* 8, 2281–2308.

Rejesus, R. M., Mohanty, S., and Balagtas, J. V., 2012. Forecasting global rice consumption, Department of Agricultural and Resource Economics, North Carolina State University. *Crit. Rev. Food Sci. Nutr.* 57, 2455–2481.

Safari, F., Zare, K., Negahdaripour, M. et al., 2019. CRISPR Cpf1 proteins: structure, function and implications for genome editing. *Cell Biosci.* 9, 36.

Sakurai, T., Kamiyoshi, A., Takei, N., Watanabe, S., Sato, M., Shindo, T., 2019. Bindel-PCR: a novel and convenient method for identifying CRISPR/Cas9-induced biallelic mutants through modified PCR using *Thermus aquaticus* DNA polymerase. *Sci Rep.* 9(1), 9923.

Sandhya, D., Jogam, P., Allini, V.R., Abbagani, S., Alok, A., 2020. The present and potential future methods for delivering CRISPR/Cas9 components in plants. *J Genet Eng Biotechnol.* 18(1), 25.

Shan, Q., Wang, Y., Li, J., Zhang, Y., Chen, K., et al., 2013. Targeted genome modification of crop plants using a CRISPR-Cas system. *Nature Biotechnol.* 31, 686–688.

Shen, C., Que, Z., Xia, Y., Tang, N., Li, D., He R., 2017. Knock out of the annexin gene OsAnn3 via CRISPR/Cas9-mediated genome editing decreased cold tolerance in rice. *J. Plant Biol.* 60, 539–547.

Sheng, X., Sun, Z., Wang, X., 2020. Improvement of the rice "easy-to-shatter" trait via CRISPR/Cas9-mediated mutagenesis of the qSH1 gene. *Front Plant Sci.* 11: 619.

Shi, J., Wang, E., Milazzo, J.P., et al., 2015. Discovery of cancer drug targets by CRISPR-Cas9 screening of protein domains. *Nat. Biotechnol.* 33, 661–667.

Siddiq, E.A., Vemireddy, L.R., 2021. Advances in Genetics and Breeding of Rice: An Overview. In: Ali, J., Wani, S.H. (eds) Rice Improvement. Springer, Cham. https://doi.org/10.1007/978-3-030-66530-2_1

Singh, N., Choudhury, D. R., Tiwari, G., Singh, A. K., Kumar, S., Srinivasan, K., Tyagi, R. K., Sharma, A. D., Singh, N. K., Singh, R., 2016. Genetic diversity trend in Indian rice varieties: an analysis using SSR markers. *BMC genetics*. 17(1), 127.

Singh, R., Kaur, N., Praba, U.P., Kaur, G., Tanin, M.J., Kumar, P., Neelam, K., Sandhu, J.S. and Vikal, Y., 2022. A Prospective Review on Selectable Marker-Free Genome Engineered Rice: Past, Present and Future Scientific Realm. *Front Genet*, *13*, p.882836.

Slaymaker, I.M., Gao, L., Zetsche, B., Scott, D.A., Yan, W.X., Zhang, F., 2016. Rationally engineered Cas9 nucleases with improved specificity. *Science*. 351(6268):84–8.

Song, S., Wang, G., Wu, H., et al., 2020. OsMFT2 is involved in the regulation of ABA signaling-mediated seed germination through interacting with OsbZIP23/66/72 in rice. *Plant J.* 103(2), 532–46.

Sun, Y., Jiao, G., Liu, Z., Zhang, X., Li, J., Guo, X., Du, W., Du, J., Francis, F., Zhao, Y., et al., 2017. Generation of High-Amylose Rice through CRISPR/Cas9-Mediated Targeted Mutagenesis of Starch Branching Enzymes. *Front. Plant Sci.* 8, 1–15.

Sun, Y.W., Zhang, X., Wu, C.Y., He, Y.B., Ma, Y.Z., Hou, H., Guo, X.P., Du, W.M., Zhao, Y.D., Xia, L.Q., 2016. Engineering herbicide-resistant rice plants through CRISPR/Cas9-mediated homologous recombination of acetolactate synthase. *Mol Plant.* 9(4), 628–631.

Svitashev, S., Schwartz, C., Lenderts, B., Young, J. K., Cigan, A.M., 2016. Genome Editing in Maize Directed by CRISPR-Cas9 Ribonucleoprotein Complexes. *Nat. Commun.* 7, 13274.

Tabassum, J., Ahmad, S., Hussain, B., Mawia, A.M., Zeb, A., Ju, L., 2021. Applications and Potential of Genome-Editing Systems in Rice Improvement: Current and Future Perspectives. *Agronomy.* 11, 1359.

Tang, L., Mao, B., Li, Y., Lv, Q., Zhang, L., Chen, C., et al., 2017. Knockout of OsNramp5 using the CRISPR/Cas9 system produces low Cd accumulating indica rice without compromising yield. *Sci. Rep.* 7:14438.

Toda, E., Koiso, N., Takebayashi, A., Ichikawa, M., Kiba, T., Osakabe, K., 2019. An Efficient DNA-And Selectable-marker-free Genome-Editing System Using Zygotes in Rice. *Nat. Plants.* 5 (4), 363–368.

Usman, B., Nawaz, G., Zhao, N., Liu, Y., Li, R., 2020a. Generation of High Yielding and Fragrant Rice (*Oryza sativa* L.) Lines by CRISPR/Cas9 Targeted Mutagenesis of Three Homoeologs of Cytochrome P450 Gene Family and OsBADH2 and Transcriptome and Proteome Profiling of Revealed Changes Triggered by Mutations. *Plants.* 9, 788.

Usman, B., Nawaz, G., Zhao, N., Liao, S., Liu, Y., Li, R., 2020b. Precise Editing of the OsPYL9 Gene by RNA-Guided Cas9 Nuclease Confers Enhanced Drought Tolerance and Grain Yield in Rice (*Oryza sativa* L.) by Regulating Circadian Rhythm and Abiotic Stress Responsive Proteins. *Int. J. Mol. Sci.* 21, 7854.

Usman, B., Nawaz, G., Zhao, N., Liao, S., Qin, B., Liu, F., Liu, Y., Li, R., 2021. Programmed editing of rice (*Oryza sativa* L.) osspl16 gene using CRISPR/Cas9 improves grain yield by modulating the expression of pyruvate enzymes and cell cycle proteins. *Int. J. Mol. Sci.* 22, 249.

Van Campenhout, C., Cabochette, P., Veillard, A.C., Laczik, M., Zelisko-Schmidt, A., Sabatel, C., Dhainaut, M., Vanhollebeke, B., Gueydan, C., Kruys, V., 2019. Guidelines for optimized gene knockout using CRISPR/Cas9. *Biotechniques.* 66, 6.

Vats, S., Kumawat, S., Brar, J., et al., 2022. Opportunity and challenges for nanotechnology application for genome editing in plants. *Plant Nano Biol.* 1, 100001.

Wada, N., Ueta, R., Osakabe, Y. et al., 2020. Precision genome editing in plants: state-of-the-art in CRISPR/Cas9-based genome engineering. *BMC Plant Biol.* 20, 234.

Walton, R.T., Christie, K.A., Whittaker, M.N., Kleinstiver, B.P., 2020. Unconstrained genome targeting with near-PAMless engineered CRISPR-Cas9 variants. *Science.* 368(6488), 290–296.

Wang, F., Wang, C., Liu, P., Lei, C., Hao, W., Gao, Y., Liu, Y.G., Zhao, K., 2016 Enhanced rice blast resistance by CRISPR/Cas9-Targeted mutagenesis of the ERF transcription factor gene OsERF922. *PLoS One.* 11, e0154027.

Wang, J., Meng, X., Hu, X., Sun, T., Li, J., Wang, K., Yu, H., 2018. xCas9 expands the scope of genome editing with reduced efficiency in rice. *Plant Biotechnol. J.* doi:10.1111/pbi.13053

Wang, W., Wang, W., Pan, Y., Tan, C., Li, H., Chen, Y., Liu, X., Wei, J., Xu, N., Han, Y., Gu, H., 2022. A new gain-of-function OsGS2/GRF4 allele generated by CRISPR/Cas9 genome editing increases rice grain size and yield. *Crop J.* 10(4),1207–1212.

Wang, Z.Y., Zheng, F.Q., Shen, G.Z., Gao, J.P., Snustad, D.P., Li, M.G., Zhang, J.L., Hong, M.M., 1995. The amylose content in rice endosperm is related to the post-transcriptional regulation of the waxy gene. *Plant J.* 7, 613–622.

Woo, J.W., Kim, J., Kwon, S.I., Corvalán, C., Cho, S.W., Kim, H.,2015. DNA-free genome editing in plants with preassembled CRISPR-Cas9 ribonucleoproteins. *Nat Biotechnol.* 33, 1162–4.

Wright, A.V., et al., 2015. Rational design of a split-Cas9 enzyme complex. *Proc. Natl Acad. Sci. USA.* 112, 2984–2989.

Xie, E., Li, Y., Tang, D., Lv, Y., Shen, Y., Cheng, Z.A., 2019. Strategy for generating rice apomixis by gene editing. *J. Integr. Plant Biol.* 61(8):911–916.

Xie, K., Yang, Y., 2013. RNA-guided genome editing in plants using a CRISPR-Cas system. *Mol. Plant.* 6 (2013), 1975–1983.

Xu, R., Yang, Y., Qin, R., Li, H., Qiu, C., Li, L., Wei, P., Yang, J., 2016. Rapid improvement of grain weight via highly efficient CRISPR/Cas9-mediated multiplex genome editing in rice. *J. Genet. Genom.* 43, 529–532.

Xu, R., Qin, R., Xie, H., Li, J., Liu, X., Zhu, M., Sun, Y., Yu, Y., Lu, P., Wei, P., 2022. Genome editing with type II-C CRISPR-Cas9 systems from Neisseria meningitidis in rice. *Plant Biotechnol J.* 20(2):350–359.

Xu, X., Xu, Z., Li, Z., Zakria, M., Zou, L., Chen, G., 2021. Increasing resistance to bacterial leaf streak in rice by editing the promoter of susceptibility gene OsSULRT3;6. *Plant Biotechnol J.* 19(6):1101–1103.

Xu, Z., Xu, X., Gong, Q., Li, Z., Li, Y., Wang, S., Yang, Y., Ma, W., Liu, L., Zhu, B., Zou, L., Chen, G., 2019a. Engineering Broad-Spectrum Bacterial Blight Resistance by Simultaneously Disrupting Variable TALE-Binding Elements of Multiple Susceptibility Genes in Rice. *Mol Plant.* 12(11):1434–1446.

Xu, F., Tang, J., Gao, S., Cheng, X., Du, L., Chu, C., 2019b. Control of rice pre-harvest sprouting by glutaredoxin-mediated abscisic acid signaling. *Plant J.* 100(5): 1036–51.

Yarra, R. and Sahoo, L., 2021. Base editing in rice: current progress, advances, limitations, and future perspectives. *Plant Cell Rep*, 40, 595–604.

Yu, Y., Zhao, Z., Shi, Y., Tian, H., Liu, L., Bian, X., Xu, Y., Zheng, X., Gan, L., Shen, Y., Wang, C., Yu, X., Wang, C., Zhang, X., Guo, X., Wang, J., Ikehashi, H., Jiang, L., Wan, J., 2016. Hybrid Sterility in Rice (*Oryza sativa* L.) Involves the Tetratricopeptide Repeat Domain Containing Protein. *Genetics*, 203(3), 1439–1451.

Yuan, G., Lu, H., De, K., Hassan, M.M., Liu, Y., Li, Y., Muchero, W., Abraham, P.E., Tuskan, G.A., Yang X., 2022. An Intein-Mediated Split-nCas9 System for Base Editing in Plants.: ACS Synth. *Biol.* 11, 2513–2517.

Yue, E., Cao, H., Liu, B., 2020. OsmiR535, a Potential Genetic Editing Target for Drought and Salinity Stress Tolerance in Oryza sativa. *Plants. 9*, 1337.

Zafar, K., Sedeek, K.E.M., Rao, G.S., Khan, M.Z., Amin, I., Kamel, R., Mukhtar, Z., Zafar, M., Mansoor, S., Mahfouz, M.M., 2020a. Genome Editing Technologies for Rice Improvement: Progress, Prospects, and Safety Concerns. *Front Genome Ed.* 4(2), 5.

Zafar, Kashaf., Khan, M.Z., Amin, I., Mukhtar, Z., Yasmin, S., Arif, M., Ejaz, K., Mansoor, S., 2020b. Precise CRISPR-Cas9 Mediated Genome Editing in Super Basmati Rice for Resistance Against Bacterial Blight by Targeting the Major Susceptibility Gene. *Front. Pl. Sci.*12(11), 575.

Zeng, X., Luo, Y., Vu, N.T.Q., Shen, S., Xia, K., Zhang, M., 2020a. CRISPR/Cas9-mediated mutation of OsSWEET14 in rice cv. Zhonghua11 confers resistance to Xanthomonas oryzae pv. oryzae without yield penalty. *BMC Plant Biol.* 20, 313.

Zeng, Y., Wen, J., Zhao, W., Wang, Q., Huang, W., 2020b. Rational Improvement of Rice Yield and Cold Tolerance by Editing the Three Genes *OsPIN5b*, *GS3*, and *OsMYB30* With the CRISPR-Cas9 System. *Front Plant Sci.* 9(10), 1663.

Zetsche, B., Volz, S.E., Zhang, F., 2015. *A split-Cas9 architecture for inducible genome editing and transcription modulation. Nat. Biotechnol.* 33(2), 139–142.

Zhang, A., Liu, Y., Wang, F., Li, T., Chen, Z., Kong, D., Bi, J., Zhang, F., Luo, X., Wang, J., et al., 2019. Enhanced rice salinity tolerance via CRISPR/Cas9-targeted mutagenesis of the OsRR22 gene. *Mol. Breed. 39*, 47.

Zhang, G., Cheng, Z., Zhang, X., Guo, X., Su, N., Jiang, L., Mao, L., Wan, J., 2011 Double repression of soluble starch synthase genes SSIIa and SSIIIa in rice (*Oryza sativa* L.) uncovers interactive effects on the physicochemical properties of starch. *Genome.* 54, 448–459.

Zhang, J., Zhang, H., Botella, J.R., Zhu, J.K., 2018a. Generation of new glutinous rice by CRISPR/Cas9-targeted mutagenesis of the Waxy gene in elite rice varieties. *J. Integr. Plant Biol.* 60, 369–375.

Zhang, Y., Massel, K., Godwin, I. D., & Gao, C., 2018b. Applications and potential of genome editing in crop improvement. *Genome biology*, 19(1), 210.

Zhang, S., Shen, J., Li, D., Cheng, Y., 2021. Strategies in the delivery of Cas9 ribonucleoprotein for CRISPR/Cas9 genome editing. *Theranostics*. 11(2):614–648.

Zhao, D.S., Li, Q.F., Zhang, C.Q., Zhang, C., Yang, Q.Q., Pan, L.X., Ren, X.Y., Lu, J., Gu, M.H., Liu, Q.Q., 2018. GS9 acts as a transcriptional activator to regulate rice grain shape and appearance quality. *Nat. Commun.* 9, 1240.

Zheng, X., Yang, S., Zhang, D., Zhong, Z., Tang, X., Deng, K., Zhou, J., Qi, Y., Zhang, Y., 2016. Effective screen of CRISPR/Cas9-induced mutants in rice by single-strand conformation polymorphism. *Plant Cell Rep.* 35(7), 1545–1554.

Zhou, J., Peng, Z., Long, J., Sosso, D., Liu, B., Eom, J.S., Huang, S., Liu, S., Vera Cruz, C., Frommer, W.B., White, F.F., Yang, B., 2015. Gene targeting by the TAL effector PthXo2 reveals cryptic resistance gene for bacterial blight of rice. *Plant J.* 82(4), 632–43.

Zhou, J., Xin, X., He, Y., Chen, H., Li, Q., Tang, X., Zhong, Z., Deng, K., Zheng, X., Akher, S.A., et al., 2019. Multiplex QTL editing of grain-related genes improves yield in elite rice varieties. *Plant Cell Rep.* 38, 475–485.

Zhou, J.H., Peng, Z., Long, J.Y., Sosso, D., Liu, B., Eom, J.S., Huang, S., Liu, S.Z., Vera Cruz, C., Frommer, W.B., White, F.F., Yang, B., 2015. Gene targeting by the TAL effector PthXo2 reveals cryptic resistance gene for bacterial blight of rice. *Plant J.* 82 (4), 632–643.

Zhu, Y., Lin, Y., Chen, S., Liu, H., Chen, Z., Fan, M., Hu, T., Mei, F., Chen, J., Chen, L., Wang, F., 2019. CRISPR/Cas9-mediated functional recovery of the recessive rc allele to develop red rice. *Plant Biotechnol J.* 17(11):2096–2105.

Zlobin, N.E., Lebedeva, M.V. and Taranov, V.V., 2020. CRISPR/Cas9 genome editing through in planta transformation. *Crit. Rev. Biotechnol*, 40(2), 153–168.